Endorsed by
University of Cambridge International Examinations

Complete Chemistry
for IGCSE

RoseMarie Gallagher
Paul Ingram

OXFORD

OXFORD
UNIVERSITY PRESS

Great Clarendon Street, Oxford OX2 6DP

Oxford University Press is a department of the University of Oxford. It furthers the University's objective of excellence in research, scholarship, and education by publishing worldwide in

Oxford New York

Athens Auckland Bangkok Bogotá Buenos Aires Calcutta Cape Town
Chennai Dar es Salaam Delhi Florence Hong Kong Istanbul Karachi
Kuala Lumpur Madrid Melbourne Mexico City Mumbai Nairobi Paris
São Paulo Singapore Taipei Tokyo Toronto Warsaw

with associated companies in Berlin Ibadan

Oxford is a registered trade mark of Oxford University Press in the UK and in certain other countries

The moral rights of the author have been asserted
Database right Oxford University Press (maker)

© RoseMarie Gallagher and Paul Ingram 2007

All rights reserved. No part of this publication may be reproduced, stored in a retrieval system, or transmitted, in any form or by any means, without the prior permission in writing of Oxford University Press, or as expressly permitted by law, or under terms agreed with the appropriate reprographics rights organisation. Enquiries concerning reproduction outside the scope of the above should be sent to the Rights Department, Oxford University Press, at the above address.

You must not circulate this book in any other binding or cover and you must impose this same condition on any acquirer

British Library Cataloguing in Publication Data
Data available

ISBN: 978-0-19-915135-6

Printed in Singapore by KHL Printing Co Pte Ltd.

10 9 8 7

Paper used in the production of this book is a natural, recyclable product made from wood grown in sustainable forests. The manufacturing process conforms to the environmental regulations of the country of origin.

Acknowledgments

The publisher would like to thank the following for their kind permission to reproduce the following photographs:

Page 6 (left) David Simson (DAS Photo), **Page 6** (right) Gair Photographic Dan/Index Stock/Photo library, **Page 9** Bowater Peter/Photo Researchers, Inc./Photo library, **Page 10**, Istock, , **Page 12** (top) David Simson (DAS Photo), **Page 12** (Centre) Prestige **Page 13** (lower) REUTERS/Bobby Yip, **Page 15** (David Simson (DAS Photo), **Page 16** (right) SHEHZAD NOORANI / Still Pictures, **Page 18** (top) Fision, **Page 20** Robert Francis/Robert Harding Picture Library Ltd/Photo library, **Page 22** (Corbis), **Page 26** (Top right and top left) OUP, **Page 26** (top centre) David Simson (DAS Photo), **Page 28** Mauritius Die Bildagentur Gmbh/Photolibrary, **Page 30** SCIENCE PHOTO LIBRARY/Photo library, **Page 31** (lower) Jtb Photo Communications Inc/Photo library, **Page 31** (Top) UKAEA Harwell, **Page 32** Getty, **Page 36** (lower) David Simson DAS Photo, **Page 43** (left) SCIENCE PHOTO LIBRARY/D Guyon, **Page 43** (right) SCIENCE PHOTO LIBRARY, **Page 48** www.suzanneplunkett.com/ images/sulphur2.jpg, **Page 49** (lower) Kevin Schafer/Peter Arnold Images Inc/Photo library, **Page 52** Andrew Lambert/University of Liverpool, **Page 53** Crafts Council, **Page 59** Science Photo Library, **Page 62** David Simson (DAS Photo), **Page 67** (left) David Simson (DAS Photo), **Page 67** (right) Jaguar, **Page 68** (lower) Bridgeman Art Library, **Page 71** David Simson (DAS Photo), **Page 74** ImageState / Alamy, **Page 78** (lower) S and R Greenhill, **Page 84** GEOFF TOMPKINSON/SCIENCE PHOTO LIBRARY/Photo library, **Page 85** (top) AJ PHOTO/HOP AMERICAIN/SCIENCE PHOTO LIBRARY/Photo library, **Page 85** (lower) David Simson (DAS Photo), **Page 90** David Simson (DAS Photo), **Page 91** (top) David Simson (DAS Photo), **Page 92** (top) David Simson (DAS Photo), **Page 92** (lower) NASA/JPL, **Page 97** USGS, **Page 99** David Simson (DAS Photo), **Page 98** (top) Science Photo Library, **Page 98** (lower) Romil Ltd, **Page 104** Régis Bossu/Sygma/Corbis, **Page 105** (top) JIM VARNEY/Science Photo Library/Photo library, **Page 116** OUP Classet, **Page 122** (top) David Simson (DAS Photo), **Page 122** (lower), INSADCO Photography / Alamy, **Page 123** (top) Jon Bower/Alamy, **Page 123** (lower) Friends of the Earth, **Page 127** (top) Maximilian Stock LTD/Phototake Inc/Photo library, **Page 127** (lower) Niall McDiarmid / Alamy, **Page 128** MARTYN F CHILLMAID/Science Photo Library, **Page 138** (top) Georg Gerster/Photo Researchers, Inc./Photo library, **Page 138** (lower) Shutterstock, **Page 139** (left) Severn Trent Water, **Page 139** (right) Weston Point Studio, **Page 141** (left) ullstein - ecopix / Still Pictures, **Page 144** (centre top) Zefa, **Page 141** (top right) David Simson (DAS Photo), **Page 144** (lower left) David Simson (DAS Photo), **Page 144** (lower right) David Simson (DAS Photo), **Page 151** top Wichita Fire Brigade Kansas, **Page 151** (lower), Holt Studios, **Page 157** (top) OUP/Classet, **Page 161** (centre) David Simson (DAS Photo), **Page 164** (right) OUP/Classet, **Page 165** Honda news, **Page 172** (right) David Simson (DAS Photo), **Page 173** Barnabys,

Page 178 Royal Mint, **Page 179** Dean Conger/CORBIS, **Page 180** (top) Thermite Welding, **Page 180** (lower) Tim Street-Porter/Beateworks/Corbis, **Page 181** David Simson (DAS Photo), **Page 184** (top) **Page 184** (lower) NASA, **Page 185** (lower right) Shutterstock, **Page 187** Ifa-Bilderteam Gmbh/Photolibrary, **Page 188** (top) Science Photo Library/Crown Copyright, **Page 188** (lower) Camera Press, **Page 190** (top centre) **Page 190** (lower right) David Simson (DAS Photo), **Page 191** (lower) Shutterstock, **Page 192** (top left) CHRISTIAN DARKIN/Science Photo Library/ Photo library, **Page 192** (top centre) Flickr, **Page 192** (top right) Len Roe, **Page 192**, (lower left) GM Motors, **Page 192** Brand X Pictures / Alamy, **Page 192** MEHAU KULYK / SCIENCE PHOTO LIBRARY, **Page 194** (top right) David Simson (DAS Photo), **Page 199** (top) JOHN MEAD/SCIENCE PHOTO LIBRARY, **Page 199** (lower) JOERG BOETHLING / Still Pictures, **Page 205** (top) Medical-on-Line / Alamy, **Page 205** (centre) Dynamic Graphics/Photo library, **Page 207** (top) OUP/Classet, **Page 212** Barnabys **Page 213** GM Motors, **Page 214** (top) OUP/Classet, **Page 214** (lower) PASQUALE SORRENTINO/Science Photo Library, **Page 216** ICI, **Page 218** (top left), David Simson (DAS Photo), **Page 218** (top centre) Shutterstock, **Page 218** (top right) Science Photo Library, **Page 218** (lower) Ron Giling / Still Pictures, **Page 222** (top) NASA, **Page 222** (lower) PHONE Labat J.M./ Still Pictures, **Page 223** (top left) Science Photo Library, **Page 223** (centre) Oxford Scientific Films, **Page 223** (top right) Tim De Waele/Corbis, **Page 227**, Friends of the Earth, **Page 228** ICI, **Page 230** Hulton Getty, **Page 231** David Simson (DAS Photo), **Page 232** (top) Flickr, **Page 232** (lower) David Simson (DAS Photo), **Page 234** David Simson (DAS Photo), **Page 239** (top left) David Simson (DAS Photo), **Page 239** (top centre) Istock, **Page 239** Index Stock Imagery/Photo library, **Page 242** Pixtal/Inmagine, **Page 243** Exxon Mobil, **Page 244** science photos / Alamy, **Page 247** Science Photo Library, **Page 248** David Simson (DAS Photo), **Page 249** Martin M Rotker/Photo Researchers, Inc. & Science Photo Library, **Page 251** Bettmann/CORBIS, **Page 256** (top right) Shutterstock, **Page 257** (centre) Istock, **Page 259** Istock, **Page 260** Dreamstime, **Page 261** (top) German Hosiery Museum, **Page 261** (lower) Shutterstock, **Page 263** (left) Roger Ressmeyer/CORBIS, **Page 264** (centre and lower) David Simson (DAS Photo), **Page 265** (lower) Symphony Environmental, **Page 269** Ray Sto, **Page 270** (top) Big Stock Photo, **Page 270** (lower) Science Photo Library.

Additional photography by RoseMarie Gallagher and Paul Ingram.

UCLES/CIE: all CIE questions are taken from the IGCSE Chemistry paper and are reproduced by permission of The University of Cambridge Local Examinations Syndicate. The University of Cambridge Local Examinations Syndicate bears no responsibility for the example answers to questions taken from its past question papers which are contained in this publication.

Introduction

If you are taking chemistry at IGSCE level, with the Cambridge International Examinations (CIE) syllabus, then this book is for you. It covers the syllabus fully. It is written in a clear direct style. It has a wealth of drawings and photos to help you. And it has been endorsed by CIE.

If you are taking the Extended curriculum, you should study all the topics in this book. For the Core curriculum you will need less detail. The CIE syllabus sets out clearly what you need to know. It is a good idea to have a look at the syllabus, and discuss it with your teacher.

We want you to understand chemistry, and do well in your exams. This book can help you do both. So make the most of it!

Work through the units Each chapter is divided into two-page units. By working through these short units, you will build up a good knowledge of the chemistry you need, and see how one topic links to another.

Use the glossary If you come across a chemical term you do not understand, look it up in the glossary at the back of the book. You can also use the glossary to test yourself, when you are revising for your exams.

Answer the questions Answering questions is a great way to get to grips with a topic. That's why this book has lots of them.

There are questions at the end of each unit, to help you understand the material in the unit. There are exam-level questions at the end of each chapter. And, at the back of the book, you will find questions from past CIE IGCSE exam papers. Work through these when you are preparing for your exams.

Answers to numerical questions are given at the back of the book. If you get stuck on a question, ask your teacher for help.

Use the contents list and index The contents list is like a map for the book. When you want to look up a big topic, start with the contents list. If you want to search for a specific detail, try the index.

And finally, enjoy! Chemistry is an exciting, important, and challenging subject. It has an impact on every part of our lives. We hope this book will help you to enjoy chemistry, and succeed in your IGCSE chemistry course.

RoseMarie Gallagher
Paul Ingram

Contents

1 States of matter
- 1.1 Everything is made of particles 6
- 1.2 Solids, liquids, and gases 8
- 1.3 Particles in solids, liquids, and gases 10
- 1.4 A closer look at gases 12
- 1.5 Mixtures, solutions and solvents 14
- 1.6 Pure substances and impurities 16
- 1.7 Separation methods (I) 18
- 1.8 Separation methods (II) 20
- 1.9 More about chromatography 22
- Questions on Chapter 1 24

2 The atom
- 2.1 Atoms, elements, and compounds 26
- 2.2 More about atoms 28
- 2.3 Isotopes and radioactivity 30
- 2.4 How electrons are arranged 32
- Questions on Chapter 2 34

3 Atoms combining
- 3.1 Why compounds form 36
- 3.2 The ionic bond 38
- 3.3 More about ions 40
- 3.4 Ionic compounds and their properties 42
- 3.5 The covalent bond 44
- 3.6 Covalent compounds 46
- 3.7 Molecular substances 48
- 3.8 Giant covalent structures 50
- 3.9 The bonding in metals 52
- 3.10 Bonding, structure and properties: a review 54
- Questions on Chapter 3 56

4 The Periodic Table
- 4.1 An overview of the Periodic Table 58
- 4.2 Group 1: the alkali metals 60
- 4.3 The patterns within Group 1 62
- 4.4 Group 7: the halogens 64
- 4.5 Group 0: the noble gases 66
- 4.6 The transition elements 68
- 4.7 Across the Periodic Table 70
- Questions on Chapter 4 72

5 The mole
- 5.1 The names and formulae of compounds 74
- 5.2 The masses of atoms, molecules. and ions 76
- 5.3 Percentage composition of a compound 78
- 5.4 The mole 80
- 5.5 The empirical formula 82
- 5.6 From empirical to final formula 84
- 5.7 The concentration of a solution 86
- Questions on Chapter 5 88

6 Chemical equations
- 6.1 Physical and chemical change 90
- 6.2 Equations for chemical reactions 92
- 6.3 Calculations from equations 94
- 6.4 Reactions involving gases 96
- 6.5 Finding % yield and % purity 98
- Questions on Chapter 6 100

7 Redox reactions
- 7.1 Different types of reaction 102
- 7.2 Oxidation and reduction 104
- 7.3 Redox and electron transfer 106
- 7.4 Changes in oxidation state 108
- 7.5 Oxidising and reducing agents 110
- Questions on Chapter 7 112

8 Acids and bases
- 8.1 Acids and alkalis 114
- 8.2 A closer look at acids and alkalis 116
- 8.3 The reactions of acids and bases 118
- 8.4 A closer look at neutralisation 120
- 8.5 Acids and bases outside the lab 122
- 8.6 Making salts 124
- 8.7 Making insoluble salts by precipitation 126
- 8.8 Finding concentrations by titration 128
- Questions on Chapter 8 130

9 Electricity and chemical change
- 9.1 Conductors and insulators 132
- 9.2 The principles of electrolysis 134
- 9.3 The electrolysis of solutions 136
- 9.4 The electrolysis of brine 138
- 9.5 Two more uses of electrolysis 140
- Questions on Chapter 9 142

10 How fast are reactions?
- 10.1 Rates of reaction 144
- 10.2 Measuring the rate of a reaction 146
- 10.3 Changing the rate of a reaction (I) 148
- 10.4 Changing the rate of a reaction (II) 150

- 10.5 Explaining rates **152**
- 10.6 Catalysts **154**
- 10.7 Enzymes: biological catalysts **156**
  Questions on Chapter 10 **158**

11 Energy changes, and reversible reactions

- 11.1 Energy changes in reactions **160**
- 11.2 Explaining energy changes **162**
- 11.3 Energy from fuels **164**
- 11.4 Reversible reactions **166**
- 11.5 Shifting the equilibrium **168**
  Questions on Chapter 11 **170**

12 The behaviour of metals

- 12.1 Metals and non-metals **172**
- 12.2 Comparing metals for reactivity (I) **174**
- 12.3 Comparing metals for reactivity (II) **176**
- 12.4 The reactivity series **178**
- 12.5 Making use of the reactivity series **180**
  Questions on Chapter 12 **182**

13 Making use of metals

- 13.1 Metals in the Earth's crust **184**
- 13.2 Extracting metals from their ores **186**
- 13.3 Extracting iron **188**
- 13.4 Extracting aluminium **190**
- 13.5 Making use of metals **192**
- 13.6 Steels and other alloys **194**
- 13.7 Corrosion **196**
- 13.8 Mining, using, and recycling metals **198**
  Questions on Chapter 13 **200**

14 Air and water

- 14.1 What is air? **202**
- 14.2 Making use of air **204**
- 14.3 Pollution alert! **206**
- 14.4 Water supply **208**
  Questions on Chapter 14 **210**

15 Non-metals: hydrogen and nitrogen

- 15.1 Hydrogen **212**
- 15.2 Nitrogen and ammonia **214**
- 15.3 Making ammonia in industry **216**
- 15.4 Fertilisers **218**
  Questions on Chapter 15 **220**

16 Non-metals: oxygen, sulphur, chlorine, and carbon

- 16.1 Oxygen **222**
- 16.2 Oxides **224**
- 16.3 Sulphur and sulphur dioxide **226**
- 16.4 Sulphuric acid **228**
- 16.5 Chlorine **230**
- 16.6 Carbon and its compounds **232**
- 16.8 Limestone **234**
  Questions on Chapter 16 **236**

17 Organic chemistry

- 17.1 Oil: a source of useful compounds **238**
- 17.2 Separating oil into fractions **240**
- 17.3 Cracking hydrocarbons **242**
- 17.4 The alkanes **244**
- 17.5 The alkenes **246**
- 17.6 The alcohols **248**
- 17.7 The manufacture of ethanol **250**
- 17.8 Carboxylic acids, and esters **252**
  Questions on Chapter 17 **254**

18 Polymers

- 18.1 Introducing polymers **256**
- 18.2 Addition polymerisation **258**
- 18.3 Condensation polymerisation **260**
- 18.4 Making use of synthetic polymers **262**
- 18.5 Plastics: here to stay? **264**
- 18.6 The macromolecules in food (I) **266**
- 18.7 The macromolecules in food (II) **268**
- 18.8 Breaking down the macromolecules **270**
  Questions on Chapter 18 **272**

Cambridge IGCSE exam questions

- Core material **274**
- Extended material **282**

Working in the lab

- Working with gases in the lab **288**
- Testing for ions in the lab **290**
- Safety first! **292**

Reference

- Answers to numerical questions **293**
- Glossary **294**
- The Periodic Table and atomic masses **298**
- Index **300**

1.1 Everything is made of particles

Made of particles

Rock, air, and water look very different. But they have one big thing in common: they are all made of very tiny pieces, far too small to see. For the moment, we will call these pieces **particles**.

In fact everything around you is made of particles – and so are you!

Particles on the move

In rock and other solids, the particles are not free to move around. But in liquids and gases, they move freely. As they move they collide with each other, and bounce off in all directions.

So the path of one particle, in a liquid or gas, looks like this:

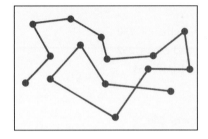

All made of particles!

The particle moves in a random way, changing direction every time it hits another particle. We call this **random motion**.

Some evidence for particles

There is evidence all around you that things are made of particles, and that they move around in liquids and gases. Look at these examples.

Evidence outside the lab

1 Cooking smells can spread through the house. This is because 'smells' are caused by gas particles mixing with, and moving through, the air. They dissolve in moisture in the lining of your nose.

2 You often see dust and smoke dancing in the air, in bright sunlight. The dust and smoke are clusters of particles. They dance around because they are being bombarded by tiny particles in the air.

Evidence in the lab

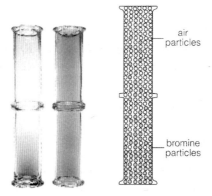

1 Place a crystal of potassium manganate(VII) in a beaker of water. The colour spreads because particles leave the crystal and mix through the water particles. The crystal **dissolves**.

2 Place an open gas jar of air upside down on an open gas jar of red-brown bromine vapour. The colour spreads upwards because particles of bromine mix through the particles of air.

Diffusion

In all the examples above, particles bounce off all directions when they collide. That's how they get mixed. This mixing process is called **diffusion**.

So what are these particles?

The smallest particle that cannot be broken down by chemical means is called an **atom**.

- In some substances, the particles are just single atoms. For example the gas argon, found in air, is made up of single argon atoms
- In many substances, the particles consist of two or more atoms joined together. These particles are called **molecules**. Water, bromine, and the gases nitrogen and oxygen in air, are made up of molecules.
- In other substances the particles consist of atoms or groups of atoms that carry a charge. These particles are called **ions**. Potassium manganate(VII) is made of ions.

You'll find out more about all these particles in later sections.

'Seeing' particles

We are now able to 'see' the particles in some solids, using very powerful microscopes. For example the image on the right shows the atoms in silicon, which is used to make computer chips. In this image, the atoms appear over 70 million times larger than they really are!

The atoms in silicon. The image was taken using a tunnelling electron microscope. The colour was added to help you see them more clearly.

Questions

1 The particles in liquids and gases show *random motion*. What does that mean, and why does it occur?
2 Why does the purple colour spread when a crystal of potassium manganate(VII) is placed in water?
3 Bromine vapour is heavier than air. Even so, it spreads upwards in the experiment above. Why?
4 a What is *diffusion*? b Use the idea of diffusion to explain how the smell of perfume travels.

1.2 Solids, liquids, and gases

What's the difference?

It is easy to tell the difference between a solid, a liquid and a gas:

A solid has a definite shape and a definite volume.

A liquid flows easily. It has a definite volume but no definite shape. It takes the shape of the container.

A gas has no definite volume or shape. It spreads out to fill its container. It is much lighter than the same volume of solid or liquid.

Water: solid, liquid and gas

Water can be a solid (ice), a liquid (water), and a gas (water vapour or steam). Its state can be changed by heating or cooling:

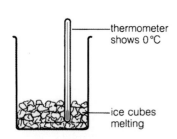

1 **Ice** slowly changes to **water**, when it is put in a warm place. This change is called **melting**. The thermometer shows 0°C until all the ice has melted, so 0°C is called its **melting point**.

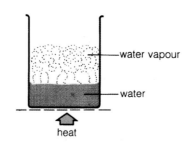

2 When the water is heated its temperature rises, and some of it changes to **water vapour**. This change is called **evaporation**. The hotter the water gets, the more quickly it evaporates.

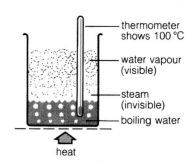

3 Soon bubbles appear in the water. It is **boiling**. The water vapour shows up as steam. The thermometer stays at 100°C while the water boils off. 100°C is the **boiling point** of water.

And when steam is cooled, the opposite changes take place:

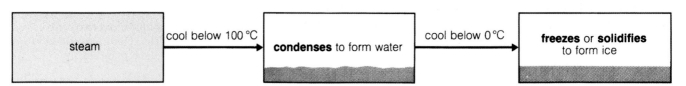

You can see that:

- condensing is the opposite of evaporating
- freezing is the opposite of melting
- the freezing point of water is the same as the melting point of ice, 0°C.

Other things can change state too

It's not just water! Nearly all substances can exist as solid, liquid and gas. Even iron and diamond can melt and boil! Some melting and boiling points are given below. Look how different they are.

Substance	Melting point/°C	Boiling point/°C
oxygen	−219	−183
ethanol	−15	78
sodium	98	890
sulphur	119	445
iron	1540	2900
diamond	3550	4832

Molten iron being poured out at an iron works. Hot – over 1540 °C!

Showing changes of state on a graph

Look at this graph. It shows how the temperature changes as a block of ice is steadily heated. First the ice melts to water, then the water gets warmer and warmer, and eventually turns to steam:

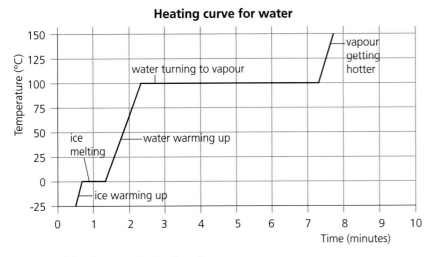

A graph like this is called a **heating curve**.

Look at the step where the ice is melting. Once melting starts, the temperature stays at 0°C until *all* the ice has melted. When the water starts to boil, the temperature stays at 100°C until *all* the water has turned to steam. So the melting and boiling points are clear and sharp.

Questions

1. Write down two properties of a solid, two of a liquid, and two of a gas.
2. Which word means the opposite of :
   a boiling?   b melting?
3. Which has a lower freezing point, oxygen or ethanol?
4. Which has a higher boiling point, oxygen or ethanol?
5. Look at the heating curve above.
   a About how long did it take for the ice to melt, once melting started?
   b How long did boiling take to complete, once it started?
   c Try to think of a reason for the difference in **a** and **b**.
6. See if you can sketch a heating curve for sodium.

1.3 Particles in solids, liquids, and gases

How the particles are arranged

Water can change from solid to liquid to gas. Its *particles* do not change. They are the same in each state. But their *arrangement* changes. The same is true for all substances.

State	How the particles are arranged	Diagram of particles
Solid 	The particles in a solid are packed tightly in a fixed pattern. There are strong forces holding them together. So they cannot leave their positions. The only movements they make are tiny vibrations to and fro.	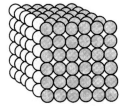
Liquid 	The particles in a liquid can move about and slide past each other. They are still close together, but are not in a fixed pattern. The forces that hold them together are weaker than in a solid.	
Gas 	The particles in a gas are far apart, and they move about very quickly. There are almost no forces holding them together. They collide with each other and bounce off in all directions.	

Changing state

Melting When a solid is heated, its particles get more energy and vibrate more. This makes the solid **expand**. At the melting point, the particles vibrate so much that they break away from their positions. The solid turns liquid.

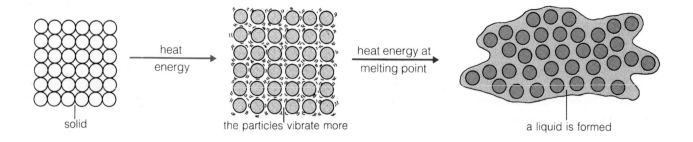

Boiling When a liquid is heated, its particles get more energy and move faster. They bump into each other more often, and bounce further apart. This makes the liquid expand. At the boiling point, the particles get enough energy to overcome the forces between them. They break away to form a gas:

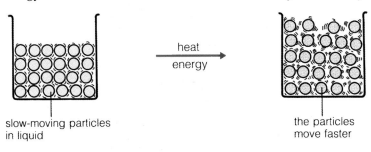
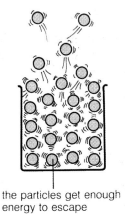

Evaporating Some particles in a liquid have more energy than others. Even well below the boiling point, some have enough energy to escape and form a gas. This is called **evaporation**. It is why puddles of rain dry up in the sun.

Condensing and solidifying When a gas is cooled, the particles lose energy. They move more and more slowly. When they bump into each other, they do not have enough energy to bounce away again. They stay close together, and a liquid forms. When the liquid is cooled, the particles slow down even more. Eventually they stop moving, except for tiny vibrations, and a solid forms.

How much heat is needed?

The amount of heat needed to melt or boil a substance is different for every substance. That's because the particles in each substance are different, with different forces between them. The stronger the forces, the more heat energy is needed to overcome them.

This table shows the amount of heat needed to melt and boil some substances. (1 mole of each substance has exactly the same number of particles.)

Substance	Amount of heat (in kilojoules) needed to ...	
	melt 1 mole of it	boil 1 mole of it
ethanol	5.0	38.7
water	6.0	41.2
sodium	2.6	97.0

Note that it always takes more heat to boil a substance than to melt it. That's because it takes a lot more energy to separate the particles completely (in boiling) than to free them from a lattice (in melting).

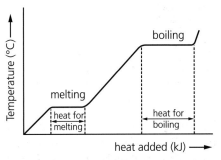

Even though you keep on adding heat, the temperature stays steady while the substance melts, and again while it boils.

Questions
1. Using the idea of particles, explain why:
   a you can pour liquids   b solids expand on heating
2. Draw a diagram to show what happens to the particles, when a liquid cools to a solid.
3. In which substance is the force between particles stronger? (Use the table above.)
   a in solid sodium, or in ice?
   b in liquid sodium, or in water?

11

1.4 A closer look at gases

What is gas pressure?

When you blow up a balloon, you fill it with air particles. They hit against the sides of the balloon and exert **pressure** on it. This pressure is what keeps the balloon inflated.

In the same way, *all* gases exert a pressure. The pressure depends on the **temperature** of the gas and the **volume** it takes up, as you'll see below.

When you heat a gas

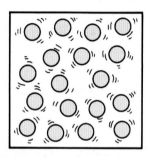
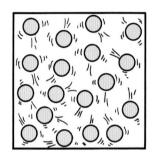

The particles in this gas are moving fast. They hit the walls of the container and exert pressure on them. If you now heat the gas ...

... the particles take in heat energy and move even faster. They hit the walls more often, and with more force. So the gas pressure increases.

The same happens with all gases:
When you heat a gas in a closed container, its pressure increases.
That is why the pressure gets very high inside a pressure cooker.

When you squeeze a gas into a smaller space

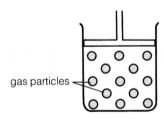
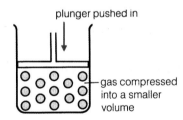

There is a lot of space between the particles in a gas. You can force them closer by pushing in the plunger ...

... like this. Now the particles are in a smaller space – so they hit the walls more often. So the gas pressure increases.

The same thing is true for all gases:
When a gas is squeezed into a smaller space, its pressure increases.

All gases can be squeezed into a smaller space, or **compressed**. If enough force is applied to the plunger above, the gas particles will get so close that the gas turns into a liquid. But liquids and solids cannot be compressed. Their particles are already very close together.

When you blow air into a balloon, the gas particles exert pressure on the balloon and inflate it. The more you blow, the greater the pressure.

In a pressure cooker, water vapour is heated to well over 100°C, so its pressure is very high. You must let a pressure cooker cool before you open it!

Cylinders of compressed air, that allow divers to breathe under water.

The diffusion of gases

On page 7 you saw that gases **diffuse** because the particles collide with other particles, and bounce off in all directions. But gases do not all diffuse at the same rate, every time. It depends on these factors:

1 The mass of the particles
The particles in hydrogen chloride gas are twice as heavy as those in ammonia gas. So which gas do you think will diffuse faster? Let's see:

- Cotton wool soaked in ammonia solution is put into one end of a long tube. It gives off ammonia gas.
- *At the same time*, cotton wool soaked in hydrochloric acid is put into the other end of the tube. It gives off hydrogen chloride gas.
- The gases diffuse along the tube. Smoke forms where they meet:

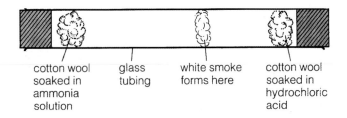

The scent of flowers travels faster in a warm room. Can you explain why?

The white smoke forms closer to the hydrochloric acid end of the tube. So the ammonia particles have travelled further than the hydrogen chloride particles – which means they have travelled *faster*.
The lower the mass of its particles, the faster a gas will diffuse.

That makes sense when you think about it. When particles collide and bounce away, the lighter particles will bounce further.

The particles in the above two gases are molecules. We use the term **relative molecular mass** for the mass of a molecule. So we can also say:
The lower its relative molecular mass, the faster a gas will diffuse.

2 The temperature
When a gas is heated, its particles take in heat energy, and move faster. So they collide with more energy, and bounce further away. And so the gas diffuses faster.
The higher the temperature, the faster a gas will diffuse.

The faster a particle is moving when it hits another, the faster and further it will bounce away. Just like snooker balls!

Questions
1. What causes the *pressure* in a gas?
2. Why does a balloon burst if you keep on blowing?
3. A gas is in a sealed container. How do you think the pressure will change if the container is cooled? Explain your answer.
4. A gas flows from one container into a larger one. What do you think will happen to its pressure? Draw diagrams to explain.
5. a Why does the scent of perfume spread?
   b Why does the scent of perfume wear off faster in warm weather than in cold?
6. Of all gases, hydrogen diffuses fastest at any given temperature. What can you tell from this?
7. Look at the glass tube above. Suppose it was warmed a little in an oven, before the experiment. Do you think that would change the result? If so, how?

1.5 Mixtures, solutions, and solvents

Mixtures

A **mixture** contains more than one substance. The substances are just mixed together, and not chemically combined. For example:

- air is a mixture of nitrogen, oxygen, and small amounts of other gases
- shampoo is a mixture of several chemicals and water.

Solutions

When you mix sugar with water, the sugar seems to disappear. That is because its particles spread all through the water particles, like this:

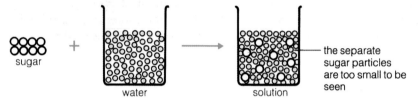

A mixture of sugar and water. This mixture is a solution.

The sugar has **dissolved** in the water, giving a solution. Sugar is the **solute**, and water is the **solvent**:

solute + solvent = solution

You can't get the sugar out again by filtering.

Not everything dissolves so easily

Now think about chalk. If you mix chalk powder with water, most of the powder eventually sinks to the bottom. You can get it out again by filtering.

Why is it so different for sugar and chalk? Because their particles are very different! How easily a substance dissolves depends on the particles in it. Look at the examples in this table:

Compound	Mass (g) dissolving in 100 g of water at 25°C
silver nitrate	241.3
calcium nitrate	102.1
sugar (glucose)	91.0
potassium nitrate	37.9
potassium sulphate	12.0
calcium hydroxide	0.113
calcium carbonate (chalk)	0.0013
silver chloride	0.0002

decreasing solubility

A mixture of chalk powder and water. This is not a solution. The tiny chalk particles do not separate and spread through the water particles. They stay in clusters big enough to see. In time, most sink to the bottom.

So silver nitrate is much more soluble than sugar – but potassium nitrate is a lot less soluble than sugar. It all depends on the particles.

Look at calcium hydroxide. It is only very slightly or **sparingly soluble** compared with the compounds above it. Its solution is called **lime water**.

Now look at the last two substances in the table. They are usually called **insoluble** since only a very tiny amount dissolves.

What's soluble, and what's not?

- The solubility of every substance is different.
- But there are some overall patterns. For example *all* sodium compounds are soluble.
- Find out more on page 126.

Helping a solute dissolve

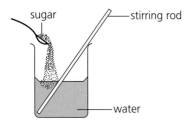
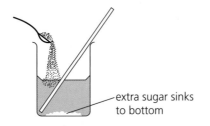
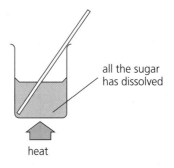

Sugar dissolves quite slowly in water at room temperature. Stirring helps. But if you keep on adding sugar …

… eventually no more of it will dissolve, no matter how hard you stir. The extra sinks to the bottom. The solution is now **saturated**.

But look what happens if you heat the solution. The extra sugar dissolves. Add more sugar and it will dissolve too, as the temperature rises.

So sugar is **more soluble** in hot water than in cold water.
If a solid is soluble in water, it usually gets more soluble as the temperature rises.

Water is not the only solvent

Water is the world's most common solvent. A solution in water is called an **aqueous solution** (from *aqua*, the Latin word for water).

But many other solvents are used in industry and about the house, to dissolve substances that are insoluble in water. Some examples are:

Solvent	It dissolves
white spirit	gloss paint
propanone (acetone)	grease, nail polish
ethanol	glues, printing inks, the scented substances that are used in perfumes and aftershaves

All three solvents above evaporate easily at room temperature – they are **volatile**. This means that glues and paints dry easily. Aftershaves feel cool because ethanol cools the skin when it evaporates.

Nail polish is insoluble in water. So she will remove it later by dissolving it in propanone.

Questions

1. **a** Are all solutions mixtures? **b** Are all mixtures solutions?
2. Explain each term in your own words. (Check the glossary?)
   a insoluble **b** solubility **c** aqueous solution
3. Look at the table on page 14.
   a Which substance in it is the most soluble?
   b About how many times more soluble is this substance than potassium sulphate, at 25 °C?
   c It gives a colourless solution. What will you see if you add 300 g of it to 100 g of water at 25 °C?
   d What will you see if you heat up the mixture in **c**?
4. Now turn to the table at the top of page 126.
   a Name two metals that have *no* insoluble salts.
   b Name one other group of salts that is *always* soluble.
5. See if you can give three examples of:
   a solids you dissolve in water, at home
   b insoluble solids you use at home.
6. Name two solvents other than water that are used in the home. What are they used for?
7. Many gases dissolve in water too. See how many examples you can think of. (Fish need one of them!)

1.6 Pure substances and impurities

What is a pure substance?

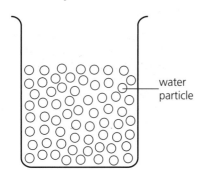 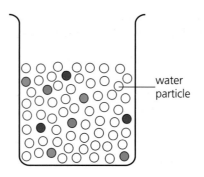 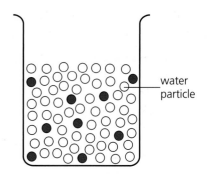

This is water. It has only water particles in it, and nothing else. So it is 100% **pure**.

This water has particles of other substances mixed with it. So it is not pure.

This water has particles of a poisonous substance in it. It is not pure – and could be deadly.

A pure substance has no particles of any other substance mixed with it.

In real life, very few substances are 100% pure. Tap water contains small amounts of many different particles (for example calcium ions and chloride ions). The particles in it are not usually harmful – and some are even good for you. Distilled water is much purer, but still not 100% pure. For example it has gas particles from the air dissolved in it.

Does purity matter?

Often, it does not matter if a substance is not pure. Most of the time we use tap water without thinking about what's in it. But sometimes purity is very important. If you are making a new medical drug, or food flavouring, you must make sure it contains nothing that could harm people.

An unwanted substance, mixed with the substance you want, is called **an impurity.**

 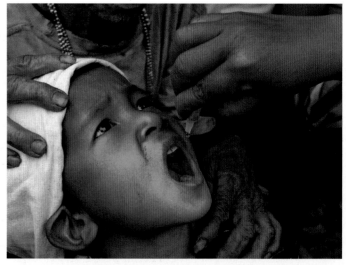

Baby foods and milk powder are tested in the factory, to make sure they contain no harmful impurities.

Vaccination against polio, by mouth. Medicines must be safe, and free of harmful impurities. So they are tested heavily.

How can you tell if a substance is pure?

Chemists use some complex methods to check purity. But there is one simple method *you* can use in the lab: **you can check melting and boiling points**.

- A pure substance has a definite, sharp, melting point and boiling point. You can look these up in data tables.
- When the substance contains an impurity, its melting point falls and its boiling point rises. And melting and boiling no longer take place sharply, but over a range of temperature.
- The more impurity present, the bigger the change in melting and boiling points, and the wider the temperature range over which the substance melts and boils.

Compare these examples:

ID check!
- Every substance has a unique pair of melting and boiling points.
- So you can use melting and boiling points to identify a substance.
- First, measure them. Then look up data tables to find out what the substance is.

This sulphur sample melts sharply at 119°C and boils at 445°C. So it is pure.

This water freezes around −0.5°C and boils around 101°C. So it's not pure.

So, melting and boiling points will give you an idea of purity. But if you are making a new medical drug, you will also do several other purity checks.

Separation: the first step in obtaining a pure substance

When you carry out a reaction to make something, you usually end up with a mixture of substances. Then you have to separate the one you want.

The table below shows some separation methods. These can give quite pure substances. For example when you filter off a solid, and rinse it really well with distilled water, you remove a lot of impurity. But it is just not possible to remove every tiny particle of impurity, in the school lab.

Method of separation	Used to separate…
filter	a solid from a liquid
centrifuge	a solid from a liquid
evaporate	a solid from its solution
crystallise	a solid from its solution
distil	a solvent from a solution
fractional distillation	liquids from each other
chromatography	different substances from a solution

There is more about these methods in the next three units.

At the end of this reaction, the beaker may contain several products, plus reactants that have not reacted. Separating them can be a challenge!

Questions
1. What does *a pure substance* mean?
2. You mix instant coffee with water, to make a cup of coffee. Is the coffee an *impurity*? Explain.
3. Explain why melting and boiling points can be used as a way to check purity.
4. Could there be impurities in a gas? Explain.

1.7 Separation methods (I)

Which method?

As you saw in the last unit, there are several ways to separate mixtures of substances. The method you choose depends on the physical state of the substances.

Separating a solid from a liquid

1 By filtering
Chalk can be separated from water by filtering, as shown here. The chalk gets trapped in the filter paper while the water passes through.

The chalk is called the **residue**. The water is the **filtrate**.

Filtering is very widely used to separate solids from liquids. For example it is used in cleaning up river water, to make it fit to drink. Can you think of a use in the kitchen?

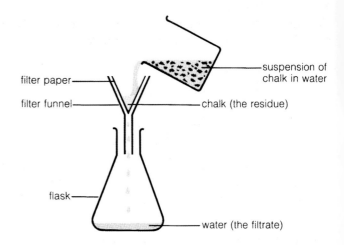

2 By centrifuging
A centrifuge is used to separate small amounts of solid and liquid. Inside the centrifuge, test tubes are spun very fast, so the solid gets flung to the bottom:

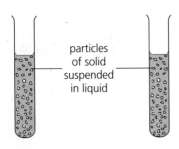 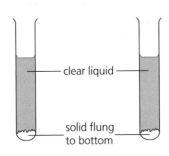

Before centrifuging, the solid is mixed all through the liquid.

After centrifuging, all the solid has collected at the bottom.

A centrifuge.

The liquid is poured out of the test tubes, or removed with a small pipette. The solid is left behind.

3 By evaporating the solvent
If the mixture is a *solution*, the solid cannot be separated by filtering or centrifuging, because its tiny particles are spread all through the solvent.

Instead, the solution is heated to evaporate the solvent. So the solid is left behind. You could use this method to obtain salt from its aqueous solution, for example:

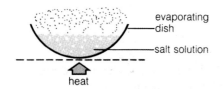 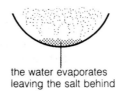

Evaporating the water from a solution of salt in water.

4 By crystallising

You can obtain many solids from their solutions by allowing them to form crystals. Copper(II) sulphate is an example:

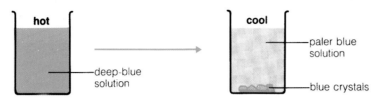

This is a saturated solution of copper(II) sulphate in water at 70 °C. If it is cooled to 20 °C . . .

. . . crystals begin to appear, because the compound is *less soluble* at 20 °C than at 70 °C.

The process is called **crystallisation**. It is carried out like this:

 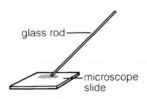 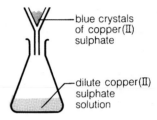

1 A solution of copper(II) sulphate is heated, to get rid of some water. As the water evaporates, the solution becomes more concentrated.

2 The solution can be checked to see if it is ready by placing one drop on a microscope slide. Crystals should form quickly on the cool glass.

3 Then the solution is left to cool and crystallize. The crystals are removed by filtering, rinsed with water and dried with filter paper.

Separating a mixture of two solids

You can separate two solids by choosing a solvent that will dissolve just one of them. Suppose the solids are salt and sand. Water dissolves salt but not sand. So you can separate them like this:

1 Add water to the mixture, and stir. The salt dissolves.
2 Filter the mixture. The sand is trapped in the filter paper, but the salt solution passes through.
3 Rinse the sand with water, and dry it in an oven.
4 Evaporate the salt solution until dry salt is left.

Water could *not* be used to separate salt and sugar, because it dissolves both. But you could use ethanol, which dissolves sugar but not salt. Ethanol is inflammable, so should be evaporated over a water bath, as shown here.

Evaporating the ethanol from a solution of sugar in ethanol, over a water bath.

Questions

1 What does this term mean? Give an example.
   a filtrate   b residue
2 You have a solution of sugar in water. You want to get the sugar from it.
   a Explain why filtering won't work.
   b Which method will you use instead?
3 Describe how you would crystallize potassium nitrate from its aqueous solution.
4 How would you separate salt and sugar? Mention any special safety precaution you would take.
5 Now see if you can think of a way to get clean sand from a mixture of sand and little bits of iron wire.

1.8 Separation methods (II)

Simple distillation

This is a way to obtain the *solvent* from a solution. The apparatus is shown on the right. It could be used to obtain water from salt water, for example. Like this:

1. The solution is heated in the flask. It boils, and steam rises into the condenser. The salt is left behind.
2. The condenser is cold, so the steam condenses to water in it.
3. The water drips into the beaker. It is called **distilled water**. It is almost pure.

You could use this method to get drinking water from sea water. Many countries in the Middle East obtain drinking water by distilling sea water in giant distillation plants.

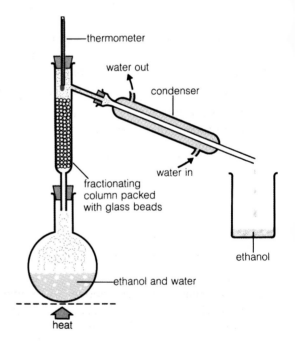

Fractional distillation

This is a way to separate *a mixture of liquids* from each other. It makes use of their different boiling points. You could use it to separate a mixture of ethanol and water, for example. The apparatus is shown on the right.

These are the steps:
1. The mixture is heated. At about 78°C, the ethanol begins to boil. Some water evaporates too, at that temperature. So a mixture of ethanol vapour and water vapour rises up the column.
2. The vapours condense on the glass beads in the column, making them hot.
3. When the beads reach about 78°C, ethanol vapour no longer condenses on them. Only the water vapour does. So water drips back into the flask, while the ethanol vapour goes into the condenser.
4. There it condenses. Pure liquid ethanol drips into the beaker.
5. Eventually, the thermometer reading rises above 78°C. This is a sign that all the ethanol has gone. So heating can be stopped.

Fractional distillation in industry

Fractional distillation is very important in industry.

It is used in the oil industry to **refine** crude oil into groups of similar compounds. The oil is heated and the vapours rise to different heights, up a tall steel fractionating column. (See page 241 for more.)

It is also used in producing **ethanol**, in industry. Some ethanol is made by fermenting sugar cane and other plant material. When fermentation is over, the fermented liquid is run off. Then it undergoes fractional distillation, to separate the ethanol from the other substances in it. The ethanol is used as a solvent, and as fuel for cars – on its own or mixed with petrol. (See page 250 for more.)

An oil refinery. Petrol is produced here, as well as many other useful chemicals.

Paper chromatography

This method can be used to separate a mixture of substances. For example, you could use it to find out how many different coloured substances there are in black ink:

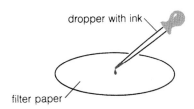

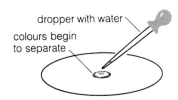

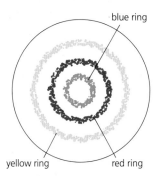

1 A drop of black ink is put at the centre of a piece of filter paper, and allowed to dry. Three or four more drops are added on the same spot.

2 Water is then dripped on to the ink spot, one drop at a time. The ink slowly spreads out and separates into rings of different colours.

3 Suppose there are three rings: yellow, red and blue. This shows that the ink contains three substances, coloured yellow, red and blue.

The substances in the ink travel across the paper at different rates. That's why they separate into rings. The filter paper showing the separate substances is called a **chromatogram**.

Paper chromatography can also be used to **identify** substances. For example, mixture X is thought to contain substances A, B, C, and D, which are all soluble in propanone. The mixture could be checked like this:

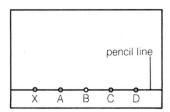

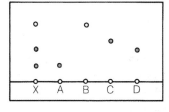

1 Concentrated solutions of X, A, B, C, and D are made up in propanone. A spot of each is placed on a line, on a sheet of filter paper, and labelled.

2 The paper is stood in a little propanone, in a covered glass tank. The solvent rises up the paper; when it's near the top, the paper is taken out again.

3 X has separated into three spots. Two are at the same height as A and B, so X must contain substances A and B. Does it also contain C and D?

The substances shown here are all coloured. You can also use paper chromatography for *colourless* substances, and to identify a substance from tables, by *measuring* how far it travels. See the next unit for more.

Questions
1 How would you obtain pure water from sea water? Draw the apparatus, and explain how the method works.
2 Why are *condensers* called that? What's the cold water for?
3 You would not use *exactly* the same apparatus you described in **1**, to separate ethanol and water. Why not?
4 Explain how fractional distillation works.
5 In the last chromatogram above, how can you tell that X does not contain substance C?
6 Look at the first chromatogram above. Can you think of a way to separate the coloured substances from the paper?

1.9 More about chromatography

How chromatography works

You already met paper chromatography. There are other kinds too. All of them depend on the interaction between:

- a non-moving or **stationary** phase (such as paper, plastic beads, or a coating on a glass plate)
- a moving or **mobile** phase. This consists of the mixture you want to separate, dissolved in a solvent. (This phase can be a liquid or a gas.)

1 Here a mixture of two substances is shown by coloured dots. The mixture is dissolved in a solvent, forming the mobile phase.

2 The two substances travel over the stationary phase at different speeds – because they have different levels of attraction to it.

3 Eventually they get completely separated from each other. Now you can collect each substance, and/or identify it.

Making use of chromatography

Chromatography is really useful. You can use it to:

- separate mixtures of substances
- purify a substance, by separating it from its impurities
- identify a substance.

Example: Identify substances in a colourless mixture

On page 21, paper chromatography was used to identify *coloured* substances. Now for a bigger challenge!

Test tubes **A – E** on the right below contain five *colourless* solutions of compounds called amino acids. **A** contains several amino acids. The others contain just one each. Your task is to identify *all* the amino acids in **A – E**.

1 Place a spot of each solution along a line drawn on slotted filter paper, as shown here. (The slots are to keep the samples separate.) Label each spot in pencil at the *top* of the paper.

A scientist using gas chromatography to identify the substances in a mixture. The sample goes into the machine on the left of the screen. From the peaks on the graph, she can tell what is in it.

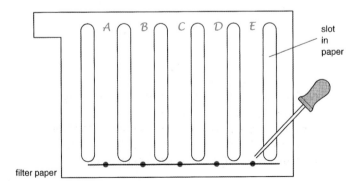

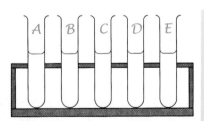

The five mystery solutions. The solvent in each is water.

2. Place a suitable solvent in the bottom of a beaker. (A mixture of water, ethanoic acid and butanol is suitable.)
3. Roll the filter paper into a cylinder and place it in the beaker. Cover the beaker.
4. The solvent rises up the paper. When it has almost reached the top, remove the paper.
5. Mark a line on it to show where the solvent reached. (You can't tell where the amino acids have reached because they are colourless.)
6. Put the paper in an oven to dry out.
7. Next spray it with a **locating agent** to make the amino acids show up. **Ninhydrin** is a good choice. (Use it in a fume cupboard!) After spraying, heat the paper in the oven for 10 minutes. The spots turn purple. So now you have a proper chromatogram. (*Chroma* means *colour*.)
8. Mark a pencil dot at the centre of each spot. Measure from the base line to each dot, and to the line showing the final solvent level.

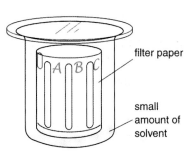

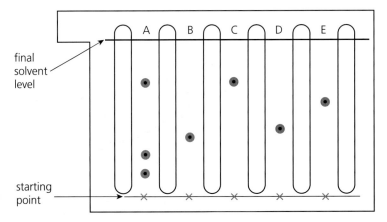

9. Now work out the R_f value for each amino acid. Like this:

$$R_f \text{ value} = \frac{\text{distance moved by amino acid}}{\text{distance moved by solvent}}$$

10. Finally, look up R_f tables to identify the amino acids. Part of an R_f table for the solvent you used is shown on the right. The method works because: **the R_f value of a compound is always the same for a given solvent, under the same conditions.**

R_f **values for amino acids** (for water/butanol/ethanoic acid as solvent)

amino acid	R_f value
cysteine	0.08
lysine	0.14
glycine	0.26
serine	0.27
alanine	0.38
proline	0.43
valine	0.60
leucine	0.73

Questions

1. Say what each term means, in chromatography:
   a. stationary phase
   b. mobile phase
2. Explain in your own words how chromatography works.
3. a. What do you think a *locating agent* is?
   b. Why is one needed in the experiment above?
4. For the chromatogram above:
   a. Were any of the amino acids in B–E also present in A? How can you tell at a glance?
   b. Using a ruler, work out the R_f values for the amino acids in A–E.
   c. Now use the R_f table above to name them.

Questions on Chapter 1

1. A large crystal of potassium manganate(VII) was placed in the bottom of a beaker of cold water and left for several hours.

   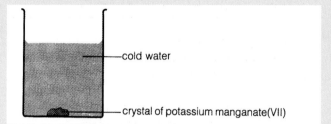

   a Describe what would be seen:
     i after five minutes   ii after several hours
   b Explain your answers using the idea of particles.
   c Name the two processes that have taken place during the experiment.

2. Use the idea of particles to explain why:
   a solids have a definite shape
   b solids cannot be poured
   c liquids fill the bottom of a container
   d you can't store gases in open containers
   e you can't squeeze a sealed plastic syringe that is completely full of water
   f a balloon expands as you blow into it.

3. The graph below is a heating curve for a pure substance. It shows how the temperature rises with time, when the solid is heated steadily until it melts, and then the liquid is heated until it boils.

   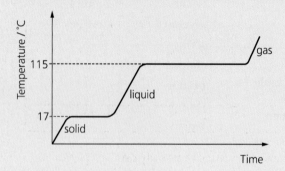

   a What is the melting point of the substance?
   b What is its boiling point?
   c What happens to the temperature while the substance changes state?
   d The graph shows that the substance takes longer to boil than to melt. Give a reason for this.
   e How can you tell that the substance is not water?
   f Sketch a rough heating curve for pure water.

4. A **cooling curve** is the opposite of a heating curve. It shows how the temperature of a substance changes with time, as it is cooled from a gas to a solid. Here is the cooling curve for one substance:

   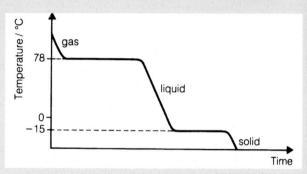

   a What is the state of the substance at room temperature (20°C)?
   b Use the list of melting and boiling points on page 9 to identify the substance.

5. Using the idea of particles explain why:
   a the smell of burnt food can travel all over the house
   b when two solids are placed on top of each other, they do not mix
   c a liquid is used in a car's braking system
   d pumping up your bike tyres quite hard gives a smooth ride
   e compressing a gas into half the volume will double its pressure
   f poisonous gases from a factory chimney can affect a large area.

6. Ammonia is an alkaline gas that turns litmus blue. It is lighter than air. A test tube of ammonia gas is placed over a test tube of air, like this:

   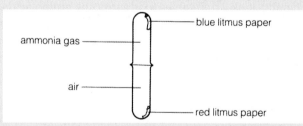

   a After a short time the red litmus paper in the lower test tube turns blue. Explain why.
   b Would it make any difference if you reversed the test tubes? Explain your answer.
   c What will you see if the test tube of air is replaced by one containing hydrogen chloride?

7 a Which of these are examples of diffusion?
   i   a helium-filled balloon rising in air
   ii a hydrogen-filled balloon deflating, due to gas passing through the skin
   iii the smell of perfume from a person standing on the other side of a room
   iv sucking a drink from a bottle, using a straw
   v   an ice lolly turning liquid when left out of the freezer
   vi all the tea in the cup changing colour when you add milk
   vii a heavy, coloured gas spreading down through a gas jar
   viii rice going soft when left in hot water
   ix a blue crystal forming a blue solution, when it is left sitting in a glass of water.
  b For *one* of the examples of diffusion, draw a diagram showing the particles before and after diffusion has taken place.

8 The rate of diffusion of a gas can be measured using this apparatus:

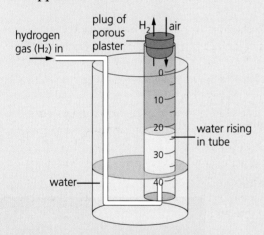

The glass tube is sealed at one end with a plug of plaster. This has tiny holes in it, just large enough to let gases pass through.
Water will rise in the tube if a gas escapes from the tube faster than air enters it. (Air is mainly nitrogen and oxygen.)
  a The water level in the tube rises, when hydrogen is the gas used. Why?
  b What does this tell you about the rate of diffusion of hydrogen compared to air?
  c Explain your answer to b using the idea of particle mass.
  d The molecules in carbon dioxide are heavier than those in nitrogen and oxygen.
    So what do you think will happen to the level of the water in the tube, if the gas in the tube is carbon dioxide? Explain your answer.

9

Gas	Formula	Relative atomic or molecular mass
methane	CH_4	16
helium	He	4
oxygen	O_2	32
nitrogen	N_2	28
chlorine	Cl_2	71

Look at the table above.
  a Which two gases will mix fastest? Explain.
  b Which gas will take least time to escape from a gas syringe?
  c Would you expect chlorine to diffuse more slowly than the gases in air? Explain.
  d An unknown gas diffuses faster than nitrogen, but more slowly than methane. What you can say about its relative molecular mass?

10 A mixture of salt and sugar has to be separated, using the solvent ethanol.
  a Which of the two substances is soluble in ethanol?
  b Draw a diagram to show how you would separate the salt.
  c How could you obtain sugar crystals from the sugar solution, *without* losing the ethanol?
  d Draw a diagram of the apparatus for c.

11 In a chromatography experiment, eight coloured substances were spotted on to a piece of filter paper. Three were the basic colours red, blue, and yellow. The other five were dyes, labelled A–E. The resulting chromatogram is shown below.

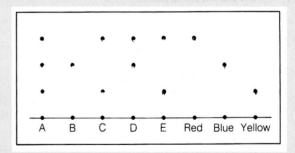

  a Which dye contains only one basic colour?
  b Which dye contains all three basic colours?
  c Which basic colour is the most soluble in propanone?

12 You have three colourless solutions. Each contains an amino acid you must identify.
Explain how to do this using chromatography. (Use the terms R_f and *locating agent* in your answer, and show that you understand what they mean.)

2.1 Atoms, elements, and compounds

Atoms

Sodium is made of tiny particles called sodium atoms.

Diamond is made of carbon atoms – different from sodium atoms.

Mercury is made of mercury atoms – different again!

Atoms are incredibly small. And mostly empty space! Each atom consists of a **nucleus** and a cloud of particles called **electrons** that whizz around the nucleus. The drawing on the right shows what a sodium atom might look like, magnified about 100 million times.

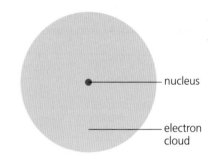

The elements

Sodium is made of sodium atoms only, so it is an **element**.
An element contains only one kind of atom.

Around 90 elements have been found in the Earth and atmosphere. Scientists have made nearly 30 others in the lab. Many of the 'artificial' elements last just a few seconds before breaking down into other elements.

Symbols for the elements

To make life easy, each element has a symbol. These are usually taken from the English or Latin name. For example the symbol for carbon is C. The symbol for potassium is K, from its Latin name *kalium*. Some elements are named after the people who discovered them.

The Periodic Table of the elements

Look at the table on the next page. It is called the **Periodic Table**. It is like a map and address book for the elements, all rolled into one.

- It gives the names and symbols for the elements.
- Look at the numbered columns. The elements in these form families or **groups**, with similar properties. If you know how one element in Group 1 behaves, for example, you can make a good guess about how the others in that group will behave.
- The rows are called periods.
- Look at the zig-zag line. It separates metals from non-metals, with the non-metals on the right. So most of the elements are metals.

Now look at the small numbers to the left of each symbol. These give vital information about the atoms of the element, as you will see later.

The elements have been 'discovered' one by one, over the centuries. This painting shows the alchemist who discovered phosphorus in the 17th century. To his amazement it glows in the dark.

The Periodic Table

[Periodic table image showing groups 1-7 and 0, with transition elements, lanthanides and actinides]

Compounds

Elements can combine with each other to form **compounds**.
A compound contains atoms of different elements joined together.
We say the atoms are **chemically combined**.
There are millions of compounds. This table shows three common ones:

Name of compound	Elements in it	How the atoms are joined up
water	hydrogen and oxygen	H–O–H
carbon dioxide	carbon and oxygen	O–C–O
ethanol	carbon, hydrogen, and oxygen	H₃C–CH₂–OH (C₂H₆O structure)

Symbols for compounds

The symbol for a compound is called its **formula**. It is made from the symbols of the elements in it. So the formula for water is H_2O.
Note that the plural of **formula** is **formulae**.

Questions

1. What is: **a** an atom? **b** an element?
2. Which element has this symbol? **a** Ca **b** Mg **c** N
3. Using the Periodic Table to help you, name three *metals* that you expect to behave in a similar way.
4. What's the difference between an element and a compound?
5. The formula for water is H_2O. What does the $_2$ tell you?
6. Using the table above, see if you can give the formula for:
   a carbon dioxide **b** ethanol

2.2 More about atoms

Protons, neutrons, and electrons

Atoms consist of a **nucleus** and a cloud of **electrons** that move round the nucleus. The nucleus is itself a cluster of two kinds of particles, **protons** and **neutrons**.

All the particles in an atom are very light. So their mass is measured in **atomic mass units**, rather than grams. Protons and electrons also have an **electric charge**:

Particle in atom	Mass	Charge
proton ●	1 unit	positive charge (1 +)
neutron ●	1 unit	none
electron •	almost nothing	negative charge (1 –)

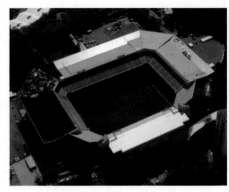

The nucleus is very tiny compared with the rest of the atom. If the atom was the size of a football stadium, the nucleus would be the size of a pea!

How the particles are arranged

The sodium atom is a good one to start with. It has **11** protons, **11** electrons, and **12** neutrons. They are arranged like this:

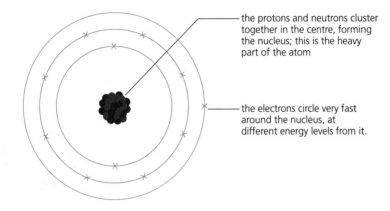

- the protons and neutrons cluster together in the centre, forming the nucleus; this is the heavy part of the atom
- the electrons circle very fast around the nucleus, at different energy levels from it.

Atomic number

A sodium atom has 11 protons. This can be used to identify it, since *only* a sodium atom has 11 protons. Every other atom has a different number of protons.
You can identify an atom by the number of protons in it.

The number of protons in an atom is called its **atomic number** or **proton number**. The atomic number of sodium is 11.

How many electrons?

The sodium atom also has 11 electrons. So it has an equal number of protons and electrons. The same is true for every sort of atom:
Every atom has an equal number of protons and electrons.
So atoms have no overall charge.

Look at the box on the right. You can see that the total charge on the electrons cancels out the total charge on the protons, for the sodium atom.

The charge on a sodium atom:
●●●● 11 protons
●●●● Each has a charge of 1 +
●●● Total charge 11 +

× × × × 11 electrons
× × × × Each has a charge of 1 –
× × × Total charge 11 –

Adding the charges: 11 +
                                    11 –
                                    ———
                                      0

The answer is zero.
The atom has no overall charge.

Nucleon number

Protons and neutrons are in the nucleus of the atom, so are called **nucleons**. The total number of protons and neutrons in an atom is called its **nucleon number**. The nucleon number for the sodium atom is 23. (11 + 12 = 23)

In the Periodic Table, sodium is shown as: $^{23}_{11}Na$.

This tells you that sodium atoms have a nucleon number of 23 and an atomic number of 11. So they must have 12 neutrons. (23 − 11 = 12)

> **Try it yourself!**
> You can describe any element in a short way like this:
>
> nucleon number
> atomic number symbol
>
> For example: $^{16}_{8}O$

The atoms of the first 20 elements

In the Periodic Table, the elements are arranged in order of increasing atomic number. Here are the first 20 elements:

Element	Symbol	Atomic number (protons)	Electrons	Neutrons	Nucleon number (protons + neutrons)
hydrogen	H	1	1	0	1
helium	He	2	2	2	4
lithium	Li	3	3	4	7
beryllium	Be	4	4	5	9
boron	B	5	5	6	11
carbon	C	6	6	6	12
nitrogen	N	7	7	7	14
oxygen	O	8	8	8	16
fluorine	F	9	9	10	19
neon	Ne	10	10	10	20
sodium	Na	11	11	12	23
magnesium	Mg	12	12	12	24
aluminium	Al	13	13	14	27
silicon	Si	14	14	14	28
phosphorus	P	15	15	16	31
sulphur	S	16	16	16	32
chlorine	Cl	17	17	18	35
argon	Ar	18	18	22	40
potassium	K	19	19	20	39
calcium	Ca	20	20	20	40

Questions
1. Name the particles that make up the atom.
2. Which particle has:
   a. a positive charge? b. no charge? c. almost no mass?
3. An atom has 9 protons. Which element is it?
4. Why do atoms have no overall charge?
5. What does this term mean?
   a. atomic number b. nucleon number
6. Name each of these atoms, and say how many protons, electrons, and neutrons it has:
   $^{12}_{6}C$ $^{16}_{8}O$ $^{24}_{12}Mg$ $^{27}_{13}Al$ $^{64}_{29}Cu$

2.3 Isotopes and radioactivity

How to identify an atom: a reminder

Only sodium atoms have 11 protons.
You can identify an atom by the number of protons in it.

Isotopes

All carbon atoms have 6 protons. But not all carbon atoms are identical. Some have more *neutrons* than others.

6 protons, 6 electrons, 6 neutrons	6 protons, 6 electrons, 7 neutrons	6 protons, 6 electrons, 8 neutrons
Most carbon atoms are like this, with **6 neutrons**. That makes **12** nucleons (protons + neutrons) in total, so it is called **carbon-12**.	But about one in every hundred carbon atoms is like this, with **7** neutrons. It has **13 nucleons** in total, so is called **carbon-13**.	And a very tiny number of carbon atoms are like this, with **8 neutrons**. It has **14 nucleons** in total, so is called **carbon-14**.

The three atoms above are called **isotopes** of carbon.
Isotopes are atoms of the same element, with different numbers of neutrons.

Most elements have isotopes. For example calcium has six, magnesium has three, iron has four, and chlorine has two.

Some isotopes are radioactive

A carbon-14 atom behaves in a strange way. It is **radioactive**. That means its nucleus is unstable. Sooner or later the atom breaks down or **decays**, giving out **radiation** in the form of rays and tiny particles, as well as a large amount of energy.

Like carbon, a number of other elements have radioactive isotopes – or **radioisotopes** – that occur naturally, and eventually decay.

But the other two isotopes of carbon (like most natural isotopes) are **non-radioactive**.

How fast do radioisotopes decay?

Radioisotopes decay at random. We can't tell whether a given atom of carbon-14 will decay in the next few seconds, or in a thousand years. But we do know how long it takes for half the radioisotopes in a sample to decay. This is called the **half-life**.

The half-life for carbon-14 is 5730 years. So if you have 100 carbon-14 atoms, 50 of them will have decayed 5730 years from now.

Half-lives vary a lot. For example:
    for radon-220   55.5 seconds
    for cobalt-60   5.26 years
    for potassium-40   1300 million years

Radiation is very dangerous, so you must handle radioisotopes with great care. This scientist is using a glove box, for safety.

Radiation can harm you

If the radiation from radioisotopes gets into your body, it will kill body cells. A large dose causes **radiation sickness**. Victims vomit a lot, and feel really tired. Their hair falls out, their gums bleed, and they die within weeks. Even small doses of radiation over a long period will cause cancer.

Making use of radioisotopes

Radioisotopes are dangerous – but they are also useful. For example:

To check for leaks Engineers can check oil and gas pipes for leaks by adding radioisotopes to the oil or gas. At a leak, the radiation is detected using an instrument called a **Geiger counter**. Radioisotopes used in this way are called **tracers**.

In cancer treatment Radioisotopes can cause cancer. But they are also used in **radiotherapy** to *cure* cancer – because the rays they give out on decay will kill cancer cells. The radioisotope cobalt-60 is usually used for this. The beam of rays is aimed exactly at its target in the body.

Checking for leaks in a gas pipe using a Geiger counter. A tracer has been added to the gas.

To find the age of old remains Living things (including us) are built largely of carbon atoms. A tiny percentage of these are carbon-14. When a living thing dies, it no longer takes in new carbon atoms. But its existing carbon-14 atoms decay over time – and we can measure the faint radiation from them.

Experts use this method to work out the age of ancient fragments of wood, bones, and cloth, that were all once living things. It is called **carbon dating**.

Using carbon dating, this Egyptian mummy was found to be around 5300 years old. The body was buried in the hot desert, which helped to preserve it. The plates and bowls had food and drink for the journey to the next world.

Questions

1. **a** What is an *isotope*?
   b Name the three isotopes of carbon, and write symbols for them.
2. Carbon-14 is *radioactive*. What does that mean?
3. What is a *radioisotope*?
4. **a** Radiation can kill us. Why?
   b So why are radioisotopes used to treat cancer?
5. We take in some carbon-14 when we breathe. Suggest a reason why it does not harm us.
6. What is the *half-life* of a radioisotope?
7. Radioisotopes can be used to check pipes for leaks.
   a Explain how this works.
   b Do you think it would be better to use a radioisotope with a long half-life, or a short one, for this? Explain.

2.4 How electrons are arranged

Electron shells

The electrons in an atom circle fast around the nucleus, at different **energy levels** from it. These energy levels are called **electron shells**.

The first shell, closest to the nucleus, is the lowest energy level. The further a shell is from the nucleus, the higher the energy level.

Electrons occupy the lowest available energy level. But they can't all crowd into the first shell, because a shell can hold only a limited number of electrons, like this:

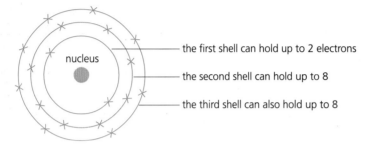

- the first shell can hold up to 2 electrons
- the second shell can hold up to 8
- the third shell can also hold up to 8

Niels Bohr, a Danish scientist, was the first person to put forward the idea of electron shells. He died in 1962.

So the electrons fill up the shells one by one, starting with the first shell. When a shell is full they start a new one.

The electron shells for the first 20 elements

This is how the shells fill for the first 20 elements of the Periodic Table:

Element	Symbol	Atomic number	1st shell	2nd shell	3rd shell	4th shell
hydrogen	H	1	1			
helium	He	2	2			
lithium	Li	3	2	1		
beryllium	Be	4	2	2		
boron	B	5	2	3		
carbon	C	6	2	4		
nitrogen	N	7	2	5		
oxygen	O	8	2	6		
fluorine	F	9	2	7		
neon	Ne	10	2	8		
sodium	Na	11	2	8	1	
magnesium	Mg	12	2	8	2	
aluminium	Al	13	2	8	3	
silicon	Si	14	2	8	4	
phosphorus	P	15	2	8	5	
sulphur	S	16	2	8	6	
chlorine	Cl	17	2	8	7	
argon	Ar	18	2	8	8	
potassium	K	19	2	8	8	1
calcium	Ca	20	2	8	8	2

Note that only three of these elements, helium, neon, and argon, have atoms with full outer shells.

Electronic configuration and the Periodic Table

Electronic configuration means the arrangement of electrons in an atom. The electronic configurations for the first 20 elements are shown below, in rows and columns, to match the Periodic Table:

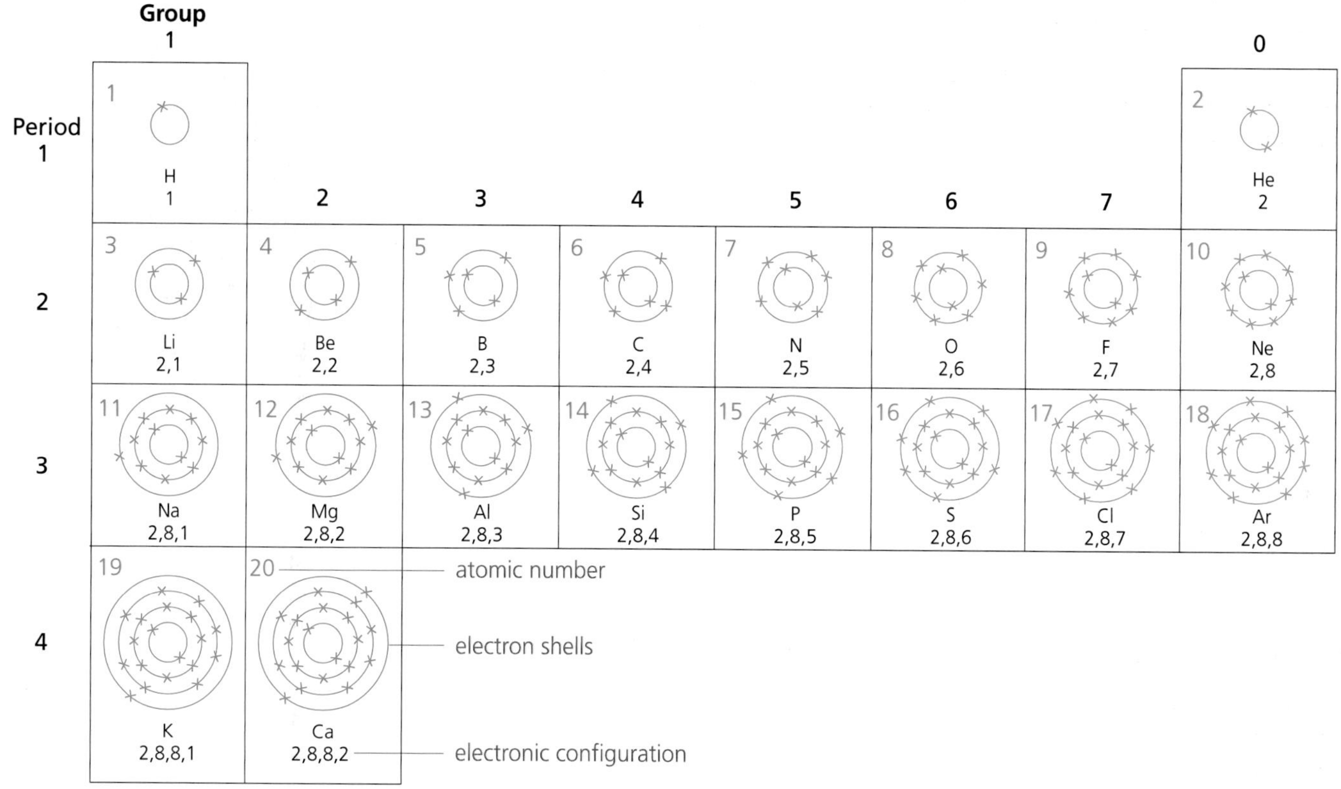

Period	Group 1	2	3	4	5	6	7	0
1	1 H 1							2 He 2
2	3 Li 2,1	4 Be 2,2	5 B 2,3	6 C 2,4	7 N 2,5	8 O 2,6	9 F 2,7	10 Ne 2,8
3	11 Na 2,8,1	12 Mg 2,8,2	13 Al 2,8,3	14 Si 2,8,4	15 P 2,8,5	16 S 2,8,6	17 Cl 2,8,7	18 Ar 2,8,8
4	19 K 2,8,8,1	20 Ca 2,8,8,2						

Labels: atomic number, electron shells, electronic configuration

Note these important points:

- The shells fill in order, the innermost one first (lowest energy level).
- All the elements in a group have the same number of electrons in their outer shells. These are called the **valency electrons**.
- The group number *is the same as* the number of outer electrons, *except* for Group 0, where the atoms have full shells.
- The period number shows how many shells there are.
- The valency electrons dictate how an element reacts. So the elements in Group I all have similar reactions. (More about this later.)
- Because of their full shells, the elements of Group 0 are unreactive.

Questions

1. An atom has 13 electrons. Draw a diagram to show how its electrons are arranged. Which element is it?
2. What does *electronic configuration* mean?
3. For sodium, the electronic configuration is 2,8,1. What is it for: **a** lithium? **b** magnesium? **c** hydrogen?
4. An atom has 5 electrons in its outer shell. To which group of the Periodic Table does it belong?
5. How many shells of electrons do atoms of Period 3 have?
6. Which of the 20 elements have full outer shells of electrons? Write the electronic configuration for each of them.

Questions on Chapter 2

1

Particle	Electrons	Protons	Neutrons
A	12	12	12
B	12	12	14
C	10	12	12
D	10	8	8
E	9	9	10

The table above describes some particles.
a Which three particles are neutral atoms?
b Which particle is a negative ion? What is the charge on this ion?
c Which particle is a positive ion? What is the charge on this ion?
d Which two particles are isotopes?
e Use the table on page 29 to identify the particles A to E.

2 In water (H_2O) the atoms are joined as shown in this diagram:

H—O—H

See if you can suggest how the atoms are joined in the following compounds:
a sulphur dioxide, SO_2
b methane, CH_4
c methanol, CH_4O
d hydrogen peroxide, H_2O_2
e ammonia, NH_3

3 The following statements are about the particles that make up the atom. For each statement write:
  p if it describes the **proton**
  e if it describes the **electron**
  n if it describes the **neutron**
a the positively-charged particle
b the particle found with the proton, in the nucleus
c the particle that can occur in different numbers, in atoms of the same element
d held in shells around the nucleus
e the negatively-charged particle
f the particle with negligible mass
g the number of these particles is found by subtracting the atomic number from the nucleon number
h the particle with no charge
i the particle with the same mass as a neutron
j the number of these particles is the atomic number

4 This set of letters represents the atoms of one element. ^y_zX
a What does this letter stand for?
  i X ii y iii z
b How many neutrons are there in the following atoms:
  i $^{107}_{47}Ag$? ii $^{63}_{29}Cu$? iii ^1_1H? iv $^{20}_{10}Ne$? v $^{238}_{92}U$?

5 Hydrogen, deuterium, and tritium are isotopes. Their structures are shown below.

hydrogen   deuterium   tritium

a Copy and complete the following key:
  ⊗ represents _____
  ○ represents _____
  x represents _____
b What are the nucleon numbers of hydrogen, deuterium, and tritium?
c Copy and complete this statement:
  Isotopes of an element always contain the same number of _____ and _____, but different numbers of _____.
d The average mass number of naturally-occurring hydrogen is 1.008. Which isotope is present in the highest proportion, in naturally-occurring hydrogen?

6 Copy and complete the following table for isotopes of some common elements.

Isotope	Name of element	Proton number	Nucleon number	Number of p	e	n
$^{16}_8O$	oxygen	8	16	8	8	8
$^{18}_8O$						
$^{12}_6C$						
$^{13}_6C$						
$^{24}_{12}Mg$						
$^{25}_{12}Mg$						
				17	17	18
				17	17	20

7 For each of the six elements aluminium (Al), boron (B), nitrogen (N), oxygen (O), phosphorus (P), and sulphur (S), write down:
  a i which period of the Periodic Table it belongs to
    ii its group number in the Periodic Table
    iii its atomic number
    iv the number of electrons in its atoms
    v its electronic configuration
    vi the number of outer electrons in its atoms
  b The outer electrons are also called the _____ electrons. What is the missing word? (7 letters!)
  c Which of the above elements would you expect to have similar properties? Why?

8 Boron has two types of atom, shown below.

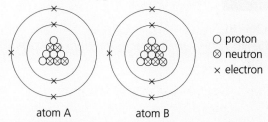

atom A   atom B

○ proton
⊗ neutron
× electron

  a What is different about these two atoms?
  b What name is given to atoms like these?
  c Describe each atom in shorthand form, as in 4.
  d What is the nucleon number of atom A?
  e i What is the difference in nucleon number for atoms A and B?
    ii Is atom B heavier, or lighter, than atom A?
  f i Give the electronic configuration for A and B.
    ii Comment on this.

9 The two metals sodium (atomic number 11) and magnesium (atomic number 12) are found next to each other in the Periodic Table.
  a Say whether this is the same, or different, for their atoms:
    i the number of electron shells
    ii the number of outer (valency) electrons
  The relative atomic mass of sodium is 23.0.
  The relative atomic mass of magnesium is 24.3.
  b Which of the two elements may exist naturally as a single isotope? Explain your answer.

10 Strontium, atomic number 38, is in the fifth period of the Periodic Table. It belongs to Group 2. Copy and complete the following.
  An atom of strontium has:
  a .... electrons
  b .... shells of electrons
  c .... electrons in its outer shell

11 This diagram represents the electronic arrangement in an atom of an element:

  a i Give the electronic configuration for the atom.
    ii What is special about this configuration?
  b Which group of the Periodic Table does the element belong to?
  c Name another element with the same number of outer-shell electrons in its atoms.

12 Gallium exists naturally as a mixture of two non-radioactive isotopes, gallium-69 and gallium-71. The atomic number of gallium is 31.
  a i How many neutrons are there in gallium-69?
    ii How many neutrons are there in gallium-71?
  Gallium also has a radioactive isotope, gallium-67. This has a half-life of 78.1 hours.
  b What is meant by the *half-life*?
  As gallium-67 decays, it releases rays called gamma rays.
  c Name two possible uses, one medical and one non-medical, for gallium-67.

13 a Copy and complete this table

Element	Proton number	Nucleon number
argon		
calcium		
chlorine		
potassium		
sulphur		

  b Which number (proton or nucleon) decides the position of an element in the Periodic Table?
  c i Write the elements from the table above in order of increasing proton number.
    ii Repeat i for the nucleon number.
    iii For which two elements are the orders different?
    iv Explain the difference you noticed in iii.
    v In the Periodic Table on page 27, find another pair of elements that shows a similar difference in order.
  d From the table above, choose the element which:
    i has atoms with a full outer electron shell.
    ii has atoms with the same number of outer-shell electrons as sodium atoms do.
    iii is in Group 6 of the Periodic Table.

3.1 Why compounds form

Most elements form compounds

Sodium and chlorine are both **elements**. When sodium is heated and placed in a jar of chlorine, it burns with a bright flame.

The result is a white solid that has to be scraped from the sides of the jar. It looks completely different from both sodium and chlorine.

The white solid is **sodium chloride**. It is formed from atoms of sodium and chlorine that are chemically combined, so it is a **compound**. The reaction can be described by a word equation like this:

sodium + chlorine → sodium chloride

The + means *reacts with*, and the → means *to form*.
Most elements react to form compounds. For example:

lithium + chlorine → lithium chloride
hydrogen + chlorine → hydrogen chloride

The Group 0 elements do not usually form compounds

The Group 0 elements are different from other elements: they do not usually form compounds. So their atoms are described as **unreactive** or **stable**. They are stable because their outer electron shells are full.
A full outer shell makes an atom stable.

Helium atom, full outer shell: *stable*

Neon atom, full outer shell: *stable*

Argon atom, full outer shell: *stable*

  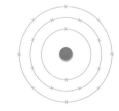

2   2,8   2,8,8

Only the atoms of the Group 0 elements have full outer shells. The atoms of all other elements have incomplete outer shells. That is why they react.
By reacting with each other, atoms can obtain full outer shells and so become stable.

The Group 0 elements are all gases. They are called **the noble gases** because they are unreactive. They glow when a current is passed through them, so they are used in lighting. For example xenon is used in lighthouse lamps.

How sodium atoms get a full outer shell

A sodium atom has just 1 electron in its outer shell. It can obtain a full outer shell by losing this electron to another atom. It becomes a **sodium ion**:

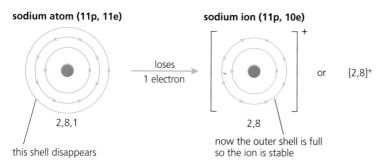

The sodium ion has 11 protons but only 10 electrons, so it has a charge of 1+, as you can see from the panel on the right.

The symbol for sodium is Na, so the symbol for the sodium ion is **Na⁺**.

The + means *1 positive charge*. Na⁺ is a **positive ion**.

The charge on a sodium ion:

charge on 11 protons	11+
charge on 10 electrons	10 −
total charge	1+

How chlorine atoms get a full outer shell

A chlorine atom has 7 electrons in its outer shell. It can reach a full shell by accepting an electron from another atom. It becomes a chloride ion:

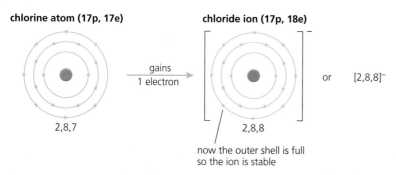

The chloride ion has a charge of 1 −, so it is a **negative ion**. Its symbol is **Cl⁻**.

Ions

An atom becomes an ion when it loses or gains electrons.
An ion is a charged particle. It is charged because it has an unequal number of protons and electrons.

The charge on a chloride ion:

charge on 17 protons	17+
charge on 18 electrons	18−
total charge	1−

Questions

1. Why are the atoms of the Group 0 elements unreactive?
2. Explain why all other atoms are reactive.
3. Draw a diagram to show how:
   a. a sodium atom obtains a full outer shell
   b. a chlorine atom obtains a full outer shell.
4. Explain why
   a. a sodium ion has a charge of 1+
   b. a chloride ion has a charge of 1 −.
5. Explain what an *ion* is, in your own words.
6. Atoms of Group 0 elements do *not* form ions. Why not?

3.2 The ionic bond

How sodium and chlorine atoms bond

A sodium atom can lose one electron, and a chlorine atom can gain one, to obtain full outer shells. So when a sodium atom and a chlorine atom react together, the sodium atom loses its electron *to the chlorine atom*, and two ions are formed.

Here, sodium electrons are shown as ● and chlorine electrons as ×, to help you see what is happening:

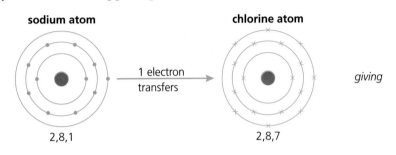
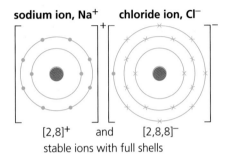

The two ions have opposite charges, so they attract each other. The force of attraction between them is strong. It is called an **ionic bond**.

How solid sodium chloride is formed

When sodium reacts with chlorine, billions and billions of sodium and chloride ions form, and are attracted to each other.

But the ions do not stay in pairs. They cluster together so that each ion is surrounded by six ions of opposite charge, as shown below. The ions are held together by strong ionic bonds.

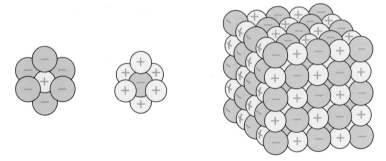

The pattern grows until a giant 3-D structure of ions is formed. It is called 'giant' because it contains many billions of ions. This giant structure is the compound **sodium chloride**, or **common salt**.

Since it is made of ions, sodium chloride is called an **ionic compound**. It contains one Na^+ ion for each Cl^- ion, so its formula is **NaCl**.

The charges in the structure add up to zero:

the charge on each sodium ion is	1+
the charge on each chloride ion is	1–
total charge	0

So the compound has no overall charge.

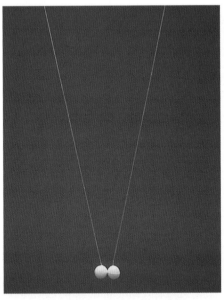

These polystyrene balls were given opposite charges. So they are attracted to each other, and cling together. The same happens with ions of opposite charge.

Other ionic compounds

Sodium is a **metal**, and chlorine is a **non-metal**. They react together to form an **ionic compound**. Other metals also react with non-metals to form ionic compounds. Below are two more examples.

Magnesium and oxygen A magnesium atom has 2 outer electrons and an oxygen atom has 6. Magnesium burns fiercely in oxygen. During the reaction, each magnesium atom loses its 2 outer electrons to an oxygen atom. Magnesium ions and oxide ions are formed:

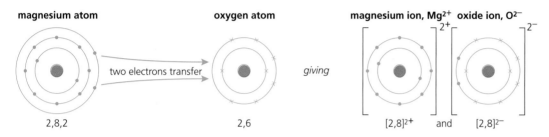

The ions attract each other because of their opposite charges. Like the sodium and chloride ions, they group to form a giant ionic structure.

The resulting compound is called **magnesium oxide**. It has one magnesium ion for each oxide ion, so its formula is **MgO**. It has no overall charge.

The charge on magnesium oxide:

charge on each magnesium ion   2+
charge on each oxide ion   2−
                    total charge   0

Magnesium and chlorine To obtain full outer shells, a magnesium atom must lose 2 electrons and a chlorine atom must gain 1 electron. So when magnesium burns in chlorine, each magnesium atom reacts with *two* chlorine atoms, to form **magnesium chloride**:

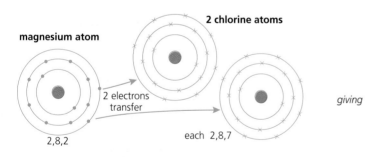

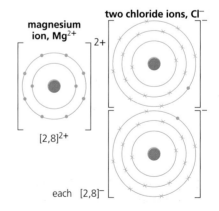

The ions form a giant ionic structure, with two chloride ions for each magnesium ion. So the formula of magnesium chloride is **MgCl$_2$**. The compound has no overall charge. Can you explain why?

Questions
1. Draw a diagram to show what happens to the electrons, when a sodium atom reacts with a chlorine atom.
2. What is an *ionic bond*?
3. Describe in your own words the structure of solid sodium chloride, and explain why its formula is NaCl.
4. Explain why:
   a. a magnesium ion has a charge of 2+
   b. the ions in magnesium oxide stay together
   c. magnesium chloride has no overall charge
   d. the formula of magnesium chloride is MgCl$_2$.

3.3 More about ions

Ions of the first twenty elements

Not every element forms ions during reactions. In fact, out of the first twenty elements in the Periodic Table, only twelve easily form ions. These ions are given below, with their names.

Group 1	2		3	4	5	6	7	0
		H^+ hydrogen						none
Li^+ lithium	Be^{2+} beryllium					O^{2-} oxide	F^- fluoride	none
Na^+ sodium	Mg^{2+} magnesium		Al^{3+} aluminium			S^{2-} sulphide	Cl^- chloride	none
K^+ potassium	Ca^{2+} calcium		transition metals					

Note that:

- hydrogen and the metals form **positive ions**, in order to gain full shells. These have the same names as the atoms.
- the non-metals form **negative ions**, and their names end in **-ide**.
- the elements in Groups 4 and 5 do not usually form ions, because their atoms would have to gain or lose several electrons, and that takes too much energy.
- Group 0 elements do not form ions; their atoms already have full shells.

The names and formulae of ionic compounds

The names   To name an ionic compound, you just put the names of the ions together, with the positive one first:

Ions in compound	Name of compound
K^+ and F^-	potassium fluoride
Ca^{2+} and S^{2-}	calcium sulphide

The formulae   The formulae of ionic compounds can be worked out using these four steps. Look at the examples that follow.

1. Write down the name of the ionic compound.
2. Write down the symbols for its ions.
3. The compound must have no overall charge, so balance the ions until the positive and negative charges add up to zero.
4. Write down the formula without the charges.

Bath salts contain Na^+ and CO_3^{2-} ions. Epsom salts contain Mg^{2+} and SO_4^{2-} ions. Can you give the names and formulae of the three main compounds in these boxes?

Example 1
1. Lithium fluoride.
2. The ions are Li^+ and F^-.
3. One Li^+ is needed for every F^-, to make the total charge zero.
4. The formula is LiF.

Example 2
1. Sodium sulphide.
2. The ions are Na^+ and S^{2-}.
3. Two Na^+ ions are needed for every S^{2-} ion, to make the total charge zero: $Na^+ Na^+ S^{2-}$.
4. The formula is Na_2S. (What does the $_2$ show?)

Some metals form more than one type of ion

Look back at the Periodic Table on page 27. Look for the block of **transition elements**. These include many common metals, such as iron and copper.

Some transition metals form only one type of ion:
- silver forms only Ag^+ ions
- zinc forms only Zn^{2+} ions.

But most transition metals can form more than one type of ion. For example, copper and iron can each form two:

Ion	Name	Example of compound
Cu^+	copper(I) ion	copper(I) oxide, Cu_2O
Cu^{2+}	copper(II) ion	copper(II) oxide, CuO
Fe^{2+}	iron(II) ion	iron(II) chloride, $FeCl_2$
Fe^{3+}	iron(III) ion	iron(III) chloride, $FeCl_3$

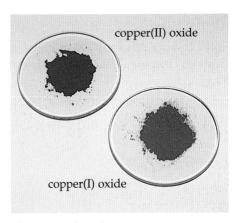

The two oxides of copper.

The (II) in a name shows that the ion has a charge of 2+. What do the (I) and (III) show?

Compound ions

All the ions you met so far have been formed from single atoms. But ions can also be formed from groups of joined atoms. These are called **compound ions**.

The most common ones are shown on the right. Remember, each is just one ion, even though it contains more than one atom.

The formulae for their compounds can be worked out as before. Some examples are shown below.

NH_4^+, the ammonium ion

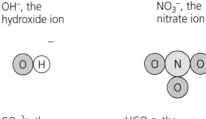

OH^-, the hydroxide ion

NO_3^-, the nitrate ion

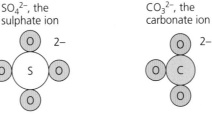

SO_4^{2-}, the sulphate ion

CO_3^{2-}, the carbonate ion

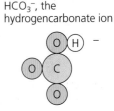

HCO_3^-, the hydrogencarbonate ion

Example 3
1. Sodium carbonate.
2. The ions are Na^+ and CO_3^{2-}.
3. Two Na^+ are needed to balance the charge on one CO_3^{2-}.
4. The formula is Na_2CO_3.

Example 4
1. Calcium nitrate.
2. The ions are Ca^{2+} and NO_3^-.
3. Two NO_3^- are needed to balance the charge on one Ca^{2+}.
4. The formula is $Ca(NO_3)_2$. Note that brackets are put round the NO_3, before the $_2$ is put in.

Questions
1. Explain why a calcium ion has a charge of 2+.
2. Why is the charge on an aluminium ion 3+?
3. Write down the symbols for the ions in:
    a potassium chloride   b calcium sulphide
    c lithium sulphide   d magnesium fluoride
4. Now work out the formula for each compound in 3.
5. Work out the formula for each compound:
    a copper(II) chloride   b iron(III) oxide
6. Write a name for each compound:
   $CuCl$, FeS, Na_2SO_4, $Mg(NO_3)_2$, NH_4NO_3, $Ca(HCO_3)_2$
7. Work out the formula for: a sodium sulphate
    b potassium hydroxide   c silver nitrate

3.4 Ionic compounds and their properties

Remember

Ionic compounds form when metals react with non-metals.
Sodium chloride (NaCl) and magnesium oxide (MgO) are examples.

The structure of ionic compounds

Sodium chloride is ordinary table salt. You have met its structure already:

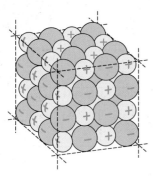

The sodium and chloride ions are arranged in a regular **lattice** like this. They are held together by strong ionic bonds.

The pattern repeats millions of times. The result is a crystal. This is a crystal of sodium chloride magnified 35 times.

The crystals look white and shiny. A box of table salt contains millions of them. There might be dozens of them in your dinner!

Sodium chloride is a typical ionic compound. In all ionic compounds, the ions are held together in a lattice by strong ionic bonds, forming a **giant structure**. Since the ions are in a regular pattern, all ionic solids are crystalline.

Note that 'giant' does not mean the lattice is huge. (Most crystals don't grow that big.) It means it contains many billions of particles.

Some ionic solids you can find in the laboratory:

sodium chloride
sodium hydroxide
copper(II) sulphate
magnesium oxide
silver nitrate

What happens when you heat an ionic compound?

It's hard to imagine salt melting – but it will if you heat it enough!

 heat energy → 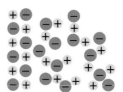 heat energy →

When you heat an ionic solid, the ions take in energy and start to vibrate to and fro in the lattice. When the temperature reaches the **melting point** …

… they vibrate so much that they break away from the lattice. Now the substance is a **liquid**, and the ions are free to move. If you keep on heating …

…some ions get enough energy to escape as a **gas**. This is called **evaporation**. When the temperature reaches the **boiling point** all the liquid boils into a gas.

If you cool sodium chloride gas, the reverse happens. The ions lose more and more energy, and eventually settle back into a lattice.

It is not only ionic solids that behave this way. You saw in Unit 1.3 that most solids are a lattice of particles, and melt and boil like this.

Left: because of its high melting point, magnesium oxide is used to line furnaces.

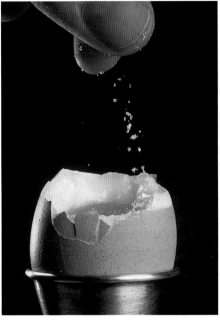

Sodium chloride is used to bring out the taste in food. Like all ionic solids it dissolves in moisture, and we taste the sodium ions. (But too much salt is bad for you.)

The properties of ionic compounds

1 Ionic compounds have high melting and boiling points.
For example:

Compound	Melting point/°C	Boiling point/°C
sodium chloride	801	1413
magnesium oxide	2852	3600

This is because the ionic bonds are very strong, so it takes a lot of heat energy to break up the lattice. That's why all ionic compounds are solid at room temperature.

Note that magnesium oxide has a far higher melting and boiling point than sodium chloride does. This is because its ions have double the charge (Mg^{2+} and O^{2-} compared with Na^+ and Cl^-), so its ionic bonds are stronger.

2 Ionic compounds are usually soluble in water.
The water molecules can attract the ions away from the lattice.
The ions can then move freely, surrounded by water molecules.

3 Ionic compounds can conduct electricity when they are melted or dissolved.
A solid ionic compound will not conduct electricity. But when it melts, the lattice breaks up and the ions are free to move. Since they are charged, this means they can conduct electricity.

The ions also move freely when the compound is dissolved in water. So solutions of ionic compounds conduct too.

Questions
1 What is a *lattice*?
2 Explain how a crystal of sodium chloride is formed.
3 A crystal of sodium chloride has flat faces. See if you can explain why.
4 Why do ionic solids have high melting points?
5 A solution of salt in water can conduct electricity. Why?
6 Explain why magnesium oxide has a higher melting point than sodium chloride does.

3.5 The covalent bond

Gaining full shells: a reminder

As you saw in Unit 3.2, atoms react in order to gain full outer shells. For example when sodium and chlorine react together, each sodium atom gives up an electron to a chlorine atom.

But that is not the only way to gain full shells. As you'll see below, atoms can also gain full shells by *sharing* electrons with each other.

Sharing electrons

When two non-metal atoms react together, *both need to gain electrons* to reach full shells. They can manage this only by sharing electrons.

We will look at non-metal elements in this unit, and at non-metal compounds in the next unit. Atoms can share only their outer (valence) electrons, so the diagrams will show only these.

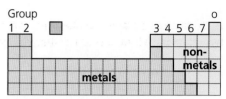

Atoms of non-metals don't *give up* electrons to gain a full shell, because they would have to lose so many. It would take too much energy to overcome the pull of the positive nucleus.

Hydrogen

A hydrogen atom has only one electron. Its shell can hold two electrons, so is not full. When two hydrogen atoms get close enough, their shells overlap and then they can share electrons. Like this:

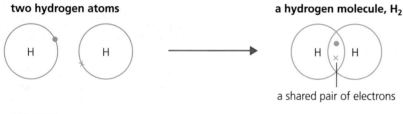

By sharing, each hydrogen atom gains a full shell of two electrons.

The bond between the atoms

Each hydrogen atom has a positive nucleus. Both nuclei attract the shared electrons – and it is this strong force of attraction that holds the two atoms together. This force of attraction is called a **covalent bond**.

Molecules

The two bonded hydrogen atoms above form a **molecule**.
A molecule is a group of atoms held together by covalent bonds.

Since it is made up of molecules, hydrogen is a **molecular** element. Its formula is H_2. The $_2$ tells you there are 2 hydrogen atoms in each molecule.

Many other non-metals are also molecular. For example:

chlorine, Cl_2 iodine, I_2 oxygen, O_2

nitrogen, N_2 sulphur, S_8 phosphorus, P_4

Elements made up of molecules containing two atoms are called **diatomic**. Hydrogen and chlorine are diatomic. Can you give three other examples?

A model of the hydrogen molecule. The molecule can also be shown as H–H. The line represents a single bond.

Chlorine

A chlorine atom needs a share in one more electron, to obtain a full shell. So two chlorine atoms bond covalently like this:

A model of the chlorine molecule. You can also show the molecule as Cl–Cl. The line represents a single bond.

Since only one pair of electrons is shared, the bond between the atoms is called a **single covalent bond**, or just a **single bond**. You can show it in a short way by a single line, like this: Cl—Cl.

Oxygen

The formula for oxygen is O_2, so each molecule must contain two atoms. Each oxygen atom has six outer electrons, so needs a share in *two* more to reach a full shell:

A model of the oxygen molecule. You can also show the molecule as O=O. The two lines represent a double bond.

Since the oxygen atoms share *two* pairs of electrons, the bond between them is called a **double bond**. You can show it like this: O=O.

Nitrogen

The formula for nitrogen is N_2, so each molecule must contain two atoms. Each nitrogen atom has five outer electrons, so it needs a share in *three* more electrons to reach a full shell:

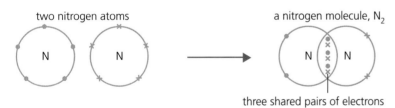

A model of the nitrogen molecule. You can also show the molecule as N≡N. The three lines represent a triple bond.

Since the nitrogen atoms share three pairs of electrons, the bond between them is called a **triple bond**. You can show it like this: N≡N.

Questions
1. a Name the bond between atoms that share electrons.
   b What holds the bonded atoms together?
2. What is a *molecule*?
3. Give five examples of molecular elements.
4. Draw a diagram to show the bonding in:
   a chlorine   b oxygen   c nitrogen
5. Now, a challenge. Hydrogen chloride (HCl) is a covalent *compound*. Draw a diagram to show the bonding in it.

3.6 Covalent compounds

Covalent compounds

In the last unit you saw that many non-metal elements exist as molecules. A huge number of compounds also exist as molecules.

In a molecular compound, atoms of *different* elements share electrons with each other. The compounds are called **covalent compounds** because of their covalent bonds. Let's look at some examples.

Methane

The formula of methane is CH_4. In each molecule, a carbon atom shares electrons with four hydrogen atoms, and all reach full shells:

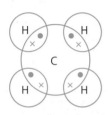

A molecule of methane.

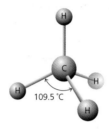

A model of the methane molecule.

Another model of the methane molecule.

The shape of the methane molecule
The methane molecule is not flat, as the first drawing above suggests. It is **tetrahedral** in shape, as the two models show. The four bonds point to the corners of a tetrahedron. The angles between the hydrogen atoms are all the same.

Why is it this shape? Because the four pairs of shared electrons repel each other, and try to get as far apart as possible.

Ammonia

Ammonia has the formula NH_3. The nitrogen atom shares electrons with three hydrogen atoms, and they all reach full shells:

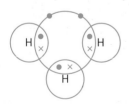

A molecule of ammonia.

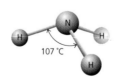

A model of the ammonia molecule.

Another model of the ammonia molecule.

Again the molecule is not flat, because of repulsion between electrons. It is **pyramidal** – shaped like a pyramid.

The hydrogen atoms in ammonia are pushed a little closer together than in methane. This is because the two non-bonding electrons in nitrogen have a stronger repelling effect than the three pairs of bonding electrons do.

Some other covalent compounds

This table shows molecules of three other covalent compounds. The basic idea is the same each time:
- the atoms share electrons, to reach full outer shells
- repulsion between electrons dictates the shape of the molecule.

Covalent compound	Description	Model
water, H$_2$O A molecule of water.	The oxygen atom shares electrons with two hydrogen atoms, and all gain full shells. Because the non-bonding electrons repel more strongly, the two pairs of bonding electrons get pushed together. So the molecule is **angular** in shape.	
hydrogen chloride, HCl 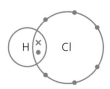 A molecule of hydrogen chloride.	The chlorine atom shares one electron with the hydrogen atom, and both gain full shells. Since it has only two atoms, the molecule can only be straight, or **linear**.	
carbon dioxide, CO$_2$ 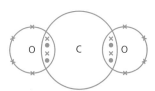 A molecule of carbon dioxide.	The carbon atom shares all four of its electrons: two with each oxygen atom. So all three atoms gain full shells. The two groups of bonding electrons repel each other, and move as far apart as possible. The result is a **linear** molecule. All the bonds are double bonds, so we can show the molecule like this: O=C=O.	

Questions

1. **a** What is a *covalent compound*?
   b Give five examples, with their formulae.
2. Draw a diagram to show the bonding in a molecule of:
   a methane **b** carbon dioxide **c** water
   then describe and explain the shape of each molecule.
3. A challenge! Methanol has the formula CH$_3$OH. The drawing on the right shows the structure of the molecule. Draw a diagram showing the bonding in methanol. Then see if you can predict the shape of the molecule.

3.7 Molecular substances

Remember

Molecules are made of non-metal atoms joined by covalent bonds.
Oxygen (O_2), water (H_2O), and ammonia (NH_3) all exist as molecules.

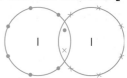

an iodine molecule, I_2

Molecular solids

Some molecular substances are solid at room temperature. Iodine is one example. The drawing on the right shows an iodine molecule. So how do these molecules form a solid? Let's see.

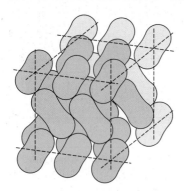

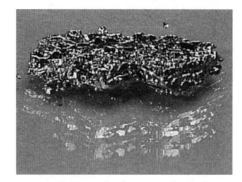

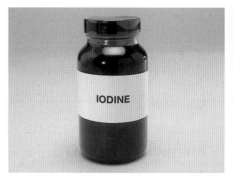

In the solid, the iodine molecules (◯) are held in a **lattice** like this, in a regular pattern. But the forces between the molecules are weak.

The pattern repeats millions of times. The result is a crystal. This photo shows a single crystal of iodine, magnified 15 times.

Iodine crystals are grey-black and shiny. A jar of iodine in the laboratory contains hundreds of thousands of crystals.

All molecular solids have a similar structure. The molecules are held in a regular pattern in a lattice. So the solids are crystalline. The forces *between* the molecules – the **intermolecular forces** – are weak.

Some properties of molecular solids

1 **Molecular solids have low melting and boiling points.** This is because the forces between the molecules are weak. In fact many molecular substances melt or even boil below room temperature, so are liquids or gases at room temperature.
2 **They do not conduct electricity.** Molecules are not charged, so molecular substances cannot conduct, even when melted.

Some molecular substances and their state at room temperature:

solids	iodine
	sulphur
liquids	bromine
	water
gases	nitrogen
	carbon dioxide

All have low melting points.

Sulphur, a yellow molecular solid, occurs naturally around volcanoes.
This shows sulphur at the Mount Baker volcano in the USA.

48

When you cool molecular liquids and gases

As you saw, many molecular substances are liquids, or gases, at room temperature. So what happens if you cool them down? The usual story! The molecules lose energy, and move more and more slowly. Eventually, at the freezing point of the substance, they form a lattice.

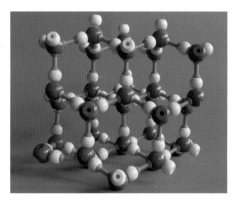

For example water is a molecular compound. It is a liquid at room temperature. But if you put it in an ice-cube tray in the freezer, it turns to a solid, ice.

This model shows the molecules within ice. The red balls represent oxygen atoms, white are hydrogen, and the purple sticks represent the weak forces between molecules.

Then, as the ice warms up again, the lattice breaks down. Even at 1 °C, the heat energy is enough to overcome the weak forces that hold the molecules in the lattice.

It's the same with all the other molecular substances that are liquids or gases at room temperature – even oxygen and nitrogen. Cool them enough and a lattice of molecules will form. Warm the lattice and it will break down again.

Penguins on the ice sheet in Antarctica: held up by the forces between the H_2O molecules in the lattice.

Questions

1. **a** What is a molecular solid?
    b A molecular solid will not conduct electricity. Why not?
2. A substance melts at 20 °C. What type of structure do you think it has? Give reasons.
3. Explain why many molecular substances are gases or liquids at room temperature, and give four examples.
4. Bromine belongs to the same family of elements as iodine. It is a liquid at room temperature. What can you say about the intermolecular forces in solid bromine, compared with those in solid iodine?
5. Describe the arrangement of the molecules in ice. How does the arrangement change as the ice warms up?

3.8 Giant covalent structures

Not all covalent solids are molecular

In all the solids in this table, the atoms are held together by covalent bonds. But compare their melting points. What do you notice?

Substance	Melting point/°C
ice	0
phosphorus	44
sulphur	115
silicon dioxide (silica)	1710
carbon (as diamond)	3550

The first three substances are molecular solids. Their molecules are held in a lattice by weak forces – so the solids melt easily, at low temperatures.

But diamond and silica are different. Their melting points show that *they* are not molecular solids with weak lattices. In fact they exist as **giant covalent structures**, or **macromolecules**.

Diamond – a giant covalent structure

Diamond is made of carbon atoms, held in a strong lattice:

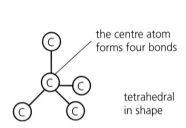

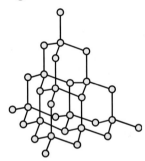

A carbon atom forms covalent bonds to *four* others, as shown above. Each outer atom then bonds to three more, and so on.

Eventually billions of carbon atoms are bonded together, in a giant covalent structure. This shows just a very tiny part of it.

The result is a single crystal of diamond. This one has been cut, shaped, and polished, to make it sparkle.

Diamond has these properties:

1. It is very hard, because each atom is held in place by four strong covalent bonds. In fact it is the hardest substance on Earth.
2. For the same reason it has a very high melting point, 3550 °C.
3. It can't conduct electricity because there are no ions or free electrons to carry the charge.

Silica is similar to diamond

Silica, SiO_2, occurs naturally as **quartz**, the main mineral in **sand**. Like diamond, it forms a giant covalent structure, as shown on the right.

Each silicon atom bonds covalently to four oxygen atoms, and each oxygen atom bonds covalently to two silicon atoms. The result is a very hard substance with a melting point of 1710 °C.

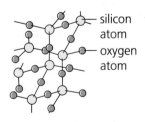

Silicon dioxide is made up of oxygen atoms ● and silicon atoms ○. Billions of them bond together like this, to give a giant structure.

Graphite – a very different giant structure

Like diamond, graphite is made only of carbon atoms. So diamond and graphite are **allotropes** of carbon – two forms of the same element.

Diamond is the hardest solid on Earth. But graphite is one of the softest! This difference is a result of their very different structures:

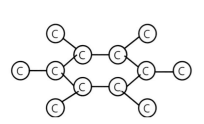

In graphite, each carbon atom forms covalent bonds to *three* others. This gives rings of *six* atoms.

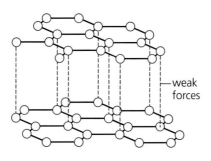

The rings form flat sheets that lie on top of each other, held together by weak forces.

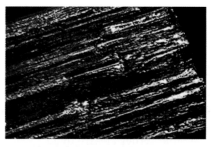

Under a microscope, you can see the layered structure of graphite quite clearly.

Graphite has these properties:
1. Unlike diamond, it is soft and slippery. That's because the sheets can slide over each other easily.
2. Unlike diamond, it is a good conductor of electricity. That's because each carbon atom has four outer electrons, but forms only three bonds. So the fourth electron is free to move through the graphite, carrying charge.

Uses of diamond, silica and graphite

Their uses reflect their properties, as this table shows.

Substance	Properties	Uses
diamond	hardest known substance does not conduct	in tools for drilling and cutting
	sparkles when cut	for jewellery
silica	hard, can scratch things	in sandpaper
	hard, lets light through	for making glass and lenses
	high melting point	in bricks for lining furnaces
graphite	soft and slippery	as a lubricant for engines and locks
	soft and dark in colour	for pencil 'lead' (mixed with clay)
	conducts electricity	for electrodes, and connecting brushes in generators

Pencil 'lead' is a mixture of graphite and clay.

Questions

1. Diamond and graphite are *allotropes* of carbon. What does that mean?
2. Why is diamond so hard?
3. Why do diamond and graphite have such very different properties? Draw diagrams to help you explain.
4. a Explain why silica has a high melting point.
   b Suggest reasons why it is lower than diamond's.
5. Diamond is not used to line furnaces. Why not?
6. You can buy solid air fresheners in shops. Are these substances giant structures, or molecular? Explain.

3.9 The bonding in metals

Clues from melting points

Compare these melting points:

Structure	Example	Melting point /°C
molecular	carbon dioxide	–56
	water	0
ionic	sodium chloride	801
	magnesium oxide	2852
giant covalent	diamond	3550
	silica	1610
metal	iron	1535
	copper	1083

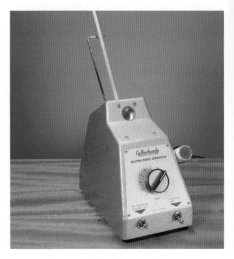

Equipment for measuring melting points in the school lab. It can heat substances up to 300 °C – so it's no good for sodium chloride!

The table shows clearly that:

- **molecular substances have low melting points**. That's because the forces between molecules in the lattice are weak.
- **giant structures such as sodium chloride and diamond have much higher melting points**. That's because the bonds between ions or atoms within giant structures are very strong.

Now look at the metals. They too have high melting points – not as high as for diamond, but much higher than for carbon dioxide or water. This gives us a clue that they too might be giant structures. And so they are, as you'll see below.

The structure of metals

In metals, the atoms are packed tightly together in a regular pattern. The tight packing allows outer electrons to separate from their atoms. The result is a lattice of ions in a 'sea' of electrons that are free to move through the lattice. Look at copper:

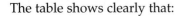
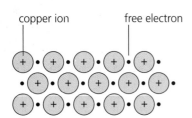

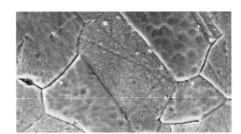

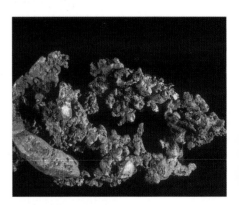

The copper ions are held together by their attraction to the electrons between them. The strong forces of attraction are called **metallic bonds**.

The regular arrangement of ions results in **crystals** of copper. This shows the crystals in a piece of copper magnified 1000 times.

The copper crystals are called grains. A lump of copper consists of millions of grains joined together. You need a microscope to see them.

It's the same with all metals. The ions sit in a regular lattice, held together by their strong attraction to the free electrons. And because the ions are in a regular lattice, metals are crystalline. (The crystals join at different angles, as you can see above.)

Some properties of metals

1 **Metals usually have high melting points.**
That's because it takes a lot of heat energy to break up the lattice, with its strong metallic bonds. Copper melts at 1083 °C, and nickel at 1455 °C. But there are exceptions. Sodium melts at only 98 °C, for example. And mercury melts at −39 °C, so it is a liquid at room temperature.

2 **Metals are malleable and ductile.**
Malleable means they can be bent and pressed into shape. *Ductile* means they can be drawn out into wires. This is because the layers can slide over each other:

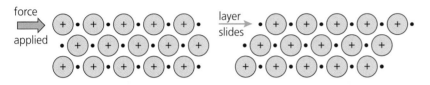

Metals: malleable, ductile, and sometimes very glamorous – like this silver bracelet.

The layers can slide without the metallic bond breaking, because the electrons are free to move too.

3 **Metals are good conductors of heat.**
That's because the free electrons take in heat energy, which makes them move faster. They quickly transfer the heat through the metal structure:

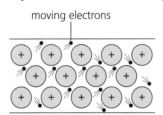

4 **Metals are good conductors of electricity.**
That's because the free electrons can move through the lattice carrying charge, when a voltage is applied across the metal. Silver is the best conductor of all the metals. Copper is next – but it is used much more than silver because it is cheaper.

These properties mean that metals are very useful substances. You can find out more about the uses of some metals, in Chapter 13.

What uses of metals can you see in this picture?

Questions

1 Describe in your own words the structure of a metal.
2 What is a *metallic bond*?
3 What does *malleable* mean?
4 Explain why metals can be drawn out into wires without breaking.
5 a Explain why metals can conduct electricity.
  b Would you expect molten metals to conduct? Give reasons for your answer.
6 Because metals are malleable, we use some of them to make saucepans. Give two other examples of uses of metals that depend on:
  a their malleability   b their ductility
  c their ability to conduct electricity
7 Mercury forms ions with a charge of 2+. It goes solid (freezes) at −39 °C. Try drawing a diagram to show the structure of solid mercury.

3.10 Bonding, structure and properties: a review

Pulling the ideas together

You have met some of the key ideas in chemistry, about atoms, and how they bond, in Chapters 2 and 3. So now is a good time to pull all these ideas together.

The electrons in atoms are arranged in shells.
Each shell can hold a certain number of electrons. Then it is full.

> For the first 20 elements: first (inner) shell holds 2 electrons, second holds 8, third also holds 8.

A full outer shell makes an atom stable.
- So if the atoms of an element already have full outer shells:
  - they do not form bonds to other atoms
  - the elements exist as single atoms (they are monatomic).
- But if the atoms do not have full outer shells:
  - they form bonds to other atoms, to gain full shells.
  - this may mean reacting with atoms of other elements.

> The elements with full outer shells are called the **noble gases**.

Atoms can bond in different ways, to gain full outer shells.
- In **ionic bonding**, atoms give electrons to, or accept them from, atoms of other elements. Examples: Na^+, Cl^-.
- In **covalent bonding**, they share electrons with each other, or with atoms of other elements. Examples: Cl_2, H_2O.
- In **metallic bonding**, the outer-shell electrons are free to move about, in a sea of electrons. Example: copper, Cu.

The type of bonding depends on how many outer-shell (valency) electrons the atom has.

- **Metal atoms** have only a small number of valency electrons.
  - So it is easiest to *lose* these, to reach a full shell.
  - So they tend to give up these electrons easily to non-metals, forming ionic compounds.
  - The valency electrons also leave the metal atoms easily, to form metallic bonds.

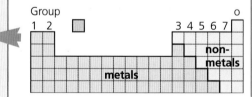

- **Non-metal atoms** have a larger number of valency electrons.
  - So it is easiest for them to *gain* electrons, to reach a full shell.
  - So they accept electrons from metal atoms, forming ionic compounds. Example: sodium chloride, NaCl.
  - They also share electrons with each other, or atoms of other non-metals, to form molecules with covalent bonds. Examples: Cl_2, H_2O.
  - Or sometimes they share electrons to form giant covalent structures. Examples: diamond, graphite, silica.

The bonding leads to a solid with a regular structure.
- The particles are held in a regular lattice, giving **crystals**.
  - In ionic solids, the ions are held in the lattice by strong ionic bonds.
  - In molecular solids, the molecules are held in the lattice by weak intermolecular forces.
  - In giant covalent structures, the atoms are in a regular pattern, with strong covalent bonds between them.
  - In metals, the metal ions are held in the lattice by metallic bonds between them and the 'sea' of electrons.
- On heating, a lattice breaks up or **melts**.
- The amount of heat needed to melt it depends on the strength of the forces between the particles in the lattice.

Graphite has an unusual giant covalent structure: flat sheets of covalently-bonded atoms, with weak forces between the sheets.

 **The bonding dictates the physical properties of the substance.**
For example:
- Ionic solids
  - do not conduct electricity (but will do if melted or dissolved)
  - have high melting points
- Molecular solids
  - do not conduct electricity
  - have low melting points
- Giant molecular structures (with the exception of graphite):
  - are very hard
  - do not conduct electricity
  - have high melting points
- Metals
  - conduct electricity and heat
  - have high melting points
  - are malleable and ductile.

Both ionic and molecular solids are **brittle**: they break up if you hammer them.

Graphite is soft since the sheets can slide over each other.

The electrons in the sea of electrons can move. And the layers of metal ions can slide over each other.

 The physical properties of the substance, in turn, largely dictate its uses.
For example:
- copper wire is used to conduct electricity
- diamond is used for cutting tools
- silicon dioxide is used to line furnaces.

Questions
1 What conclusions can you draw about each element?
  a It is malleable, and reacts with chlorine on heating.
  b It does not react with any other elements.
  c It conducts electricity when solid and melted.
  d It is a gas at room temperature.

2 What conclusions can you draw about each substance?
  a It is brittle and conducts electricity when melted.
  b It is used to make bells.
  c Its atoms have six valency electrons.
  d It is used to make bricks for building homes.

Questions on Chapter 3

1. This question is about the ionic bond formed between the metal lithium (atomic number 3) and the non-metal fluorine (atomic number 9).
   a. How many electrons are there in a lithium atom? Draw a diagram to show its electron structure. (You can show the nucleus as a dark circle at the centre.)
   b. How does a metal atom obtain a full outer shell of electrons?
   c. Draw the structure of a lithium ion, and write the symbol for the ion.
   d. How many electrons does a fluorine atom have? Draw a diagram to show its electron structure.
   e. How does a non-metal atom become a negative ion?
   f. Draw the structure of a fluoride ion, and write a symbol for the ion.
   g. Draw a diagram to show what happens when a lithium atom reacts with a fluorine atom.
   h. Draw the arrangement of ions in the compound that forms when lithium reacts with fluorine.
   i. Write the name and formula for the compound in **h**.
   j. Which type of structure does the compound in **h** have?
   k. Give two properties of this compound.

2. The following show the electronic arrangement for two elements, aluminium and nitrogen. These elements react to form an ionic compound called aluminium nitride.

   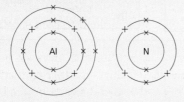

   a. Answer these questions for an aluminium atom.
      i. Does it gain or lose electrons, to form an ion?
      ii. How many electrons are transferred?
      iii. Is the ion formed positive, or negative?
      iv. What charge does the ion have?
   b. Now repeat **a**, but for a nitrogen atom.
   c. i. Write the electronic configurations for the ions formed by the two atoms.
      ii. What do you notice about these configurations? Explain it.
   d. Name another non-metal that would form an ionic compound with aluminium, in a similar way to nitrogen.

3. a. The electronic configuration of a neon atom is (2,8). What is special about the outer shell of a neon atom?
   b. The electronic configuration of a calcium atom is (2,8,8,2). What must happen to a calcium atom for it to achieve a noble gas structure?
   c. Draw a diagram of an oxygen atom, showing its eight protons (p), eight neutrons (n), and eight electrons (e).
   d. What happens to the outer-shell electrons of a calcium atom, when it reacts with an oxygen atom?
   e. Name the compound that is formed when calcium and oxygen react together. Which type of bonding does it contain?
   f. Write the formula for the compound in **e**.
   g. Does this compound have a high, or low, melting point? Explain your answer.

4. Na^+ O_2 Al CH_4 N I^-
   a. From the list above, select:
      i. two atoms   ii. two molecules   iii. two ions
   b. What do the following symbols represent?
      i. Na^+ ii. I^-
   c. Name the compound formed from Na^+ and I^- ions, and write a formula for it.

5. a. Write down a formula for each of the following:
      i. a nitrate ion
      ii. a sulphate ion
      iii. a carbonate ion
      iv. a hydroxide ion
   b. The metal strontium forms ions with the symbol Sr^{2+}. Write down the formula for each of the following:
      i. strontium oxide
      ii. strontium chloride
      iii. strontium nitrate
      iv. strontium sulphate

6. This represents a molecule of a certain gas:

   a. What is the gas? What is its formula?
   b. Which type of bonding holds the atoms together?
   c. Name another compound with this type of bonding.
   d. What do the symbols • and × represent?

7 This table gives some properties for substances A to G.

Substance	Melting point /°C	Electrical conductivity solid	Electrical conductivity liquid	Solubility in water
A	−112	poor	poor	insoluble
B	680	poor	good	soluble
C	−70	poor	poor	insoluble
D	1495	good	good	insoluble
E	610	poor	good	soluble
F	1610	poor	poor	insoluble
G	660	good	good	insoluble

a Which of the substances are metals? Give reasons for your choice.
b Which of the substances are ionic compounds? Give reasons for your choice.
c Two of the substances have very low melting points, compared with the rest. Explain why these could *not* be ionic compounds.
d Two of the substances are molecular. Which two are they?
e i Which substance is a giant covalent structure?
  ii What other name is used to describe this type of structure? (Hint: starts with *m*.)
f Name the type of bonding found in:
  i B   ii C   iii D   iv E
g What accounts for the conductivity property of:
  i D?   ii E?

8 Silicon lies directly below carbon in Group 4 of the Periodic Table. The following table lists the melting and boiling points for silicon, carbon (diamond), and their oxides.
a In which state are the two *elements* at room temperature (20 °C)?

Substance	Symbol or formula	Melting point /°C	Boiling point /°C
carbon	C	3730	4530
silicon	Si	1410	2400
carbon dioxide	CO_2	(turns to gas at −78 °C)	
silicon dioxide	SiO_2	1610	2230

b Which type of structure does carbon (diamond) have: giant covalent, or molecular?
c Which type of structure do you expect silicon to have? Give reasons.
d In which state are the two oxides, at room temperature?
e Which type of structure has carbon dioxide?
f Does silicon dioxide have the same structure as carbon dioxide? What is your evidence?

9 Hydrogen bromide is a compound of the two elements hydrogen and bromine. It has a melting point of −87 °C and a boiling point of −67 °C. Bromine is in Group 7 of the Periodic Table.
a Is hydrogen bromide a solid, a liquid, or a gas at room temperature (20 °C)?
b Is it made of molecules, or does it have a giant structure? How can you tell?
c i Which type of bond is formed between the hydrogen and bromine atoms, in hydrogen bromide?
  ii Draw a diagram of the bonding between the atoms, showing only the outer electrons.
d Write a formula for hydrogen bromide.
e Name two other compounds with bonding similar to that in hydrogen bromide.
f Write formulae for these two compounds.

10 The following diagram represents the lattice structure of the compound zinc sulphide:

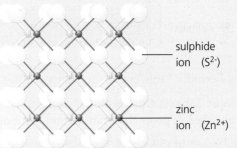

a Which does the diagram represent: a giant structure, or a molecular structure?
b Which type of bonding does zinc sulphide have?
c Look carefully at the structure. How many:
  i sulphur ions are joined to each zinc ion?
  ii zinc ions are joined to each sulphur ion?
d i From c, deduce the formula of zinc sulphide.
  ii Is this formula consistent with the charges on the two ions? Explain your answer.
e Name another metal and non-metal that would give a compound with a similar formula.

11 a Use the structures of diamond and graphite to explain why:
  i graphite is used for the 'lead' in pencils
  ii diamonds are used in cutting tools
  iii graphite is used in electrical circuits.
b Give reasons why:
  i copper is used in electrical wiring
  ii steel is used for domestic radiators.
c Ethanol is used as the solvent in perfume and aftershave, because it evaporates easily. What does that tell you about the bonding in it?

4.1 An overview of the Periodic Table

What is the Periodic Table?

Period	Group 1	Group 2											Group 3	Group 4	Group 5	Group 6	Group 7	Group 0
1					$^{1}_{1}$H hydrogen													$^{4}_{2}$He helium
2	$^{7}_{3}$Li lithium	$^{9}_{4}$Be beryllium											$^{11}_{5}$B boron	$^{12}_{6}$C carbon	$^{14}_{7}$N nitrogen	$^{16}_{8}$O oxygen	$^{19}_{9}$F fluorine	$^{20}_{10}$Ne noen
3	$^{23}_{11}$Na sodium	$^{24}_{12}$Mg magnesium	The transition elements										$^{27}_{13}$Al aluminium	$^{28}_{14}$Si silicon	$^{31}_{15}$P phosphorus	$^{32}_{16}$S sulphur	$^{35.5}_{17}$Cl chlorine	$^{40}_{18}$Ar argon
4	$^{39}_{19}$K potassium	$^{40}_{20}$Ca calcium	$^{45}_{21}$Sc scandium	$^{48}_{22}$Ti titanium	$^{51}_{23}$V vanadium	$^{52}_{24}$Cr chromium	$^{55}_{25}$Mn manganese	$^{56}_{26}$Fe iron	$^{59}_{27}$Co cobalt	$^{59}_{28}$Ni nickel	$^{64}_{29}$Cu copper	$^{65}_{30}$Zn zinc	$^{70}_{31}$Ga gallium	$^{73}_{32}$Ge germanium	$^{75}_{33}$As arsenic	$^{79}_{34}$Se selenium	$^{80}_{35}$Br bromine	$^{84}_{36}$Kr krypton
5	$^{85}_{37}$Rb rubidium	$^{88}_{38}$Sr strontium	$^{89}_{39}$Y yurium	$^{91}_{40}$Zr zirconium	$^{93}_{41}$Nb niobium	$^{96}_{42}$Mo molybdenum	$^{99}_{43}$Tc technetium	$^{101}_{44}$Ru ruthenium	$^{103}_{45}$Rh rhodium	$^{106}_{46}$Pd palladium	$^{108}_{47}$Ag silver	$^{112}_{48}$Cd cadmium	$^{115}_{49}$In indium	$^{119}_{50}$Sn tin	$^{122}_{51}$Sb antimony	$^{128}_{52}$Te tellurium	$^{127}_{53}$I iodine	$^{131}_{54}$Xe xenon
6	$^{133}_{55}$Cs caesium	$^{137}_{56}$Ba barium	$^{139}_{57}$La lanthanium	$^{178.5}_{72}$Hf hafnium	$^{181}_{73}$Ta tantalum	$^{184}_{74}$W tungsten	$^{186}_{75}$Re rhenium	$^{190}_{76}$Os osmium	$^{192}_{77}$Ir iridium	$^{195}_{78}$Pt platinum	$^{197}_{79}$Au gold	$^{201}_{80}$Hg mercury	$^{204}_{81}$Tl thallium	$^{207}_{82}$Pb lead	$^{209}_{83}$Bi bismuth	$^{210}_{84}$Po polonium	$^{210}_{85}$At astatine	$^{222}_{86}$Rn radon
7	$^{223}_{87}$Fr francium	$^{226}_{88}$Ra radium	$^{227}_{89}$Ac actinuim															

Lanthanides	$^{140}_{58}$Ce cerium	$^{141}_{59}$Pr praseodymium	$^{144}_{60}$Nd neodymium	$^{147}_{61}$Pm promethium	$^{150}_{62}$Sm samarium	$^{152}_{63}$Eu europium	$^{157}_{64}$Gd gadolinium	$^{159}_{65}$Tb terbium	$^{162}_{66}$Dy dysprosium	$^{165}_{67}$Ho holmium	$^{167}_{68}$Er erbium	$^{169}_{69}$Tm thutium	$^{173}_{70}$Yb ytterbium	$^{175}_{71}$Lu lutetium
Actinides	$^{232}_{90}$Th thorium	$^{231}_{91}$Pa protactinium	$^{238}_{92}$U uranium	$^{237}_{93}$Np neptunium	$^{244}_{94}$Pu plutonium	$^{243}_{95}$Am americium	$^{247}_{96}$Cm curium	$^{247}_{97}$Bk berkelium	$^{251}_{98}$Cf californium	$^{252}_{99}$Es einsteinium	$^{257}_{100}$Fm fermium	$^{258}_{101}$Md mendelevium	$^{259}_{102}$No nobelium	$^{262}_{103}$Lw lawrencium

You met the Periodic Table briefly in Chapter 2. Let's review its key points.

- The Periodic Table is a list of all the elements, **in order of increasing atomic number**. It gives their names and symbols.
- When arranged in this order, the elements show **periodicity**: elements with similar properties appear at regular intervals. So the list is set out with these similar elements together, in the numbered columns.
- The columns are called **groups**, and numbered 0 to 7. Note where group 0 is.
- The rows are called **periods**. They too are numbered 0 to 7.
- The heavy zig-zag line above separates **metals** from **non-metals**, with the non-metals to the right.

More about the groups

- The group number tells you how many electrons there are in the outer shell of the atoms. The atoms of Group 1 elements have one outer-shell electron. The atoms of Group 2 elements have two, and so on.
- The outer-shell electrons are also called **valency electrons**. And their number dictates how the elements behave.
- So all the elements in a group have similar properties.
- The atoms of the Group 0 elements have *full outer shells*. This makes them unreactive.
- Some of the groups have special names:

Group 1 – **the alkali metals**
Group 2 – **the alkaline earth metals**
Group 7 – **the halogens**
Group 0 – **the noble gases**.

Those numbers in the table
Remember, the numbers beside the symbols in the Periodic Table tell you about the atoms:

nucleon number
symbol
atomic number

- The nucleon number is the number of protons + neutrons.
- The atomic number is the number of protons.

These numbers are for the main isotope of each element.

More about the periods

The period numbers give information about the number of electron shells in the atoms. In the elements of Period 2, the atoms have two electron shells. In Period 3 they have three. (But the shell structure gets more complex from Period 4 onwards.)

Metals versus non-metals

Look again at the table. The metals are to the left of the zig-zag line, and the non-metals to the right. Over three-quarters of the elements are metals. Only two groups are completely non-metal. Which two?

Metals and non-metals have very different properties. For example metals conduct electricity, but non-metals do not. (The non-metal graphite, a form of carbon, is an exception.) For more about their properties see Unit 12.1.

Hydrogen

Hydrogen sits alone in the table. That is because it has one outer electron, like the Group 1 metals – but unlike them it is a gas, and usually reacts like a non-metal.

The transition elements

The long block in the middle of the Periodic Table contains only metals – the *transition metals*. There is more about these in Unit 4.6.

Some artificial elements

Not all elements occur naturally. Some are artificial, created in the lab. Most of these are in the lowest block of the Periodic Table. They include the elements neptunium (Np) to lawrencium (Lr) in the very bottom row. All the isotopes of the artificial elements are **radioactive**. Their atoms break down over time, giving out radiation.

Trends in the Periodic Table

The elements in each numbered group show trends in their properties. For example as you go down Group 1, the elements become *more* reactive. As you go down Group 7, they become *less* reactive. We discuss trends within groups, and across a period, in the rest of this chapter.

A world-famous structure, made from iron. Find iron in the Periodic Table. Which block is it in?

Dmitri Mendeleev, the Russian chemist who published the first Periodic Table in 1869. The table has grown a lot since then, with many more elements added.

Questions

1. Use the Periodic Table to find the names of:
   a three metals in common use around you
   b two non-metals that you breathe in.
2. Using *only* the Periodic Table to help you, write down everything you can about:
   a the element nitrogen   b the element magnesium
   You can include details about atomic structure.
3. Name three elements that are likely to react in a similar way to: a sodium   b fluorine
4. Which is likely to be more reactive, oxygen or krypton? Why?
5. Which element is named after:
   a Europe?   b Albert Einstein?   c America?
6. Why do chemists consider the Periodic Table so useful?

4.2 Group 1: the alkali metals

What are they?

The alkali metals are the six metals of Group 1 in the Periodic Table – lithium, sodium, potassium, rubidium, caesium, and francium. Only the first three are safe enough to keep in the school lab:

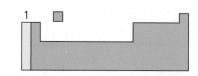

lithium   Li (2,1)   sodium   Na (2,8,1)   potassium   K (2,8,8,1)

The other three are violently reactive – and francium is radioactive.

Their physical properties

The alkali metals are *not* typical metals.

- Like all metals, they are good conductors of heat and electricity.
- But they are softer than most other metals. The samples shown above were cut with a knife.
- They are 'lighter' than most other metals – they have **low density**. The three metals above float in water (and immediately react with it).
- They have low melting and boiling points, compared with most metals.

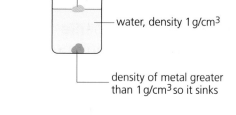

density of metal less than 1 g/cm^3 so it floats

water, density 1 g/cm^3

density of metal greater than 1 g/cm^3 so it sinks

Their chemical properties

The alkali metals are the most reactive metals. Look at these examples.

Reaction with water
They react violently with water, giving hydrogen and a hydroxide.

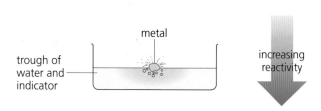

trough of water and indicator

metal

increasing reactivity

What you see

- lithium floats and fizzes
- sodium shoots across the water
- potassium melts with the heat of the reaction, and the hydrogen catches fire

For sodium the reaction is:

sodium + water → sodium hydroxide + hydrogen

The hydrogen bubbles off. The hydroxide is alkaline so the indicator changes colour.

All the alkali metals react vigorously with water, releasing hydrogen gas and forming hydroxides. The hydroxides give alkaline solutions.

Reaction with non-metals

1 **With chlorine** If you put heated alkali metals into jars of chlorine gas, they will burst into flame. They burn brightly, giving white solids called **chlorides**. The reaction with sodium is:

   sodium + chlorine → sodium chloride

   Sodium chloride is also known as common salt. (It is the salt we put on food.) It dissolves in water to give a colourless solution.

2 **With oxygen** The heated metals also burst into flame in gas jars of oxygen, and burn fiercely to form white solids called **oxides**. Note the different flame colours when the metals burn:

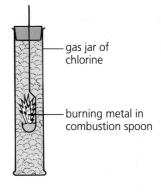

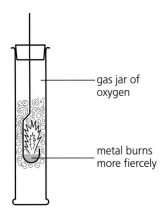

What you see
- a red flame for lithium
- a yellow flame for sodium
- a lilac flame for potassium

The reaction with sodium is:

   sodium + oxygen → sodium oxide

The oxide dissolves in water to give a colourless alkaline solution.

They form ionic compounds

The compounds formed in these reactions are **ionic**. Sodium chloride is made of sodium ions (Na^+) and chloride ions (Cl^-). Sodium oxide is made of sodium ions and oxide ions (O^{2-}). So the sodium atoms have become sodium ions with a charge of 1+. It's the same for all the alkali metals:

The alkali metals form ionic compounds, in which the metal ion has a charge of 1+. The compounds are white solids, which dissolve in water to give colourless solutions.

Because it reacts with oxygen and water, sodium is stored under oil.

Questions
1 Name the Group 1 elements and give their symbols.
2 What is the other name for the Group 1 elements? Why are they called that?
3 Which best describes the Group 1 metals:
   a soft or hard?
   b high density or low density?
   c high melting point or low melting point?
   d reactive or unreactive, with water?
4 a Write a word equation for the reaction of lithium with water.
   b The resulting solution affects litmus. Why?
5 a What forms when potassium reacts with chlorine?
   b What colour is this compound?
   c What will you *see* when you dissolve it in water?
   d Do you think the solution will conduct electricity? Give a reason for your answer.

4.3 The patterns within Group 1

The trends in physical properties

The Group 1 metals are a family. But like all families, each member of the group is a little different. Look at this table:

Metal	Symbol	This metal is silvery and ...	Density/g/cm^3	Melts at/°C	Boils at/°C
lithium	Li	soft	0.53	181	1342
sodium	Na	a little softer	0.97	98	883
potassium	K	softer still	0.86	63	760
rubidium	Rb	even softer	1.53	39	686
caesium	Cs	the softest of the five	1.88	29	669

Note the overall increase or decrease for each property, as you go down the table. This kind of pattern is called a **trend**. You can summarize the trends like this:

lithium
sodium   softness   density   melting points   boiling points
potassium increases   increases decrease   decrease
rubidium
caesium

Why they have similar chemical properties

You saw in Unit 4.2 that the elements in Group 1 react in a similar way with water, chlorine, and oxygen. Why is this? The answer is simple: *because they all have the same number of valency (outer-shell) electrons.*

2,1

2,8,1

2,8,8,1

Atoms with the same number of valency electrons react in a similar way.

The trend in reactivity

Compare the reactions of the Group 1 family with water:

Metal	What you see
lithium	a lot of fizz around the floating metal
sodium	it shoots around on the surface of water
potassium	it melts and the hydrogen bursts into flames
rubidium	sparks fly everywhere
caesium	a violent explosion

reactivity increases

The increase in violence shows the metals are getting more **reactive**.
Reactivity increases as you go down Group 1.

Strong family similarities – just like the alkali metals.

A pellet of sodium reacting with water. It shoots around on the surface. The purple colour is due to an indicator, and shows that an alkaline solution is forming.

And this shows a pellet of caesium reacting with water. Notice the difference! Caesium is further down Group 1 than sodium, and a lot more reactive. Why?

Explaining the trend in reactivity

Why do the Group 1 metals get more reactive, as you go down the group? Because their atoms get larger!

Atoms get larger down the group because they add on electron shells. Compare the shells in lithium and potassium atoms:

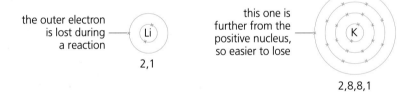

The atoms react in order to obtain a full outer shell. They do this by losing the outer electron. The more shells there are, the further the outer electron is from the positive nucleus – so the easier it is to lose.

And the easier it is to lose an electron, the more reactive the metal is. So potassium is a lot more reactive than lithium.

Questions

1. The Group 1 metals show a *trend* in melting points.
   a What does that mean?
   b Describe two other physical trends for the group.
2. Find one measurement that does *not* fit the trend exactly, in the table on the opposite page.
3. The Group 1 metals all have similar chemical properties. Why is this?
4. When a Group 1 metal reacts, what happens to the outer shell electron of its atoms?
5. a Which holds its outer electron more strongly: a lithium atom, or a sodium atom? Explain why you think so.
   b Sodium is more reactive than lithium. Why?
6. The Group 2 metals also lose electrons when they react. What can you predict about reactivity down Group 2?

4.4 Group 7: the halogens

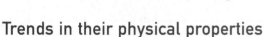

A non-metal group

Group 7 is a non-metal group. It contains the elements fluorine, chlorine, bromine, and iodine. They all have similar properties. For example they all:

- form coloured gases. Fluorine is a pale yellow gas and chlorine is a green gas. Bromine forms a red vapour, and iodine a purple vapour.
- are poisonous.
- are brittle and crumbly in their solid form, and do not conduct electricity. (These are typical properties of non-metals.)
- form diatomic molecules (containing two atoms). For example, Cl_2.

Trends in their physical properties

As usual, the elements show trends in their properties. Look at these:

Halogen	Formula	At room temperature (20°C) the element is …	Melts at … /°C	Boils at … /°C
fluorine	F_2	a yellow gas	–220	–188
chlorine	Cl_2	a green gas	–101	–35
bromine	Br_2	a red liquid	–7	59
iodine	I_2	a black solid	114	184

(size and mass of atoms increase ↓; density increases ↓; melting and boiling points increase ↓)

Density increases down the group as the mass of the atoms increases. Melting and boiling points increase too, showing that more energy is needed to overcome the attraction between the molecules.

Trends in their chemical properties

The halogens are among the most reactive elements in the Periodic Table. For example they react with metals to form compounds called **halides**. This table compares their reaction with iron:

Halogen	Reaction with iron wool	The product	Its appearance
fluorine	Iron wool bursts into flame as fluorine passes over it – without any heating!	iron(III) fluoride, FeF_3	pale green solid
chlorine	Hot iron wool glows brightly when chlorine passes over it.	iron(III) chloride, $FeCl_3$	yellow solid
bromine	Hot iron wool glows, but less brightly, when bromine vapour passes over it.	iron(III) bromide, $FeBr_3$	red-brown solid
iodine	Hot iron wool shows a faint red glow when iodine vapour passes over it.	iron(III) iodide, FeI_3	black solid

(reactivity increases ↑)

So the halogens all react in a similar way – and show a trend in reactivity. But this time, reactivity increases as you go *up* the group.
Reactivity increases as you go up Group 7. This is the opposite of the trend in Group 1.

Now look at the products in the table. What patterns do you notice?

Why do the halogens have similar properties?

The halogens have similar properties because their atoms all have the same number of valency (outer-shell) electrons: 7.

Why are they so reactive?

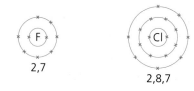

Halogen atoms: 7 valency electrons, so just one electron short of a full shell.

They are reactive because their atoms are only one electron short of a full shell. A halogen atom can gain an electron by accepting an electron from another atom, *or* sharing an electron with it.

In reacting with **metals**, a halogen atom *accepts* an electron from the metal atom. The metal atom becomes a positive ion, and the halogen atom a negative ion. For example when fluorine reacts with sodium, Na⁺ and F⁻ ions form. The result is an **ionic solid** with a giant structure, as for sodium chloride on page 38.

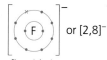

a fluoride ion

With hydrogen and other **non-metals**, the halogen atoms *share* electrons, forming **molecules** with covalent bonds. Halogen atoms also form covalent bonds with each other, to give diatomic molecules, like the fluorine molecule shown on the right.

a fluorine molecule

A halogen atom bonds to other atoms to gain, or share, an extra electron.

Explaining the trend in reactivity

When they react with metals, halogen atoms gain an electron. The smaller the atom the easier it is to attract this electron – so the more reactive the element will be:

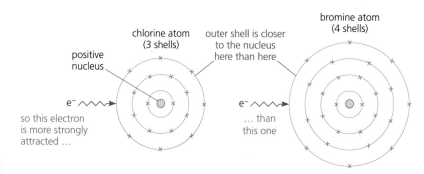

So the trend in reactivity is opposite to the trend in the size of the atoms, as you saw on page 64.

Questions
1. What is the other name for Group 7?
2. The elements in Group 7 have similar properties. Why?
3. Describe the appearance of the halogens.
4. **a** Summarise the trends in density, and melting and boiling points, for the Group 7 elements.
   b Are these trends the same as for Group 1? (Check on page 62.)
5. **a** Why are the halogens so reactive?
   b Why does reactivity *decrease* down the group?
6. The fifth element in Group 7 is called astatine. Do you expect it to be:
   a a gas, a liquid, or a solid? Give your reason.
   b coloured or colourless?
   c harmful or harmless?

4.5 Group 0: the noble gases

The noble gases

This group of non-metals contains the elements helium, neon, argon, krypton and xenon. These elements are all:

- non-metals
- colourless gases, which occur naturally in air
- **monatomic** – they exist as single atoms
- unreactive. This is their most striking property. They do not normally react with anything. That is why they are called **noble**.

Why are they unreactive?

As you have seen, atoms react with each other to gain full outer shells of electrons. But the atoms of the noble gases have full shells already:

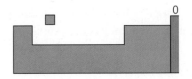

Noble gas	% in air
helium	only very tiny traces
argon	just under 1 %
neon	0.002 %
krypton	0.0001 %
xenon	less than 0.0001%

Helium is so light that most if it escapes from the atmosphere. But it is the most abundant element in the universe!

Helium is obtained from natural gas, in which it is an impurity. The other noble gases are obtained from the fractional distillation of liquid air.

a helium atom   a neon atom   an argon atom

So the atoms have no need to react. The full shells make them stable. **The noble gases are unreactive, and monatomic, because their atoms already have full outer electron shells.**

Trends in their physical properties

The noble gases are all unreactive – but they are not all identical. Like all the groups, they show trends. Look at this table.

Noble gas	Its atoms		A balloon full of this gas …		Boiling point /°C	
helium	$^{4}_{2}$He		rises quickly into the air		−269	
neon	$^{20}_{10}$Ne	the atoms increase in size and mass	rises slowly	the density of the gases increases	−246	the boiling points increase
argon	$^{40}_{18}$Ar		falls slowly		−186	
krypton	$^{84}_{36}$Kr		falls quickly		−152	
xenon	$^{131}_{54}$Xe		falls very quickly		−107	

The gases grow denser (or 'heavier') down the group, because the mass of the atoms increases. The increase in boiling points is a sign of increasing attraction between atoms. It gets harder to separate them to form a gas.

Compare these trends with those for the Group 7 non-metals on page 64. What do you notice?

These advertising signs use neon.

The headlights on this car are filled with xenon.

Uses of the noble gases

The noble gases are unreactive or **inert**, which makes them safe to use. They also glow when a current is passed through them at low pressure. These properties lead to many uses.

- Helium is used to fill balloons and airships, because it is much lighter than air – and will not catch fire.

- Argon is used to provide an inert atmosphere. For example it is used:
  - as a filler in ordinary tungsten light bulbs. (If air were used, the oxygen in it would make the tungsten filament burn away.)
  - to protect metals that are being welded. It won't react with the hot metals (unlike the oxygen in air).

- Neon is used in advertising signs. It glows red, but the colour can be changed by mixing it with other gases.

- Krypton is used in lasers – for example for eye surgery – and in car headlamps.

- Xenon gives a light like bright daylight, but with a blue tinge. It is used in lighthouse lamps, lights for hospital operating theatres, and car headlamps.

Helium is lighter than air, and inert – which makes it ideal for balloons.

Questions

1. Why do the members of Group 0 have similar properties?
2. Explain why the noble gases are unreactive.
3. a What are the trends in density and boiling point for the noble gases?
   b Are these trends the same as for:
      i Group 1?   ii Group 7? (Check pages 62 and 64.)
4. Explain why the noble gases are widely used, and give one use for each.
5. The sixth element in Group 0 is radon (Rn). Would you expect it to be:
   a a gas, a liquid, or a solid, at room temperature?
   b heavier, or lighter, than xenon?
   c chemically reactive?

67

4.6 The transition elements

What are they?

The **transition elements** are the block of 30 elements in the middle of the Periodic Table. They are all **metals**, and include most of the metals you find in everyday use, such as iron, tin, copper, and silver.

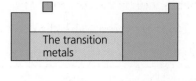

Their physical properties

Here are three of the transition metals:

Iron: the most widely used; grey with a metallic lustre (shine).

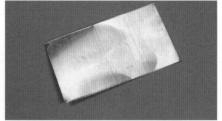

Copper: reddish with a metallic lustre.

Nickel: silvery with a metallic lustre.

Here is some data for them, with sodium for comparison:

Metal	Symbol	Density / g/cm^3	Melts at / °C
iron	Fe	7.9	1535
copper	Cu	8.9	1083
nickel	Ni	8.9	1455
sodium	Na	0.97	98

The transition metals share these physical properties:

- **hard, tough and strong**. They are not soft like the Group 1 metals.
- **high melting points**. Look at the values in the table. But mercury is an exception. It is a liquid at room temperature. (It melts at –39°C.)
- **malleable** (can be hammered into different shapes) and **ductile** (can be drawn out into wires).
- **good conductors of heat and electricity**. Of all metals, silver is the best conductor of electricity, and copper is next.
- **high density**. Unlike sodium, they sink in water, since their density is greater than the density of water (which is 1 g/cm^3).

Some transition metals	
iron	copper
nickel	zinc
silver	gold
platinum	mercury
chromium	cadmium

Their chemical properties

1. **They are much less reactive than the metals of Group 1.**
   For example copper does not react with water, or catch fire in air, unlike sodium. Their low reactivity means they do not corrode readily in air or water. But iron is an exception – it rusts very easily, and we spend a fortune every year on rust prevention.

2. **They show no clear trend in reactivity, unlike the metals of Group 1.**
   But those next to each other in the Periodic Table do tend to be similar.

Transition metal compounds are used in pottery glazes, because they are coloured.

3 **Most transition metals form coloured compounds.** In contrast, the Group 1 metals form white compounds.

4 **Most can form ions with different charges.** Compare these:

Metal	Charge on ions	Examples
Group 1 metals	always 1+	sodium: Na^+
Group 2 metals	always 2+	magnesium: Mg^{2+}
Group 3 metals	always 3+	aluminium: Al^{3+}
Transition metals	variable	copper: Cu^+, Cu^{2+}
		iron: Fe^{2+}, Fe^{3+}

So we say the transition metals have **variable valency**.

5 **They can form more than one compound with another element.** This is because of their variable valency. For example:

copper(I) oxide, CuO   copper(II) oxide, Cu_2O
iron(II) oxide, FeO   iron(III) oxide, Fe_2O_3

The Roman number in brackets tells you the number of electrons the metal atom has lost. This number is called its **oxidation state**.

6 **Most transition metals can form complex ions.** For example, if you add ammonia to a solution containing copper(II) ions you first get a pale blue precipitate of copper(II) hydroxide. But if you add more ammonia, the precipitate dissolves again, giving a deep blue solution.

That's because each copper ion has now attracted four ammonia molecules and two water molecules, forming a large soluble **complex ion**. This ion gives the solution its deep blue colour. Its formula is $[Cu(H_2O)_2(NH_3)_4]^{2+}$.

Salts of transition metals

- The oxides and hydroxides of transition metals are bases.
- You can make salts of the transition metals by reacting these bases with acids.

Uses of the transition metals

- The hard, strong transition metals are used in structures such as bridges, railway lines, buildings, and cars. Iron is the most widely used of all. It is usually used as **steels**, which are **alloys** of iron. (In alloys, small amounts of other substances are mixed with a metal, to improve its properties.)
- Many transition metals are used in making alloys. For example, chromium and nickel are mixed with iron to make **stainless steel**.
- Transition metals are used as conductors of heat and electricity. For example, steel is used for radiators, and copper for electricity cables.
- Many transition metals and their compounds acts as **catalysts**. These are substances that speed up reactions, but are not themselves chemically changed. Iron is used as a catalyst in making ammonia (page 216).

Iron rods are being used to strengthen a building.

Questions

1 Name five transition metals.
2 Which best describes the transition metals, overall:
   a soft or hard?   b high density or low density?
   c high melting point or low melting point?
   d reactive or unreactive, with water?
3 What is unusual about mercury?
4 Most paints contain compounds of transition elements. Why do you think this is?
5 Suggest reasons why copper is used in hot water pipes, while iron is not.

4.7 Across the Periodic Table

Trends across Period 3

As you saw, there are trends within groups in the Periodic Table. There are also trends as you move across a period. Look at this table for Period 3:

Group	1	2	3	4	5	6	7	0
Element	sodium	magnesium	aluminium	silicon	phosphorus	sulphur	chlorine	argon
Valency electrons	1	2	3	4	5	6	7	8
Element is a ...	metal	metal	metal	metalloid	non-metal	non-metal	non-metal	non-metal
Reactivity	high →			low →			high	unreactive
Melting point (°C)	98	649	660	1410	590	119	−101	−189
Boiling point (°C)	883	1107	2467	2355	(ignites)	445	−35	−186
Oxide is ...	basic		amphoteric			acidic	−	−
Typical compound	NaCl	MgCl$_2$	AlCl$_3$	SiCl$_4$	PH$_3$	H$_2$S	HCl	−
Valency shown in that compound	1	2	3	4	3	2	1	0

Notice these trends, as you move across the period:

1. The number of valency (outer-shell) electrons increases by 1 each time – like the group number. By argon (Group 0) the shell is full.
2. The elements go from metals to non-metals. Silicon is in between, like a metal in some ways and a non-metal in others. It is called a semi-metal or **metalloid**.
3. The metal atoms have few outer-shell electrons, so they *lose* them during reactions, to obtain full shells. Sodium atoms lose 1 electron, and aluminium atoms lose 3 (like their group numbers).
   But the non-metal atoms are closer to full shells. So they react to *gain* or *share* electrons. Group 7 atoms need to gain or share just 1 electron to reach a full shell, Group 6 atoms need 2, and so on.
4. Reactivity *decreases* as you move across the metals. Aluminium is less reactive than magnesium, which is less reactive than sodium. That is because the more electrons an atom has to lose, the more difficult this becomes. (More energy is needed to overcome the pull of the positive nucleus.)
5. But reactivity tends to *increase* as you move across the non-metals (apart from Group 0). So chlorine is more reactive than sulphur. This is because the fewer electrons an atom needs to reach a full shell, the easier it is to attract them.

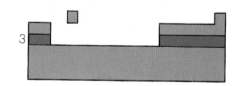

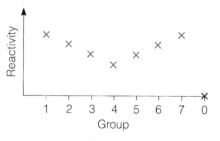

The trend in reactivity across a period

6 The melting and boiling points increase to the middle of the period, then decrease again. They are lowest on the right. (Only chlorine and argon are gases at room temperature.)

7 The oxides of the metals are **basic**, which means they react with acids to form salts. Those of the non-metals are **acidic** – they react with alkalis to form salts. But aluminium oxide is in between: it reacts with both acids and alkalis to form salts. So it is called an **amphoteric oxide**. (There is more about oxides in Unit 16.2.)

The elements in Period 2 show similar trends.

Valency

Now look at the last two rows in the table. One shows a typical compound of each element. The other shows the **valency** of the element in that compound.

The valency of an element is the number of electrons its atoms lose, gain or share, to form a compound.

Sodium always loses 1 electron to form a compound. So it has a valency of 1. Chlorine shares or gains 1, so it also has a valency of 1. Valency rises to 4 in the middle of the period, then falls again. It is zero for the noble gases.

Note that *valency* is not the same as *the number of valency electrons*. But:

- the valency does match the number of valency electrons, up to Group 4
- the valency matches the charge on the ion, where an element forms ions.

The change from metals to non-metals

As you saw, the change from metals to non-metals across Period 3 is not clear-cut. Silicon behaves like a metal in some ways, and a non-metal in others. That is why it is called a metalloid.

Several other elements in the Periodic Table are metalloids. They lie along the zig-zag line that separates metals from non-metals. Look on the right above.

Metals can conduct electricity. The metalloids can do so too, under certain conditions. So they are called **semi-conductors**. This property leads to the use of germanium and silicon in computer chips, and silicon and boron in PV (photovoltaic) cells for solar power.

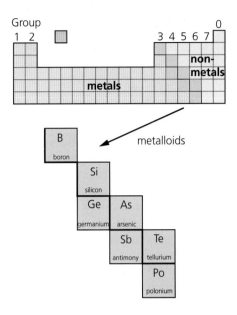

Plenty of silicon! It occurs naturally in rocks and sand as silica (silicon oxide). To extract it the silica is heated with carbon (coke) in a furnace.

Questions

1 a Describe how the number of valency electrons changes with group number, across the Periodic Table.
  b Describe the change from metal to non-metal properties, across Period 3.
2 How does the reactivity of the metals change as you move across a period? Why?
3 What does *valency* mean? Give three examples.
4 What is a *metalloid*? Give three examples.
5 What is a *semi-conductor*? Name one.

6 a A challenge! Make a table like the one opposite, but for *Period 2*. For each element the table should show:
  i   the group number
  ii the name of the element
  iii the number of valency electrons it has
  iv a typical compound
  v   the valency shown in that compound.
  b Now try to predict the melting and boiling points for the elements in the period.

Questions on Chapter 4

1 This extract from the Periodic Table shows the symbols for the first 20 elements.

					H												He
Li	Be											B	C	N	O	F	Ne
Na	Mg											Al	Si	P	S	Cl	Ar
K	Ca																

Look at the row from lithium (Li) to neon (Ne).
a What is this row of the Periodic Table called?
b Which element in it is the least reactive? Why?
c Which one shows the highest valency? Why?
Look at the column of elements from beryllium (Be) to calcium (Ca).
d What is this column of the table called?
e Which is the most reactive element in it? Explain why, in terms of electronic structure.
f Describe how the atomic structures of the first 20 elements relate to their positions in the Periodic Table. Draw diagrams to illustrate some examples.

2 This question is about **elements** from these families: alkali metals, alkaline earth metals, transition metals, halogens, noble gases.
A is a soft, silvery metal that reacts violently with water.
B is a gas at room temperature. It reacts violently with other elements, without heating.
C is a gas that sinks in air. It does not normally react with anything.
D is a hard solid at room temperature, and forms coloured compounds.
E conducts electricity, and reacts slowly with water. During the reaction, its atoms give up two electrons each.
F is a reactive liquid that shows a valency of 1 in its reactions.
G is a hard solid that conducts electricity, can be beaten into shape, and rusts easily.
a For each element above, say which of the listed families it belongs to.
b **i** Comment on the position of elements **A**, **B**, and **C** within their families.
   ii Describe the valence (outer) shell of electrons for each of the elements **A**, **B**, and **C**.
c Explain why the arrangement of electrons in their atoms makes some elements very reactive, and others unreactive.
d Name elements that fit descriptions **A** to **G**.
e Which of the above elements may be useful as catalysts?

3 The elements of Group 7 have similar properties.
a What name is given to this group of elements?
b Reactions with iron can be used to show their similar chemical properties.
   i Say how the reactions differ when chlorine, bromine, and iodine, react with iron.
   ii Describe the trend in the reactivity of these three halogens.
c **i** Suggest how fluorine would react with iron.
   ii Name the compound formed in this reaction.

4 Argon has atomic number 18, one more than chlorine. Explain why:
a chlorine is reactive, but argon is unreactive
b chlorine forms diatomic molecules, while argon is monatomic
c both chlorine and argon are gases at room temperature.

5 This diagram shows some of the elements in Group 7 of the Periodic Table.

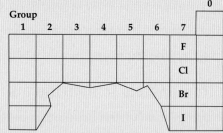

a What are the elements in this group called?
b Chlorine reacts explosively with hydrogen. The word equation for the reaction is:
   hydrogen + chlorine → hydrogen chloride

The reaction requires sunlight, but not heat.
   i How would you expect fluorine to react with hydrogen?
   ii Write the word equation for the reaction.
c **i** How might bromine react with hydrogen?
   ii Write the word equation for that reaction.
d *Explain* the differences in reactivity that you have described above.

6 **a** Strontium lies below calcium in Group 2, and next to rubidium in Group 1. Explain why it is:
   i more reactive than calcium
   ii less reactive than rubidium.
b Barium lies below strontium in Group 2. What can you conclude about its chemical properties?
c Krypton lies below argon in Group 0. What can you conclude about its physical properties?

7 Before the development of the modern Periodic Table, a German scientist called Döbereiner noticed that some elements could be grouped into sets of three, with similar properties.
He called these sets of three elements **triads**.
In the modern Periodic Table, the elements in each triad now appear in the same group.
  a Copy and complete the following triads by inserting the missing element.
   chlorine (Cl) .... iodine (I)
   lithium (Li) .... potassium (K)
   calcium (Ca) .... barium (Ba)
  b To which group of the modern Periodic Table does each triad belong?
  c What name is now given to those groups?
  d Why do elements in the same group have similar properties?

8 Rubidium is an alkali metal. It lies below potassium in Group 1. Here is data for Group 1:

Element	Atomic number	Melting point / °C	Boiling point / °C	Chemical reactivity
lithium	3	180	1330	quite reactive
sodium	11	98	890	reactive
potassium	19	64	760	very reactive
rubidium	37	?	?	?
caesium	55	29	690	violently reactive

  a Describe the trends in melting point, boiling point, and reactivity, as you go down the group.
  b Now using these trends, predict the missing data for rubidium.
  c *Explain* the trend in the reactivity of the Group 1 elements.
  d In a rubidium atom:
   i how many electron shells are there?
   ii how many electrons are there?
   iii how many valency electrons are there?

9 Identify these non-metal elements:
  a a colourless gas, used in balloons and airships
  b a poisonous green gas
  c a colourless gas that glows with a red light in advertising signs
  d a red liquid
  e a yellow gas which is so reactive that it is not allowed in school labs
  f a black solid that forms a purple vapour when heated gently.

10 The elements of Group 0 are called the noble gases. They are all monatomic gases.
  a Name four of the noble gases.
  b i What is meant by *monatomic*?
   ii Explain *why* the noble gases, unlike all other gaseous elements, are monatomic.
  When reactive non-metals react, they are said to become *like noble gases*.
  c i Explain what this statement means.
   ii What can you conclude about the reactivity of Group 7 *ions*?

11 The Periodic Table is the result of hard work by many scientists, in many countries, over hundreds of years. They helped to develop it by discovering, and investigating, new elements.

The Russian chemist Mendeleev was the first person to produce a table like the one we use today. He put all the elements he knew about into his table. But he realized that gaps should be left for elements not yet discovered. He even predicted the properties of some of these.

Mendeleev published his Periodic Table in 1869. This shows Groups 1 and 7 in his table.

	Group 1	Group 7
Period 1	H	
Period 2	Li	F
Period 3	Na	Cl
Period 4	K Cu	Mn Br
Period 5	Rb Ag	I

Use the modern Periodic Table (page 58) to help you answer these questions.

  a What does Period 2 mean?
  b i How does Group 1 in the modern Periodic Table differ from the Group 1 shown above?
   ii The arrangement in the modern table is more appropriate for this group. Explain why.
   iii What are the Group 1 elements called?
  c i What are the Group 7 elements called?
   ii Mn is out of place in Group 7. Why?
  d Mendeleev left gaps in several places in his table. Why did he do this?
  e There was no group to the right of Group 7 in Mendeleev's table. Suggest a reason.

5.1 The names and formulae of compounds

The names of compounds

Many compounds contain just two elements. If you know which elements they are, you can usually name the compound. You just have to follow a few rules.

- When the compound contains a metal and a non-metal:
  - the name of the metal is always given first
  - and then the name of the non-metal, but ending with - *ide*.
  Examples: sodium chloride, magnesium oxide, iron sulphide.

- When the compound is made of two non-metals:
  - if one is hydrogen, that is named first
  - otherwise the one with the lower group number comes first
  - and then the name of the other non-metal, ending with - *ide*.
  Examples: hydrogen chloride, carbon dioxide.

But some compounds have 'everyday' names that give you no clue about the elements in them. Water, methane, and ammonia are examples. You just have to remember their formulae!

That very common compound, water. Your body is full of it. What elements does it contain?

Finding formulae from the structure of compounds

Every compound has a formula as well as a name. The formula is made up of the symbols for the elements, and often has numbers too.

In Chapter 3, you saw how the formula of a compound is related to its structure. For example:

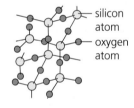

Sodium chloride forms a giant structure with one sodium ion for every chloride ion. So its formula is NaCl.

Water is made up of molecules in which two hydrogen atoms are bonded to an oxygen atom. So its formula is H_2O.

Silicon dioxide (silica) forms a giant structure in which there are two oxygen atoms for every silicon atom. So its formula is SiO_2.

Note the difference:

- in giant structures like sodium chloride and silicon dioxide, the formula tells you the *ratio* of the ions or atoms in the compound.
- in a molecular compound, the formula tells you *exactly* how many atoms are bonded together in each molecule.

Valency

But you don't need to draw the structure of a compound to work out its formula. You can work it out quickly if you know the valency of the elements:

The valency of an element is the number of electrons its atoms lose, gain or share, to form a compound.

Look at this table:

Elements	In forming a compound, the atoms ...	So the valency of the element is ...	Examples of compounds formed (those in green are covalent, with shared electrons)
Group 1	lose 1 electron	1	sodium chloride, NaCl
Group 2	lose 2 electrons	2	magnesium chloride, $MgCl_2$
Group 3	lose 3 electrons	3	aluminium chloride, $AlCl_3$
Group 4	share 4 electrons	4	methane, CH_4
Group 5	gain or share 3 electrons	3	ammonia, NH_3
Group 6	gain or share 2 electrons	2	magnesium oxide, MgO; water, H_2O
Group 7	gain or share 1 electron	1	sodium chloride, NaCl; hydrogen chloride, HCl
Group 0	(already have a full shell)	0	none
hydrogen	lose or share 1 electron	1	hydrogen bromide, HBr
transition metals	can lose different numbers of electrons	variable	iron(II) chloride, $FeCl_2$; iron(III) chloride, $FeCl_3$ copper(I) chloride, CuCl; copper(II) chloride, $CuCl_2$

Writing formulae using valencies

This is how to write the formula of a compound, using valencies:
1 Write down the valencies of the two elements.
2 Write down their symbols, in the same order as the elements in the name.
3 Add numbers after the symbols if you need to, to balance the valencies.

Example 1 What is the formula of hydrogen sulphide?
1 Valencies: hydrogen, 1; sulphur (Group 6), 2
2 HS   (valencies not balanced)
3 The formula is **H_2S**   (2 × 1 and 2, so the valencies are now balanced)

Example 2 What is the formula of aluminium oxide?
1 Valencies: aluminium (Group 3), 3; oxygen (Group 6), 2
2 AlO   (valencies not balanced)
3 The formula is **Al_2O_3**   (2 × 3 and 3 × 2, so the valencies are now balanced)

Writing formulae by balancing charges

In an ionic compound, the total charge is zero. So you can also work out the formula of an ionic compound by balancing the charges on its ions. To find out how to do this, turn to Unit 3.3.

Questions
The Periodic Table on page 58 will help you with these.
1 See if you can write a more scientific name for water.
2 Give a name for a compound containing just these elements:
   a sodium and fluorine   b fluorine and hydrogen
   c sulphur and hydrogen   d bromine and beryllium
3 Why does silica have the formula SiO_2?
4 Decide whether this formula is correct. If it is not correct, write it correctly.
   a HBr_2 b ClNa   c Cl_3Ca d Ba_2O
5 Write the correct formula for barium iodide.
6 See if you can give a name and formula for a compound that forms when phosphorus reacts with chlorine.

5.2 The masses of atoms, molecules, and ions

Relative atomic mass

A single atom weighs hardly anything. You can't use scales to weigh it.

But scientists do need a way to compare the masses of atoms. So this is what they did.

First, they chose an atom of carbon-12 to be the standard atom. They fixed its mass as exactly 12 units. (It has 6 protons and 6 neutrons, as shown on the right. They ignored the electrons.)

6 protons
6 electrons
6 neutrons

An atom of carbon-12. It is the main isotope of carbon. (See page 30.)

Then they compared all the other atoms with this standard atom, in a machine called a mass spectrometer, and found values for their masses. Like this:

This is the standard atom, $^{12}_{6}C$. Its mass is exactly 12.

This magnesium atom is twice as heavy as the standard atom, so its mass must be 24.

This hydrogen atom is $\frac{1}{12}$ as heavy as the standard atom, so its mass must be 1.

The mass of an atom found by comparing it with the $^{12}_{6}C$ atom is called its **relative atomic mass** or A_r.
So the A_r of hydrogen is 1, and the A_r of magnesium is 24.

A_r and isotopes

As you saw on page 30, the atoms of an element are not always identical. Some may have extra neutrons. Different atoms of the same element are called **isotopes**. Scientists found that chlorine has two isotopes:

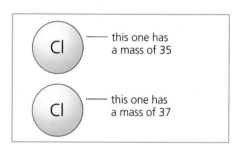

The two isotopes of chlorine.

Name	Protons	Neutrons	Nucleon number	% of chlorine atoms like this
chlorine-35	17	18	35	75%
chlorine-37	17	20	37	25%

We need to take all the natural isotopes of an element into account, to work out the relative atomic mass. This is the formula to use:

relative atomic mass (A_r) of an element =
  (% × nucleon number for the first isotope)
+ (% × nucleon number for the second isotope)
  and so on, for all its natural isotopes

The calculation for chlorine is given on the right. It shows that the relative atomic mass of chlorine is **35.5**.

The relative atomic mass (A_r) of an element is the average mass of its isotopes compared to an atom of carbon-12.

For most elements, A_r is very close to a whole number. It is usually rounded off to a whole number, to make calculations easier.

How to work out the relative atomic mass (A_r) of chlorine

Using the formula on the left, the relative atomic mass of chlorine ...

= 75% × 35 + 25% × 37

= $\frac{75}{100}$ × 35 + $\frac{25}{100}$ × 37
(changing % to fractions)

= 26.25 + 9.25

= 35.5

So the relative atomic mass for chlorine is **35.5**.

A_r values for some common elements

Element	Symbol	A_r	Element	Symbol	A_r
hydrogen	H	1	chlorine	Cl	35.5
carbon	C	12	potassium	K	39
nitrogen	N	14	calcium	Ca	40
oxygen	O	16	iron	Fe	56
sodium	Na	23	copper	Cu	64
magnesium	Mg	24	zinc	Zn	65
sulphur	S	32	iodine	I	127

Finding the mass of an ion

Mass of sodium atom = 23, so mass of sodium ion = 23 since a sodium ion is just a sodium atom minus an electron (which has negligible mass).

An ion has the same mass as the atom from which it is made.

Finding the masses of molecules and ions

Using A_r values, it is easy to work out the mass of any molecule or group of ions. Read the text panel on the right above, then look at these examples:

  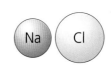

mass of ion = mass of its atom

Hydrogen gas is made of molecules. Each molecule contains 2 hydrogen atoms, so its mass is 2. (2 × 1 = 2)

The formula of water is H_2O. A water molecule contains 2 hydrogen atoms and 1 oxygen atom, so its mass is 18. (2 × 1 + 16 = 18)

Sodium chloride (NaCl) forms a giant structure with 1 sodium ion for every chloride ion. So the mass of a 'unit' of sodium chloride is 58.5. (23 + 35.5 = 58.5)

If the substance is made of molecules, its mass found in this way is called the **relative molecular mass**, or M_r. So the M_r for hydrogen is 2, and for water 18.

But if the substance is made of *ions*, its mass is called the **relative formula mass**. The short form for this is also M_r. So the M_r for NaCl is 58.5.

This table gives two more M_r values.

Substance	Formula	Atoms in formula	A_r of atoms	M_r
ammonia	NH_3	1N 3H	N = 14 H = 1	1 × 14 = 14 3 × 1 = 3 Total = **17**
magnesium nitrate	$Mg(NO_3)_2$	1Mg 2N 6O	Mg = 24 N = 14 O = 16	1 × 24 = 24 2 × 14 = 28 6 × 16 = 96 Total = **148**

Questions

1. What is the relative atomic mass of an element? How is it found? Why is it called *relative*?
2. What is the A_r of the iodide ion, I^-?
3. The relative molecular mass and formula mass are both called M_r for short. What is the difference between them?
4. Work out the M_r for each of these, and say whether it is the relative molecular mass or the relative formula mass:
   a. oxygen, O_2 b. iodine, I_2 c. methane, CH_4
   d. chlorine, Cl_2 e. butane, C_4H_{10} f. ethanol, C_2H_5OH
   g. ammonium sulphate $(NH_4)_2SO_4$

5.3 Percentage composition of a compound

Percentage composition

The percentage composition of a compound tells you *which* elements are in the compound and *how much* of each there is, as a percentage of the total mass.

Calculating the percentage of an element by mass

Methane is a compound of carbon and hydrogen, with the formula CH_4. We can show a methane molecule like this:

The mass of a carbon atom is 12, and the mass of each hydrogen atom is 1, so the relative molecular mass of methane is 16.

You can find what fraction of the total mass is carbon, and what fraction is hydrogen, like this:

$$\text{Mass of carbon as fraction of total mass} = \frac{\text{mass of carbon}}{\text{total mass}}$$

$$= \frac{12}{16} \text{ or } \frac{3}{4}$$

$$\text{Mass of hydrogen as fraction of total mass} = \frac{\text{mass of hydrogen}}{\text{total mass}}$$

$$= \frac{4}{16} \text{ or } \frac{1}{4}$$

These fractions are usually written as percentages. To change a fraction to a percentage, you just multiply it by 100:

$$\frac{3}{4} \times 100 = \frac{300}{4} = 75 \text{ per cent or } 75\% \qquad \frac{1}{4} \times 100 = \frac{100}{4} = 25\%$$

So 75% of the mass of methane is carbon, and 25% is hydrogen.

We say that the **percentage composition** of methane is 75% carbon, 25% hydrogen.

The composition of a compound does not change

Methane forms when bacteria break down organic material, such as the remains of dead plants and animals. It occurs as natural gas. You also find it in ponds, compost heaps, and sewage.

But no matter where you find it, a methane molecule *always* contains one carbon atom and four hydrogen atoms. So its percentage composition is *always* 75% carbon, 25% hydrogen. This follows one of the basic laws of chemistry:

Every pure sample of a given compound always has exactly the same composition.

The famous scientist John Dalton (1766–1844) and a pupil, collecting methane from a pond.

The 'natural gas' piped to homes in some places is also methane. It has exactly the same composition as the gas John Dalton got from ponds.

Calculating the percentage composition of a compound

Here is how to calculate the percentage of an element in a compound:

1. Write down the formula of the compound.
2. Using A_r values, work out its molecular or formula mass (M_r).
3. Write the mass of the element as a fraction of the total.
4. Multiply the fraction by 100, to give a percentage.

Example 1 The gas sulphur dioxide forms when fossil fuels are burned. It is a pollutant – it dissolves in rain to give acid rain. Its formula is SO_2. Calculate the percentage of oxygen in it.

The M_r of the compound is 64, as shown on the right.
Mass of oxygen in the formula = 32
Mass of oxygen as a fraction of the total $= \dfrac{32}{64}$

Mass of oxygen as a percentage of the total $= \dfrac{32}{64} \times 100 = 50\%$
So the compound is **50% oxygen**.
This means it is also 50% sulphur (100% – 50% = 50%).

To find the M_r for sulphur dioxide

A_r : S = 32, O = 16.

The formula has 1 S and 2 O, so the M_r is:

1 S   = 32
2 O = 2 × 16 = 32
       Total = 64

Example 2 Fertilisers contain nitrogen, which plants need to make them grow. The compound ammonium nitrate is used as a fertiliser. It has the formula NH_4NO_3. Calculate:

i the percentage of nitrogen in ammonium nitrate
ii the mass of nitrogen in a 20-kg bag of the fertiliser.

i The M_r of the compound is 80, as shown on the right.
The element we are interested in is nitrogen.

Mass of nitrogen in the formula   = 28
Mass of nitrogen as a fraction of the total $= \dfrac{28}{80}$
Mass of nitrogen as a percentage of the total $= \dfrac{28}{80} \times 100 = 35\%$
So the fertiliser is **35% nitrogen**.

ii The fertiliser is 35% nitrogen.

So the mass of nitrogen in a 20-kg bag $= \dfrac{35}{100} \times 20$ kg $= 7$ kg
So the bag contains **7 kg** of nitrogen.

To find the M_r for ammonium nitrate

A_r : N = 14, H = 1, O = 16.

The formula has 2 N, 4 H and 3 O, so the M_r is:

2 N = 2 × 14 = 28
4 H = 4 × 1 = 4
3 O = 3 × 16 = 48
       Total = 80

Questions

1. A compound contains just oxygen and sulphur. It is 40% sulphur. What percentage is oxygen?
2. Find the percentage of:
   i hydrogen   ii oxygen
   in ammonium nitrate, NH_4NO_3.
3. Calculate the percentage of copper in copper(II) oxide, CuO. (The A_r values are: Cu = 64, O = 16.)
4. 12 g of copper is converted into copper(II) oxide.
   a What mass of copper(II) oxide is formed?
   b What mass of oxygen does it contain?

5.4 The mole

What is a mole?

In Unit 5.2 you read about relative atomic mass (A_r), and relative molecular and formula masses (M_r). These terms are very important to a chemist – and it is very important that you understand them! For this reason:

If you work out the A_r or M_r of a substance, and then weigh out that number of grams of the substance, you can say how many atoms or molecules it contains.

For example, the A_r of carbon is 12. The photo on the right shows exactly 12 grams of carbon.

So we know the heap contains 602 000 000 000 000 000 000 000 carbon atoms!

We call this a **mole** of atoms.

That huge number is called **Avogadro's number** or **Avogadro's constant** after the Italian scientist who proposed it.

It is usually written in a short way as $\mathbf{6.02 \times 10^{23}}$. (The 10^{23} shows that you must move the decimal point 23 places to the right to get the full number.)

Some more examples of moles

Sodium is made of single sodium atoms. Its symbol is Na. Its A_r is 23.	Iodine is made of iodine molecules. Its formula is I_2. Its M_r is 254.	Water is made of water molecules. Its formula is H_2O. Its M_r is 18.
 This is **23 grams** of sodium. It contains 6.02×10^{23} sodium atoms, or **1 mole** of sodium atoms.	 This is **254 grams** of iodine. It contains 6.02×10^{23} iodine molecules, or **1 mole** of iodine molecules.	 The beaker contains **18 grams** of water, or 6.02×10^{23} water molecules, or **1 mole** of water molecules.

From these examples you should see that:

**One mole of a substance is 6.02×10^{23} particles of the substance.
It is obtained by weighing out the A_r or M_r of the substance in grams.**

Finding the mass of a mole

You can find the mass of one mole of any substance by these steps:
1 Write down the symbol or formula of the substance.
2 Find out its A_r or M_r.
3 Express that mass in grams (g).

This table shows three more examples:

Substance	Symbol or formula	A_r	M_r	Mass of 1 mole
helium	He	He = 4	exists as single atoms	4 grams
oxygen	O_2	O = 16	2 × 16 = 32	32 grams
ethanol	C_2H_5OH	C = 12 H = 1 O = 16	2 × 12 = 24 6 × 1 = 6 1 × 16 = 16 46	46 grams

Some calculations on the mole

Example 1 Calculate the mass of:
a 0.5 moles of bromine atoms   b 0.5 moles of bromine molecules.

a The A_r of bromine is 80, so 1 mole of bromine atoms has a mass of 80 g.
   Therefore 0.5 moles of bromine atoms has a mass of 0.5 × 80 g, or 40 g.

b A bromine *molecule* contains 2 atoms, so its M_r is 160. Therefore 0.5 moles of bromine molecules has a mass of 0.5 × 160 g, or 80 g.

So, to find the mass of a given number of moles:
mass = mass of 1 mole × number of moles

Example 2 How many moles of oxygen molecules are in 64 g of oxygen?

The M_r of oxygen is 32, so 32 g of it is 1 mole.
Therefore 64 g is $\frac{64}{32}$ moles, or 2 moles.

So, to find the number of moles in a given mass:

$$\text{number of moles} = \frac{\text{mass}}{\text{mass of 1 mole}}$$

Questions

1 How many atoms are in 1 mole of atoms?
2 How many molecules are in 1 mole of molecules?
3 What name is given to the number 6.02×10^{23}?
4 Find the mass of 1 mole of:
   a hydrogen atoms   b iodine atoms
   c chlorine atoms   d chlorine molecules
5 Find the mass of 2 moles of:
   a oxygen atoms   b oxygen molecules
6 Find the mass of 3 moles of ethanol, C_2H_5OH.
7 How many moles of molecules are there in:
   a 18 grams of hydrogen, H_2?
   b 54 grams of water?
8 Sodium chloride is made up of Na^+ and Cl^- ions.
   a How many sodium ions are there in 58.5 g of sodium chloride? (A_r: Na = 23; Cl = 35.5.)
   b What is the mass of 1 mole of chloride ions?

5.5 The empirical formula

What a formula tells you about moles and masses

The formula of carbon dioxide is **CO₂**. Some molecules of it are shown on the right. You can see that:

| 1 carbon atom | combines with | 2 oxygen atoms | so |

| 1 mole of carbon atoms | combines with | 2 moles of oxygen atoms |

A_r: C = 12, O = 16

Moles can be changed to grams, using A_r and M_r. So we can write:

| 12 g of carbon | combines with | 32 g of oxygen |

In the same way:
6 g of carbon combines with 16 g of oxygen
24 kg of carbon combines with 64 kg of oxygen, and so on.
The masses of substances that combine are *always in the same ratio*.

Therefore, from the formula of a compound, you can tell:

- **how many moles of the different atoms combine**
- **how many grams of the different elements combine.**

Finding the empirical formula

From the formula of a compound you can tell what masses of the elements combine. But you can also do things the other way round.
If you know what masses combine, you can work out the formula.
These are the steps:

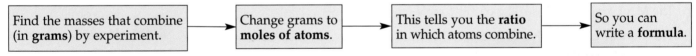

| Find the masses that combine (in **grams**) by experiment. | → | Change grams to **moles of atoms**. | → | This tells you the **ratio** in which atoms combine. | → | So you can write a **formula**. |

A formula found by this method is called the **empirical formula**.
The empirical formula shows the simplest ratio in which atoms combine.

Example 1 32 grams of sulphur combine with 32 grams of oxygen to form an oxide of sulphur. What is its empirical formula?

Draw up a table like this:

Elements that combine	sulphur	oxygen
Masses that combine	32 g	32 g
Relative atomic masses (A_r)	32	16
Moles of atoms that combine	32/32 = 1	32/16 = 2
Ratio in which atoms combine	1:2	
Empirical formula	**SO₂**	

So the empirical formula of the oxide that forms is **SO₂**.

Sulphur combines with oxygen when it burns.

Example 2 An experiment shows that compound Y is 80% carbon and 20% hydrogen. What is its empirical formula?

Y is 80% carbon and 20% hydrogen. So 100 g of Y contains 80 g of carbon and 20 g of hydrogen. Draw up a table like this:

Elements that combine	carbon	hydrogen
Masses that combine	80 g	20 g
Relative atomic masses (A_r)	12	1
Moles of atoms that combine	80/12 = 6.67	20/1 = 20
Ratio in which atoms combine	6.67 : 20 or 1 : 3 in its simplest form	
Empirical formula	CH_3	

So the empirical formula of Y is **CH_3**.

But we can tell right away that the *molecular* formula for Y must be different. (A carbon atom does not bond to just 3 hydrogen atoms.) You will learn how to find the molecular formula from the empirical formula in the next unit.

An experiment to find the empirical formula

To work out the empirical formula, you need to know the masses of elements that combine. *The only way to do this is by experiment.*

For example, magnesium combines with oxygen to form magnesium oxide. The masses that combine can be found like this:

1 Weigh a crucible and lid, empty. Then add a coil of magnesium ribbon and weigh it again, to find the mass of the magnesium.
2 Heat the crucible. Raise the lid carefully at intervals to let oxygen in. The magnesium burns brightly.
3 When burning is complete, let the crucible cool (still with its lid on). Then weigh it again. The increase in mass is due to oxygen.

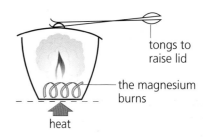

Weighing showed that 2.4 g of magnesium combined with 1.6 g of oxygen. Draw up a table again:

Elements that combine	magnesium	oxygen
Masses that combine	2.4 g	1.6 g
Relative atomic masses (A_r)	24	16
Moles of atoms that combine	2.4/24 = 0.1	1.6/16 = 0.1
Ratio in which atoms combine	1 : 1	
Empirical formula	MgO	

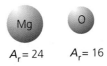

So the empirical formula for the oxide is **MgO**.

Questions

1 a How many atoms of hydrogen combine with one carbon atom to form methane, CH_4?
  b How many grams of hydrogen combine with 12 grams of carbon to form methane?
2 What does the word *empirical* mean? (Check the glossary?)
3 56 g of iron combine with 32 g of sulphur to form iron sulphide. Find its empirical formula. (A_r: Fe = 56, S = 32.)
4 An oxide of sulphur is 40% sulphur and 60% oxygen. What is its empirical formula?

5.6 From empirical to final formula

The formula of an ionic compound

You saw in the last unit that the empirical formula shows the *simplest ratio* in which atoms combine.

The diagram on the right shows the structure of sodium chloride, again. Sodium and chlorine atoms combine in a ratio of 1:1 to to form this compound. So its empirical formula is NaCl.

The formula of an ionic compound is always the same as its empirical formula.

In the experiment on page 83, the empirical formula for magnesium oxide was found to be MgO. So the formula for magnesium oxide is also **MgO**.

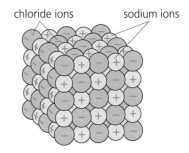

The structure of sodium chloride.

The formula of a molecular compound

The gas ethane is one of the alkane family of compounds. An ethane molecule is drawn on the right. It contains only hydrogen and carbon atoms, so ethane is a **hydrocarbon**.

From the drawing you can see that the ratio of carbon to hydrogen atoms in ethane is 2:6. The simplest ratio is therefore 1:3. So the *empirical* formula of ethane is CH_3. (It is compound Y on page 83.)
But its *molecular* formula is C_2H_6.

The molecular formula shows the actual numbers of atoms that combine to form a molecule.

The molecular formula is more useful than the empirical formula because it gives you more information. For some molecular compounds, both formulae are the same. For others they are different.

Compare the two types of formulae for the alkanes in the table on the right. What do you notice?

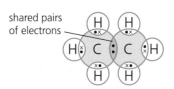

An ethane molecule.

Name of alkane	Molecular formula	Empirical formula
methane	CH_4	CH_4
ethane	C_2H_6	CH_3
propane	C_3H_8	C_3H_8
butane	C_4H_{10}	C_2H_5
pentane	C_5H_{12}	C_5H_{12}
hexane	C_6H_{14}	C_3H_7

How to find the molecular formula

To find the molecular formula for an unknown compound, you need to know:

- the **empirical formula**. This is found by experiment, as in the example on page 83.
- the **relative molecular mass** of the compound (M_r). This is also found by experiment, for example using a mass spectrometer.

Once you have these two pieces of information you can work out the molecular formula.

> To find the molecular formula:
>
> i Calculate $\dfrac{M_r}{\text{empirical mass}}$ for the substance. This gives a number, n.
>
> ii Multiply the numbers in the empirical formula by n.

Let's look at two examples.

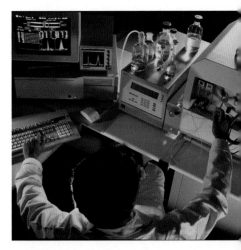

A mass spectrometer, for finding relative molecular mass. It compares the mass of a molecule with the mass of a carbon-12 atom.

Calculating the molecular formula

Example 1 A molecular compound has the empirical formula HO. Its relative molecular mass is 34. What is its molecular formula?

(A_r: H = 1, O = 16.)

For the empirical formula HO, the empirical mass = 17. But M_r = 34.

So $\dfrac{M_r}{\text{empirical mass}} = \dfrac{34}{17} = 2$

So the molecular formula is 2 × HO, or **H_2O_2**.
The compound is hydrogen peroxide.

Example 2 Octane is a hydrocarbon. It is 84.2% carbon and 15.8% hydrogen by mass. Its M_r is 114. What is its molecular formula?

1 **First find the empirical formula for the compound.**
84.2 g of carbon combines with 15.8 g of hydrogen.
Changing masses to moles:
$\dfrac{84.2}{12}$ moles of carbon atoms combine with $\dfrac{15.8}{1}$ moles of hydrogen atoms, or
7.02 moles of carbon atoms combine with 15.8 moles of hydrogen atoms, so
1 mole of carbon atoms combines with $\dfrac{15.8}{7.02}$ or **2.25 moles** of hydrogen atoms.
So atoms combine in the ratio of 1 : 2.25 or **4 : 9**. (You must give the ratio as whole numbers, since only whole atoms can combine.)
The empirical formula of octane is therefore **C_4H_9**.

2 **Then use M_r to find the molecular formula.**
For the empirical formula (C_4H_9), the empirical mass = 57.
But M_r = 114.

So $\dfrac{M_r}{\text{empirical mass}} = \dfrac{114}{57} = 2$

So the molecular formula of octane = 2 × C_4H_9 = **C_8H_{18}**.

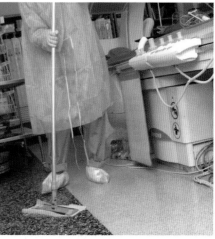

Using hydrogen peroxide solution to clean a hospital floor. Hydrogen peroxide acts as a bleach, and kills germs.

Octane is a flammable liquid, and one of the main ingredients in petrol.

Questions

1. In the ionic compound magnesium chloride, magnesium and chlorine atoms combine in the ratio 1:2. What is the formula of magnesium chloride?
2. In the ionic compound aluminium fluoride, aluminium and fluorine atoms combine in the ratio 1:3. What is the formula of aluminium fluoride?
3. What is the difference between an empirical formula and a molecular formula? Can they ever be the same?
4. What is the empirical formula of benzene, C_6H_6?
5. A compound has the empirical formula CH_2. Its M_r is 28. What is its molecular formula?
6. A hydrocarbon is 84% carbon, by mass. Its relative molecular mass is 100. Find:
   a its empirical formula b its molecular formula
7. An oxide of phosphorus has an M_r value of 220. It is 56.4% phosphorus. Find its molecular formula.

5.7 The concentration of a solution

What does 'concentration' mean?

A

Solution A contains 2.5 grams of copper(II) sulphate in 1 dm³ of water. So its concentration is **2.5 grams/dm³**.

B

Solution B contains 25 grams of the same compound in 1 dm³ of water. So its concentration is **25 grams/dm³**.

C

Solution C contains 125 grams of the compound in 0.5 dm³ of water. So its concentration is **250 grams/dm³**.

The concentration of a solution is the amount of solute, in grams or moles, that is dissolved in 1 dm³ of solution.

Finding the concentration in moles

Example Find the concentrations of A and C above, in moles per dm³.

You must first change the mass of the solute to moles.
The formula mass of copper(II) sulphate is 250, as shown on the right.
So 1 mole of the compound has a mass of 250 grams.

Solution A It has 2.5 grams of the compound in 1 dm³ of solution.

2.5 grams = $\frac{2.5}{250}$ moles = 0.01 moles

so its concentration is **0.01 moles per dm³**.

Moles per dm³ is often shortened just to M, so the concentration of solution A can be written as **0.01 M**.

Solution C It has 250 grams of the compound in 1 dm³ of solution.
250 grams = 1 mole
so its concentration is **1 mole per dm³**, or **1 M** for short.

A solution that contains 1 mole of solute per dm³ is often called a **molar solution**. So C is a molar solution.

> The formula of copper(II) sulphate is $CuSO_4.5H_2O$. The formula contains 1 Cu, 1 S, 9 O, and 10 H, so the formula mass is:
>
> 1 Cu = 1 × 64 = 64
> 1 S = 1 × 32 = 32
> 9 O = 9 × 16 = 144
> 10 H = 10 × 1 = 10
> Total = 250

In general, to find the concentration of a solution in moles per dm³:

concentration (mol/dm³) = $\frac{\text{amount of solute (mol)}}{\text{volume of solution (dm}^3\text{)}}$

Use the equation above to check that the last column in this table is correct:

Amount of solute (mol)	Volume of solution (dm³)	Concentration of solution (mol/dm³)
1.0	1.0	1.0
0.2	0.1	2.0
0.5	0.2	2.5
1.5	0.3	5.0

Remember
- 1 dm³ = 1 litre
  = 1000 cm³
  = 1000 ml
- All these mean the same thing:
  moles per dm³
  mol/dm³
  mol dm⁻³
  moles per litre

Finding the amount of solute in a solution

If you know the concentration of a solution, and its volume:
- you can work out how much solute it contains, in moles. Just rearrange the equation from the last page:
  amount of solute (mol) = concentration (mol/dm³) × volume (dm³)
- you can then convert moles to grams by multiplying the number of moles by M_r.

Use the calculation triangle

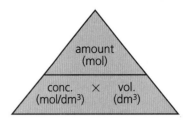

Cover one to see how to calculate it. So, cover *concentration* and you'll see that it is *amount* divided by *volume*. To draw this triangle, just remember *alligators chew visitors*!

Sample calculations

The table shows different volumes of solutions, with different concentrations. Check that you understand the calculations. Are the results correct?

	Solution A sodium hydroxide (NaOH) 2 dm³	Solution B sodium thiosulphate $Na_2S_2O_3$ 250 cm³	Solution C lead nitrate $Pb(NO_3)_2$ 100 cm³	Solution D silver nitrate $AgNO_3$ 25 cm³
concentration (mol/dm³)	1	2	0.1	0.05
amount of solute present (moles)	1 × 2 = 2	2 × $\frac{250}{1000}$ = 0.5	0.1 × $\frac{100}{1000}$ = 0.01	0.05 × $\frac{25}{1000}$ = 0.00125
M_r	40	158	331	170
mass of solute present (g)	80	79	3.31	0.2125

Questions

1. How many moles of solute are in:
   a. 500 cm³ of solution, of concentration 2 mol/dm³?
   b. 2 litres of solution, of concentration 0.5 mol/dm³?
2. What is the concentration of a solution containing:
   a. 4 moles in 2 dm³ of solution?
   b. 0.3 moles in 200 cm³ of solution?
3. Different solutions of salt X are made up. What volume of:
   a. a 4 mol/dm³ solution contains 2 moles of X?
   b. a 6 mol/dm³ solution contains 0.03 moles of X?
4. The M_r of sodium hydroxide is 40. How many grams of sodium hydroxide are there in:
   a. 500 cm³ of a molar solution?
   b. 25 cm³ of a 0.5 M solution?
5. What is the concentration in moles per litre of:
   a. a sodium carbonate solution containing 53 g of the salt (Na_2CO_3) in 1 litre?
   b. a copper(II) sulphate solution containing 62.5 g of the salt ($CuSO_4.5H_2O$) in 1 litre?

Questions on Chapter 5

Relative atomic masses (A_r) are given on page 299. Use the approximate values in the table there.

1. How many grams are there in this?
   a. 1 mole of copper atoms
   b. 1.5 moles of sulphur atoms
   c. 2 moles of magnesium atoms
   d. 5 moles of carbon atoms
   e. 10 moles of chlorine atoms
   f. 0.1 moles of nitrogen atoms
   g. 0.2 moles of neon atoms

2. How many grams are there in this?
   a. 1 mole of hydrogen molecules, H_2
   b. 2 moles of hydrogen molecules, H_2
   c. 1 mole of oxygen molecules, O_2
   d. 0.5 moles of chlorine molecules, Cl_2
   e. 2 moles of phosphorus molecules, P_4
   f. 4 moles of sulphur molecules, S_8
   g. 3 moles of ozone molecules, O_3

3. Find how many moles of atoms there are, in:
   a. 32 g of sulphur
   b. 48 g of magnesium
   c. 23 g of sodium
   d. 14 g of lithium
   e. 1.4 g of lithium
   f. 3.1 g of phosphorus
   g. 6.4 g of oxygen
   h. 5.4 g of aluminium
   i. 2 g of hydrogen

4. Which of the two amounts in each pair contains more atoms?
   a. 80 g of sulphur, 80 g of calcium
   b. 80 g of sulphur, 80 g of oxygen
   c. 1 mole of sulphur atoms, 8 moles of chlorine atoms
   d. 1 mole of sulphur atoms, 1 mole of oxygen molecules

5. How many grams are there in this?
   a. 5 moles of water
   b. 1 mole of hydrated copper(II) sulphate, $CuSO_4.5H_2O$
   c. 2 moles of ammonia, NH_3
   d. 0.3 moles of calcium carbonate, $CaCO_3$
   e. $\frac{1}{5}$ mole of magnesium oxide, MgO
   f. 0.1 moles of sodium thiosulphate, $Na_2S_2O_3$
   g. 2 moles of iron(III) chloride, $FeCl_3$

6. 1 mole of sodium carbonate (Na_2CO_3) contains 2 moles of sodium atoms, 1 mole of carbon atoms, and 3 moles of oxygen atoms.
   In the same way, write down the number of moles of each atom present in 1 mole of:
   a. lead oxide, Pb_3O_4
   b. ammonium nitrate, NH_4NO_3
   c. calcium hydroxide, $Ca(OH)_2$
   d. ethanoic acid, CH_3COOH
   e. hydrated iron(II) sulphate, $FeSO_4.7H_2O$

7. The formula of calcium oxide is CaO. The A_r values are: Ca = 40, O = 16. Complete the following statements:
   a. 1 mole of Ca (. . . g) and 1 mole of O (. . . g) combine to form . . . . mole of CaO (. . . g).
   b. 4.0 g of calcium and . . . . g of oxygen combine to form . . . . g of calcium oxide.
   c. When 0.4 g of calcium reacts with oxygen, the increase in mass is . . . . g.
   d. If 6 moles of CaO were decomposed to calcium and oxygen, ... moles of Ca and . . . . moles of O_2 would be obtained.
   e. The percentage by mass of calcium in calcium oxide is . . . . %.

8. 110 g of manganese was extracted from 174 g of manganese oxide. (A_r: Mn = 55, O = 16.)
   a. What mass of oxygen is there in 174 g of manganese oxide?
   b. How many moles of oxygen atoms is this?
   c. How many moles of manganese atoms are there in 110 g of manganese?
   d. What is the empirical formula of manganese oxide?
   e. What mass of manganese would be obtained from 1000 g of manganese oxide?

9. 27 g of aluminium burns in chlorine to form 133.5 g of aluminium chloride. (A_r: Al = 27, Cl = 35.5.)
   a. What mass of chlorine is present in 133.5 g of aluminium chloride?
   b. How many moles of chlorine atoms is this?
   c. How many moles of aluminium atoms are present in 27 g of aluminium?
   d. Use your answers for parts b and c to find the simplest formula of aluminium chloride.
   e. 1 dm^3 of an aqueous solution is made using 13.35 g of aluminium chloride. What is its concentration in moles per dm^3?

10 Copy and complete the following table.
(A_r: H = 1, C = 12, N = 14, O = 16.)

Compound	Molar mass g/mol	Empirical formula	Molecular formula
hydrazine	32	NH_2	
cyanogen	52	CN	
nitrogen oxide	92	NO_2	
glucose	180	CH_2O	

11 Hydrocarbons A and B both contain 85.7% carbon. Their molar masses are 42 and 84 g respectively.
   a Which elements does a hydrocarbon contain?
   b Calculate the empirical formulae of A and B. (A_r: H = 1, C = 12.)
   c Calculate the molecular formulae of A and B.

12 Zinc phosphide is made by heating zinc with phosphorus. 9.75 g of zinc combines with 3.1 g of phosphorus.
   a Find the empirical formula for the compound. (A_r: Zn = 65, P = 31.)
   b Calculate the percentage of phosphorus in it.

13 Phosphorus forms two oxides, which have the empirical formulae P_2O_3 and P_2O_5.
   a Which oxide contains the higher percentage of phosphorus? (A_r: P = 31, O = 16.)
   b What mass of phosphorus will combine with 1 mole of oxygen molecules (O_2) to form P_2O_3?
   c What is the molecular formula of the oxide that has a formula mass of 284?
   d Suggest a molecular formula for the other oxide.

14 Which of the two solutions in each pair contains more moles of solute?
   a 1 dm² of 1 M sodium chloride (NaCl), 1 dm³ of 2 M sodium chloride
   b 500 cm³ of 1 M sodium chloride, 1 dm³ of 1 M sodium chloride
   c 1 dm³ of 0.1 M sodium chloride, 100 cm³ of 2 M sodium chloride
   d 250 cm³ of 2 M sodium chloride, 1 dm³ of 1 M sodium hydroxide (NaOH).

15 You have to prepare some 2 M solutions, with 10 g of solute in each.
What volume of solution will you prepare, for each solute below?
(H = 1, Li = 7, N = 14, O = 16, Mg = 24, S = 32.)
   a lithium sulphate, Li_2SO_4
   b magnesium sulphate, $MgSO_4$
   c ammonium nitrate, NH_4NO_3

16 An oxide of copper can be converted to copper by heating it in a stream of hydrogen, in the apparatus shown below:

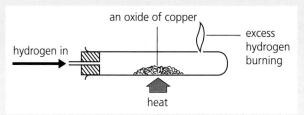

Hydrogen was allowed to pass through the test tube for some time before it was lit. The test tube was heated until all the oxide was converted to copper.
The apparatus was then allowed to cool, with hydrogen still passing through, before it was dismantled.
   a Write a word equation for the reaction.
   b Why was hydrogen allowed to pass through the apparatus for some time, before it was lit? Why was it kept flowing, during cooling?

The experiment was repeated by different groups of students. Each used a different mass of oxide. Here are the results:

Group	Mass of copper oxide/g	Mass of copper produced/g	Mass of oxygen lost/g
1	0.62	0.55	0.07
2	0.90	0.80	0.10
3	1.12	1.00	0.12
4	1.69	1.50	
5	1.80	1.60	

   c Work out the missing figures for the table.
   d On graph paper, plot the mass of copper against the mass of oxygen. (Show the mass of copper along the x-axis and oxygen along the y-axis.) Then draw the line of best fit through the origin and set of points.
   e From the graph, find the mass of oxygen that would combine with 1.28 g of copper.
   f Calculate the mass of oxygen that would combine with 128 g of copper. (Remember, 1.28 × 100 = 128.)
   g How many moles of copper atoms are there in 128 g of copper? (A_r: Cu = 64.)
   h How many moles of oxygen atoms combine with 128 g of copper? (A_r: O = 16.)
   i What is the simplest formula for this oxide of copper?
   j What is another term for *the simplest formula*, found in this way?

6.1 Physical and chemical change

A substance can be changed by heating it, adding water to it, mixing another substance with it, and so on. The change that takes place will be either a **chemical** change or a **physical** one.

Chemical change

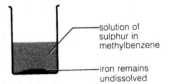

1 Some yellow sulphur and black iron filings are mixed together on filter paper.

2 The mixture is easily separated again, by using a magnet to attract the iron . . .

3 . . . or by dissolving the sulphur in methylbenzene (a solvent).

4 But when the mixture is *heated*, the yellow specks of sulphur disappear. A black solid forms.

5 This black solid is not at all like the mixture. It is not affected by a magnet . . .

6 . . . and none of it dissolves in methylbenzene.

The black solid that formed in step 4 is clearly a new chemical substance. So a chemical change has taken place.
In a chemical change, a new chemical substance is produced.

The difference between a mixture and a compound

In step 1 above, iron and sulphur particles are mixed closely together. But they are not bonded to each other. The iron particles still behave like iron, and the sulphur particles like sulphur.

But in step 4, the iron and sulphur have reacted together. They have formed ions, which bond to form the ionic compound iron sulphide. The magnet and solvent now have no effect.

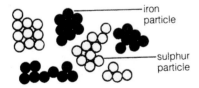
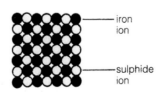
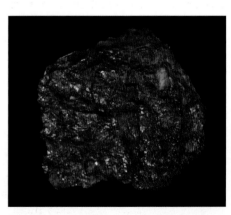

Mixture Clusters of iron and sulphur particles mixed together – but not chemically bonded.

Compound Iron ions and sulphide ions bonded together, forming iron(II) sulphide.

The compound iron(II) sulphide occurs in the earth as pyrite.

The signs of a chemical change

A chemical change is usually called a **chemical reaction**. You can tell when a chemical reaction has taken place, by these signs:

1 **One or more new chemical substances are formed.**
   The new substances usually look different from the starting substances.
   For example:

iron	+	sulphur	→	iron(II) sulphide
(black filings)		(yellow powder)		(black solid)

2 **Energy is taken in or given out, during the reaction.**
   Energy was needed to start the reaction between iron and sulphur, in the form of heat from a Bunsen. But the reaction gave out heat once it began.

   A change that gives out heat energy is called **exothermic**.
   A change that takes in heat energy is **endothermic**.

   So the reaction between iron and sulphur is exothermic.
   The reactions that take place when you fry an egg are endothermic.

3 **The change is usually difficult to reverse.**
   You would need to carry out several reactions to get the iron and sulphur back from iron sulphide. (But it can be done!)

Fireworks contain magnesium and other substances. When they burn, the reactions give out energy in the form of heat, light, and sound.

Physical change

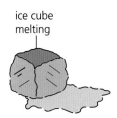
ice cube melting

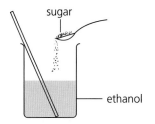
sugar
ethanol

The reactions that take place when you fry an egg are endothermic.

Ice turns to water at 0°C. It is easy to change the water back to ice again, by cooling it.

Sugar dissolves in ethanol. You can separate the two again by distilling the solution.

No new chemical substances are formed in these changes. Although ice and water *look* different, they are both made only of water molecules.
If no new chemical substance is formed, a change is a physical change.

Physical changes can be exothermic or endothermic. But, unlike chemical changes, they are usually easy to reverse.

Questions
1 Explain the difference between a *mixture* of iron and sulphur and the *compound* iron sulphide.
2 What are the signs of a chemical change?
3 What is an *exothermic* reaction?
4 Name one physical change that is *endothermic*.
5 Is the change chemical or physical? Give reasons.
   a a glass bottle breaking
   b butter and sugar being made into toffee
   c cotton being woven to make sheets
   d coal burning in air.

6.2 Equations for chemical reactions

Equations for two sample reactions

The reaction between carbon and oxygen When carbon is heated in oxygen, they react together, and carbon dioxide is formed. Carbon and oxygen are the **reactants**. Carbon dioxide is the **product** of the reaction.

You could show the reaction using a diagram, like this:

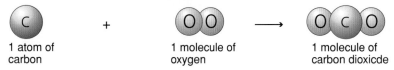

1 atom of carbon   1 molecule of oxygen   1 molecule of carbon dioxicde

or in a shorter way, using symbols and numbers, like this:

$$C + O_2 \longrightarrow CO_2$$

This short way to describe the reaction is called a **chemical equation**.

The reaction between hydrogen and oxygen When hydrogen and oxygen react together, the product is water. The diagram is:

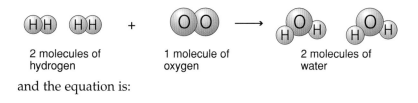

2 molecules of hydrogen   1 molecule of oxygen   2 molecules of water

and the equation is:

$$2H_2 + O_2 \longrightarrow 2H_2O$$

Why there is a 2 in front of H_2 and H_2O, in the equation?
Now look at the number of atoms on each side of the equation:

On the left:	On the right:
4 hydrogen atoms	4 hydrogen atoms
2 oxygen atoms	2 oxygen atoms

The number of each type of atoms is the same on both sides of the equation. This is because atoms do not *disappear* during a reaction – they are just *rearranged*, as shown in the diagram.

When the number of each type of atom is the same on both sides, an equation is **balanced.** If it is not balanced, it is not correct.

Adding state symbols

Reactants and products may be solids, liquids, gases or solutions. You can show their states by adding **state symbols** to the equations:

(s) for solid   (l) for liquid
(g) for gas   (aq) for aqueous solution (solution in water)

For the two reactions above, the equations with state symbols are:

$$C\,(s) + O_2\,(g) \longrightarrow CO_2\,(g)$$
$$2H_2\,(g) + O_2\,(g) \longrightarrow 2H_2O\,(l)$$

The main reaction when coal burns in plenty of air is: $C + O_2 \rightarrow CO_2$.

The reaction between hydrogen and oxygen gives out so much energy that it is used to power rockets. The hydrogen and oxygen are carried as liquids in the fuel tanks.

How to write the equation for a reaction

These are the steps to follow, when writing an equation:

1. Write the equation in words.
2. Now write it using symbols. Make sure all the formulae are correct.
3. Check that the equation is balanced, for each type of atom in turn. *Make sure you do not change any formulae.*
4. Add the state symbols.

Example 1 Calcium burns in chlorine to form calcium chloride, a solid. Write an equation for the reaction, using the steps above.

1. calcium + chlorine ⟶ calcium chloride
2. $Ca + Cl_2 \longrightarrow CaCl_2$
3. Ca: 1 atom on the left and 1 atom on the right.
   Cl: 2 atoms on the left and 2 atoms on the right.
   The equation is balanced.
4. $Ca\,(s) + Cl_2\,(g) \longrightarrow CaCl_2\,(s)$

Example 2 Hydrogen chloride is formed by burning hydrogen in chlorine. Write an equation for the reaction.

1. hydrogen + chlorine ⟶ hydrogen chloride
2. $H_2 + Cl_2 \longrightarrow HCl$
3. H: 2 atoms on the left and 1 atom on the right.
   Cl: 2 atoms on the left and 1 atom on the right.
   The equation is *not* balanced. It needs another molecule of hydrogen chloride on the right. So a 2 is put *in front of* the HCl.
   $H_2 + Cl_2 \longrightarrow 2HCl$
   The equation is now balanced. Do you agree?
4. $H_2\,(g) + Cl_2\,(g) \longrightarrow 2HCl\,(g)$

Example 3 Magnesium burns in oxygen to form magnesium oxide, a white solid. Write an equation for the reaction.

1. magnesium + oxygen ⟶ magnesium oxide
2. $Mg + O_2 \longrightarrow MgO$
3. Mg: 1 atom on the left and 1 atom on the right.
   O: 2 atoms on the left and 1 atom on the right.
   The equation is *not* balanced. Try this:
   $Mg + O_2 \longrightarrow 2MgO$ (Note, the 2 goes *in front of* the MgO.)
   Another magnesium atom is now needed on the left:
   $2Mg + O_2 \longrightarrow 2MgO$
   The equation is balanced.
4. $2Mg\,(s) + O_2\,(g) \longrightarrow 2MgO\,(s)$

Magnesium burning in oxygen.

Questions

1. What do + and ⟶ mean, in an equation?
2. Balance the following equations:
   a. $Na\,(s) + Cl_2\,(g) \longrightarrow NaCl\,(s)$
   b. $H_2\,(g) + I_2\,(g) \longrightarrow HI\,(g)$
   c. $Na\,(s) + H_2O\,(l) \longrightarrow NaOH\,(aq) + H_2\,(g)$
   d. $NH_3\,(g) \longrightarrow N_2\,(g) + H_2\,(g)$
   e. $C\,(s) + CO_2\,(g) \longrightarrow CO\,(g)$
   f. $Al\,(s) + O_2\,(g) \longrightarrow Al_2O_3\,(s)$
3. Aluminium burns in chlorine to form aluminium chloride, $AlCl_3$, a solid. Write an equation for the reaction.

6.3 Calculations from equations

What an equation tells you

When carbon burns in oxygen, the reaction can be shown as:

 + ⟶

1 atom of carbon   1 molecule of oxygen   1 molecule of carbon dioxide

or in a short way, using an equation:

$$C\,(s) + O_2\,(g) \longrightarrow CO_2\,(g)$$

This equation tells you that:

| 1 carbon atom | reacts with | 1 molecule of oxygen | to give | 1 molecule of carbon dioxide |

Now suppose there was 1 *mole* of carbon atoms. These would react with 1 *mole* of oxygen molecules:

| 1 mole of carbon atoms | reacts with | 1 mole of oxygen molecules | to give | 1 mole of carbon dioxide molecules |

Moles can be changed to grams, using A_r and M_r. (See Unit 5.4.)
The A_r values are: C = 12, O = 16.
So the M_r values are: O_2 = 32, CO_2 = (12 + 32) = 44, and we can write:

| 12 g of carbon | reacts with | 32 g of oxygen | to give | 44 g of carbon dioxide |

This also means that:

| 6 g of carbon | reacts with | 16 g of oxygen | to give | 22 g of carbon dioxide |

and so on. The masses of the substances taking part in the reaction *are always in the same ratio*.

You can find out the same kind of information from any equation.

From the equation for a reaction you can tell:
- **how many moles of each substance take part**
- **how many grams of each substance take part.**

Does the mass change during a reaction?

Now look what happens to the total mass, during the above reaction:

    mass of carbon and oxygen at the start: 12 g + 32 g = **44 g**
    mass of carbon dioxide at the end: **44 g**

The total mass has not changed, during the reaction. This is because no atoms have disappeared. They have just been rearranged.

That is one of the basic laws of chemistry:
The total mass remains unchanged, during a chemical reaction.

Calculating masses from equations

These are the steps to follow:
1. Write the balanced equation for the reaction. (It gives *moles*.)
2. Write down the A_r or M_r for each substance that takes part.
3. Using A_r or M_r, change the moles in the equation to *grams*.
4. Once you know the masses for the equation, you can find any *actual* mass.

Example Hydrogen burns in oxygen to form water. What mass of oxygen is needed to burn 1 gram of hydrogen, and what mass of water is obtained?

1. The equation for the reaction is: $2H_2 (g) + O_2 (g) \longrightarrow 2H_2O (l)$
2. A_r: H = 1, O = 16. M_r: H_2 = 2, O_2 = 32, H_2O = 18.
3. So, for the equation, the amounts in grams are:

$2H_2 (g)$	+	$O_2 (g)$	$\longrightarrow$	$2H_2O (l)$	
2 × 2 g		32 g		2 × 18 g	or
4 g		32 g		36 g	

4. But you start with only 1 g of hydrogen, so the *actual* masses are:

1 g		32/4 g		36/4 g	or
1 g		8 g		9 g	

So 1 g of hydrogen needs **8 g** of oxygen to burn, and gives **9 g** of water.

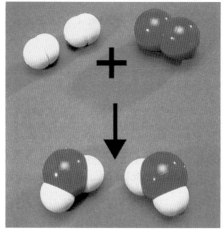

These models show how the atoms are rearranged during the reaction between hydrogen and oxygen. The equation is $2H_2 + O_2 \rightarrow 2H_2O$.

Working out equations from the masses that react

If you know the actual masses that react, you can then work out the equation for the reaction, by changing masses to moles.

Example Iron reacts with copper(II) sulphate ($CuSO_4$) solution to give copper and iron sulphate solution. The formula for the iron sulphate could be either $FeSO_4$ or $Fe_2(SO_4)_3$. In an experiment, 1.4 g of iron gave 1.6 g of copper. Write the correct equation for the reaction.

1. A_r: Fe = 56, Cu = 64.
2. Change the masses to moles of atoms:

    $\frac{1.4}{56}$ moles of iron atoms produce $\frac{1.6}{64}$ moles of copper atoms, or

    0.025 moles of iron atoms produce 0.025 moles of copper atoms, so 1 mole of iron atoms produces 1 mole of copper atoms.
3. So the equation for the reaction must be:
    $Fe + CuSO_4 \longrightarrow Cu + FeSO_4$
4. Add the state symbols to complete it:
    $Fe (s) + CuSO_4 (aq) \longrightarrow Cu (s) + FeSO_4 (aq)$

Iron wool reacting with copper(II) sulphate solution. Iron is more reactive than copper so displaces it from solution. A deep pink deposit of copper forms.

Questions

1. The reaction between magnesium and oxygen is:
   $2Mg (s) + O_2 (g) \longrightarrow 2MgO (s)$
   a. Write a word equation for the reaction.
   b. How many moles of magnesium atoms react with 1 mole of oxygen molecules?
   c. The A_r values are: Mg = 24, O = 16.
      How many grams of oxygen react with:
      i 48 g of magnesium?   ii 12 g of magnesium?

2. Copper(II) carbonate breaks down on heating, like this:
   $CuCO_3 (s) \longrightarrow CuO (s) + CO_2 (g)$
   a. Write a word equation for the reaction.
   b. Find the mass of 1 mole of each substance in the reaction. (A_r: Cu = 64, C = 12, O = 16.)
   c. When 31 g of copper(II) carbonate is used:
      i how many grams of carbon dioxide form?
      ii what mass of solid remains after heating?

6.4 Reactions involving gases

A closer look at some gases

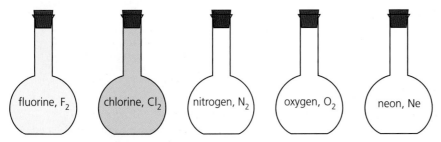

Imagine five very large flasks, each with a volume of 24 dm³. Each is filled with a different gas. Each gas is at room temperature and pressure, or **rtp**.

(We take **room temperature and pressure** as the standard conditions for comparing gases; rtp is 20 °C and 1 atmosphere.)

If you weighed the gas in the five flasks, you would discover something amazing. There is exactly 1 mole of each gas!

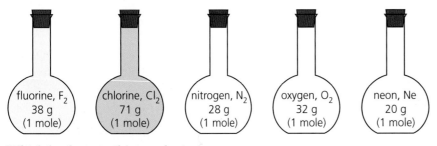

Which leads us to this conclusion:
1 mole of every gas occupies the same volume, at the same temperature and pressure. At room temperature and pressure, this volume is 24 dm³.

This was discovered by the Italian scientist Avogadro, in 1811. So it is often called **Avogadro's Law**. It does not matter whether the gas exists as atoms or molecules, or whether its atoms are large or small. The law still holds.

The volume occupied by 1 mole of a gas is called its **molar volume**.
The molar volume of a gas is 24 dm³ at rtp.

> **Remember**
> 24 dm³ = 24 litres
>   = 24 000 cm³
>
> Imagine a ball about 36 cm in diameter. Its volume is about 24 dm³.

Another way to look at it

Look at these two gas jars.
A is full of nitrogen dioxide, NO_2.
B is full of oxygen, O_2. Both gases exist as molecules.

The two gas jars have identical volumes. And the two gases are at the same temperature and pressure.

You can't see the gas molecules, let alone count them. But, from Avogadro's Law, you can say that the number of molecules is the same in each jar.

Gas volumes from moles and grams

Avogadro's Law makes it easy to work out the volumes of gases.

Example 1 What volume does 0.25 moles of a gas occupy at rtp?
1 mole occupies 24 dm³ so
0.25 moles occupies 0.25 × 24 dm³ = 6 dm³
so 0.25 moles of any gas occupies **6 dm³** (or **6000 cm³**) at rtp.

Example 2 What volume does 22 g of carbon dioxide occupy at rtp?
M_r of carbon dioxide = 44, so
44 g = 1 mole, so
22 g = 0.5 mole
so the volume occupied = 0.5 × 24 dm³ = **12 dm³**.

Use the calculation triangle
Cover the one you want to find – and you'll see how to calculate it.

(Triangle: Volume at rtp (dm³) / no of moles × 24 dm³)

Gas volumes from equations

From the equation for a reaction, you can tell how many *moles* of a gas take part. Using Avogadro's Law, you can also work out its *volume*. In these examples, all volumes are measured at rtp.

Example 1 What volume of hydrogen will react with 24 dm³ of oxygen to form water?

1. The equation for the reaction is: $2H_2\,(g) + O_2\,(g) \longrightarrow 2H_2O\,(l)$
2. So 2 volumes of hydrogen react with 1 of oxygen, or 2 × 24 dm³ of hydrogen react with 24 dm³ of oxygen.
48 dm³ of hydrogen will react.

Example 2 When sulphur burns in air it forms sulphur dioxide. What volume of this gas is produced when 1 g of sulphur burns? (A_r: S = 32.)

1. The equation for the reaction is: $S\,(s) + O_2\,(g) \longrightarrow SO_2\,(g)$
2. 32 g of sulphur atoms = 1 mole of sulphur atoms, so 1 g = $\frac{1}{32}$ mole or 0.03125 moles of sulphur atoms.
3. 1 mole of sulphur atoms gives 1 mole of sulphur dioxide molecules so 0.03125 moles of sulphur atoms gives 0.03125 moles of sulphur dioxide molecules.
4. 1 mole of sulphur dioxide molecules has a volume of 24 dm³ at rtp so 0.03125 moles has a volume of 0.03125 × 24 dm³ at rtp, or 0.75 dm³. So **0.75 dm³** (or **750 cm³**) of sulphur dioxide are produced.

Sulphur dioxide is one of the gases given out in volcanic eruptions. These scientists are collecting gas samples on the slopes of an active volcano.

Questions
(A_r: O = 16, N = 14, H = 1, C = 12.)

1. What does *rtp* mean? What values does it have?
2. What does *molar volume* mean, for a gas?
3. What is the molar volume of neon gas at rtp?
4. For any gas, calculate the volume at rtp of:
    a 7 moles   b 0.5 moles   c 0.001 moles
5. Calculate the volume at rtp of:
    a 16 g of oxygen (O_2)   b 1.7 g of ammonia (NH_3)
6. You burn 6 grams of carbon in plenty of air:
    $C\,(s) + O_2\,(g) \longrightarrow CO_2\,(g)$
    a What volume of gas will form (at rtp)?
    b What volume of oxygen will be used up?
7. If you burn the carbon in limited air, the reaction is different: $2C\,(s) + O_2\,(g) \longrightarrow 2CO\,(g)$
    a What volume of gas will form this time?
    b What volume of oxygen will be used up?

6.5 Finding % yield and % purity

Yield and purity

The **yield** is the amount of product you obtain from a reaction. Suppose you own a factory that makes paint or fertilisers. You will want the highest yield possible, for the lowest cost!

Now imagine your factory makes medical drugs, or flavouring for foods. The yield will still be important – but the **purity** of the product may be even more important. Because impurities could harm people.

In this unit you'll learn how to calculate the % yield from a reaction, and the % purity of the product obtained.

Working in a chemical factory: the yield is always important.

Finding the % yield

You can work out % yield like this:

$$\% \text{ yield} = \frac{\text{actual mass obtained}}{\text{calculated mass}} \times 100\%$$

Example The medical drug aspirin is made from salicyclic acid. 1 mole of salicylic acid gives 1 mole of aspirin:

$$\underset{\text{salicylic acid}}{C_7H_6O_3} \xrightarrow{\text{chemicals}} \underset{\text{aspirin}}{C_9H_8O_4}$$

In a trial, 100.0 grams of salicylic acid gave 121.2 grams of aspirin. What was the % yield?

1. A_r : C = 12, H = 1, O = 16.
   So M_r : salicyclic acid = 138, aspirin = 180.

2. 138 g of salicyclic acid = 1 mole
   so 100 g = $\frac{100}{138}$ mole = 0.725 moles

3. 1 mole of salicylic acid gives 1 mole of aspirin
   so 0.725 moles give 0.725 moles of aspirin
   or 0.725 × 180 g = 130.5 g
   So 130.5 g is the **calculated mass** for the reaction.

4. But the **actual mass** obtained in the trial was 121.2 g.
   So % yield = $\frac{121.2}{130.5}$ g × 100 = **92.9%**

This is a high yield – so it is worth continuing with those trials.

Finding the % purity

When you make something in a chemical reaction, and separate it from the final mixture, it will not be pure. It will still have small amounts of other substances mixed with it. The impurities could be small amounts of unreacted starting substances. Or small amounts of another product that you don't want.

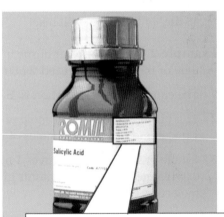

REFERENCE DATA

$C_6H_4(OH)COOH$ MW 138.12 EC[200-712
SPECIFICATION
Assay >99.5%
Loss on drying < 0.3%
Chloride < 0.05%
Sulphate < 0.02%
Heavy metals as Pb < 0.001%

Labels on chemicals usually give % purity. Some chemicals are purified until they have only 'parts per trillion' of an impurity. This proportion of impurity is like one leaf in a forest!

You can work out the % purity of the product you obtained like this:

% purity of a product = $\dfrac{\text{mass of the pure product}}{\text{mass of the impure product obtained}} \times 100\%$

Example 1 The aspirin from the trial on page 98 was not pure. 121.2 g of solid was obtained, but analysis showed that only 109.2 g of it was aspirin. The rest was impurities such as unreacted salicylic acid.

% purity of the aspirin obtained = $\dfrac{109.2}{121.2}$ g × 100% = **90.0%**

This level of purity is *not* acceptable for a medical drug. So the aspirin was purified by recrystallizing it several times. Each time, the product was a little purer. Finally an 80 g sample was obtained, containing 79.8 g of aspirin.

% purity of this aspirin = $\dfrac{79.8}{80}$ g × 100% = **99.75%**

The aspirin is now more acceptable for medical use. But making it purer has reduced the yield from 121.2 g to 80 g. So the % yield is now only **61.3%**.

Example 2 Chalk is almost pure calcium carbonate. You can work out its purity by measuring how much carbon dioxide it gives off. 10 g of chalk was reacted with an excess of dilute hydrochloric acid. 2280 cm³ of carbon dioxide gas was collected at room temperature and pressure (rtp).

The equation for the reaction is:
$CaCO_3$ (s) + 2HCl (aq) ⟶ $CaCl_2$ (aq) + H_2O (l) + CO_2 (g)

1 M_r of $CaCO_3$ = 100   (A_r: Ca = 40, C = 12, O = 16.)
2 1 mole of $CaCO_3$ gives 1 mole of CO_2 and
   1 mole of gas has a volume of 24 000 cm³ at rtp.
3 So 24 000 cm³ of gas is produced by 100 g of calcium carbonate
   and 2280 cm³ is produced by $\dfrac{2280}{24\,000}$ × 100 g or **9.5 g**.

So there is 9.5 g of calcium carbonate in the 10 g of chalk.

So the % purity of the chalk = $\dfrac{9.5}{10}$ g × 100 = **95%**.

> **Purity check!**
> You can check the purity of any product by measuring its melting and boiling points, and comparing them with tables of values. (You can buy these tables.)
> - Impurities lower the melting point and raise the boiling point.
> - The more impurity present, the greater the change.

Chalk cliffs in Normandy, France. Chalk, like limestone, is calcium carbonate.

Questions

1 Define the term: **a** % yield **b** % purity
2 100 g of aspirin was obtained from 100 g of salicylic acid. What was the % yield?
3 17 kg of aluminium was produced from 51 kg of aluminium oxide (Al_2O_3) by electrolysis. What was the percentage yield? (A_r: Al = 27, O = 16.)
4 Some sea water is evaporated. The sea salt obtained is found to be 86% sodium chloride. How much sodium chloride could be obtained from 200 g of this salt?
5 A 5.0 g sample of dry ice (solid carbon dioxide) turned into 2400 cm³ of carbon dioxide gas at rtp. What was the percentage purity of the dry ice? (M_r of CO_2 = 44.)

Questions on Chapter 6

1. Is it a physical change, or a chemical change? Give reasons for your answers.
   a. ice melting
   b. iron rusting
   c. petrol burning
   d. candle wax melting
   e. a candle burning
   f. wet hair drying
   g. milk souring
   h. perfume evaporating
   i. a lump of roll sulphur being crushed
   j. copper being obtained from copper(II) oxide
   k. clothes being ironed
   l. custard being made
   m. a cigarette being smoked
   n. copper(II) sulphate crystallizing from solution

2. Write a chemical equation for each of the following – without state symbols. For example:

   would be written as: $2H_2 + O_2 \rightarrow 2H_2O$

   a. H-H + Cl-Cl → H-Cl H-Cl
   b. N≡N + H-H H-H H-H → NH₃ NH₃
   c. P-P + Cl-Cl Cl-Cl Cl-Cl → PCl₃ PCl₃
   d. S with O O + O-O → SO₃ SO₃

3. Write equations for the following reactions:
   a. 1 mole of copper atoms combines with 1 mole of sulphur atoms to form 1 mole of copper(II) sulphide, CuS.
   b. 3 moles of lead atoms combine with 2 moles of oxygen molecules to form 1 mole of lead oxide, Pb_3O_4.
   c. 1 mole of ethanol molecules, C_2H_5OH, burns in 3 moles of oxygen molecules to form 2 moles of carbon dioxide molecules and 3 moles of water molecules.
   d. 1 mole of iron(III) oxide, Fe_2O_3, reacts with 3 moles of hydrogen molecules to form 2 moles of iron atoms and 3 moles of water molecules.

4. Balance these equations:
   a. $H_2\ (g) + Br_2\ (g) \longrightarrow HBr\ (g)$
   b. $Cl_2\ (g) + KBr\ (aq) \longrightarrow KCl\ (aq) + Br_2\ (aq)$
   c. $C_2H_4\ (g) + O_2\ (g) \longrightarrow CO_2\ (g) + H_2O\ (l)$
   d. $Zn\ (l) + Fe_2O_3\ (l) \longrightarrow Fe\ (l) + ZnO\ (s)$
   e. $NH_3\ (g) + O_2\ (g) + H_2O\ (l) \longrightarrow HNO_3\ (l)$
   f. $Pb(NO_3)_2\ (s) \longrightarrow PbO_2\ (s) + NO_2\ (g) + O_2\ (g)$
   g. $Al\ (s) + HCl\ (aq) \longrightarrow AlCl_3\ (aq) + H_2\ (g)$
   h. $C_2H_5OH\ (l) + O_2\ (g) \longrightarrow CO_2\ (g) + H_2O\ (l)$

5. Mercury(II) oxide breaks down into mercury and oxygen when heated, like this:

   $2HgO\ (s) \longrightarrow 2Hg\ (l) + O_2\ (g)$

   a. Calculate the mass of 1 mole of mercury(II) oxide. (A_r: O = 16, Hg = 201)
   b. How much mercury and oxygen should be obtained from 21.7 g of mercury(II) oxide?
   c. When the experiment was carried out, only 19.0 g of mercury was actually collected. Calculate the % yield of mercury for this experiment.

6. Iron(II) sulphide is formed when iron and sulphur react together:

   $Fe\ (s) + S\ (s) \longrightarrow FeS\ (s)$

   a. How many grams of sulphur will react with 56 g of iron? (See A_r values on page 299.)
   b. If 7 g of iron and 10 g of sulphur are used, which substance is in excess?
   c. If 7 g of iron and 10 g of sulphur are used, name the substances present when the reaction is complete, and find the mass of each.
   d. What mass of iron will react completely with 10 g of sulphur?

7. Iron is obtained from iron(III) oxide in this reaction:

   $Fe_2O_3\ (s) + 3CO\ (g) \longrightarrow 2Fe\ (s) + 3CO_2\ (g)$

   a. Write a word equation for the reaction.
   b. What is the formula mass of iron(III) oxide? (A_r: Fe = 56, O = 16.)
   c. How many moles of Fe_2O_3 are there in 320 kg of iron(III) oxide? (1 kg = 1000 g.)
   d. How many moles of Fe are obtained from 1 mole of Fe_2O_3?
   e. From c and d, find how many moles of iron atoms are obtained from 320 kg of iron(III) oxide.
   f. How much iron (in kg) is obtained from 320 kg of iron(III) oxide?

8 What is the volume at rtp, in dm^3 and cm^3, of:
a 2 moles of hydrogen, H_2?
b 0.5 moles of carbon?
c 0.01 moles of nitrogen, N_2?
d 0.3 moles of oxygen, O_2?
e 0.1 moles of neon, Ne?
f 0.003 moles of ammonia, NH_3?

9 The volumes of some gases were measured at rtp. The readings were:
  i hydrogen, $6 \, dm^3$
  ii oxygen, $3 \, dm^3$
  iii neon, $2400 \, cm^3$
  iv carbon monoxide, $600 \, dm^3$
  v carbon dioxide, $1.2 \, dm^3$
  vi sulphur dioxide, $480 \, cm^3$
a What does rtp mean?
b Calculate the number of moles of particles of each gas present. Say if they are atoms or molecules.
c Now calculate the number of grams of each gas present. (See A_r values on page 299.)

10 Sodium hydrogen carbonate, $NaHCO_3$, breaks down on heating, like this:

$2NaHCO_3 \, (s) \longrightarrow$
$\quad Na_2CO_3 \, (s) + H_2O \, (l) + CO_2 \, (g)$

a Write a word equation for the reaction.
b i How many moles of sodium hydrogen carbonate are in the equation?
  ii What is the mass of this amount?
    (A_r: Na = 23, H = 1, C = 12, O = 16.)
c i How many moles of carbon dioxide are there in the equation?
  ii What is the volume of this amount at rtp?
d What volume at rtp of carbon dioxide will be obtained from the decomposition of:
  i 84 g of sodium hydrogen carbonate?
  ii 8.4 g of sodium hydrogen carbonate?

11 When calcium carbonate is heated strongly, this chemical change occurs:

$CaCO_3 \, (s) \longrightarrow CaO \, (s) + CO_2 \, (g)$

a Write a word equation for the change.
b How many moles of $CaCO_3$ are there in 50 g of calcium carbonate? (A_r: Ca = 40, C = 12, O = 16.)
c i What mass of calcium oxide is obtained from the thermal decomposition of 50 g of calcium carbonate, assuming a 40% yield?
  ii What mass of carbon dioxide will be given off at the same time?
  iii What volume will this gas occupy at rtp?

12 Nitrogen monoxide reacts with oxygen to form nitrogen dioxide. The equation is:

$2NO \, (g) + O_2 \, (g) \longrightarrow 2NO_2 \, (g)$

a How many moles of oxygen molecules react with 1 mole of nitrogen monoxide molecules?
b What volume of oxygen will react with $50 \, cm^3$ of nitrogen monoxide?
c Using the volumes in **b**, what is:
  i the total volume of the two reactants?
  ii the volume of nitrogen dioxide formed?

13 Nitroglycerine is used as an explosive. The equation for the explosion reaction is:

$4C_3H_5(NO_3)_3 \, (l) \longrightarrow$
$\quad 12CO_2 \, (g) + 10H_2O \, (l) + 6N_2 \, (g) + O_2 \, (g)$

a How many moles does the equation show for:
  i nitroglycerine? ii gas molecules produced?
b How many moles of gas molecules are obtained from 1 mole of nitroglycerine?
c What is the total volume of gas (at rtp) obtained from 1 mole of nitroglycerine?
d What is the mass of 1 mole of nitroglycerine? (A_r values: H = 1, C = 12, N = 14, O = 16.)
e What will be the total volume of gas (at rtp) from exploding 1 kg of nitroglycerine?
f Using your answers above, try to explain *why* nitroglycerine is used as an explosive.

14 A 5g sample of impure magnesium carbonate is reacted with an excess of hydrochloric acid:

$MgCO_3 \, (s) + 2HCl \, (aq) \longrightarrow$
$\quad MgCl_2 \, (aq) + H_2O \, (l) + CO_2 \, (g)$

$1250 \, cm^3$ of carbon dioxide is collected at rtp.
a How many moles of CO_2 are produced?
b What mass of pure magnesium carbonate would give this volume of carbon dioxide? (A_r values: C = 12, O = 16, Mg = 26.)
c Calculate the % purity of the sample of magnesium carbonate.

15 2 g (an excess) of iron is added to $50 \, cm^3$ of 0.5 M sulphuric acid. When the reaction is over, the reaction mixture is filtered. The mass of the unreacted iron is found to be 0.6g. (Fe = 56.)
a What mass of iron took part in the reaction?
b How many moles of iron atoms took part?
c How many moles of sulphuric acid reacted?
d Write the equation for the reaction, and deduce the charge on the iron ion that formed.
e What volume of hydrogen (calculated at rtp) bubbled off during the reaction?

7.1 Different types of reaction

Reactions everywhere

Thousands of reactions are going on all around you – and inside you. Here we review some different types of reaction you meet in the school lab. Then we show that there is a pattern to them.

Some different types of reaction

1 **Combination** or **synthesis**   This is where two or more substances react to form *just one compound*.

   Example   Iron reacts with sulphur to form iron sulphide:

   iron + sulphur ⟶ iron sulphide
   $Fe(s) + S(s) \longrightarrow FeS(s)$

   You have to heat the mixture to get the reaction started. But once started it gives out heat, and the mixture glows.

The reaction between iron and sulphur: they combine together.

2 **Decomposition**   Here a substance breaks down into simpler substances. This can be brought about by heat, light, electricity, or enzymes.

   Example 1   Silver chloride is a white solid. If you leave it standing in daylight it goes dark grey, because light causes it to break down to silver and chlorine:

   silver chloride $\xrightarrow{light}$ silver + chlorine
   $2AgCl(s) \xrightarrow{light} 2Ag(s) + Cl_2(g)$

   Silver bromide and silver iodide decompose in the same way. These reactions are used in photographic film, as you will see on page 127.

   Example 2   Ionic substances can be broken down by passing electricity through them. The process is called **electrolysis**.

   For example if you connect two graphite rods to a battery and stand them in molten lithium chloride, beads of silvery lithium form at one rod, and bubbles of chlorine gas at the other:

   lithium chloride $\xrightarrow{electricity}$ lithium + chlorine
   $2LiCl(l) \xrightarrow{electricity} 2Li(l) + Cl_2(g)$

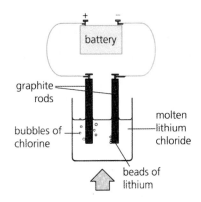

The decomposition of molten lithium chloride, by electrolysis.

   Electrolysis is used to extract aluminium from its ore, and to get chlorine from sodium chloride (salt). You can find out more about it in Chapter 9.

3 **Combustion**   This is where a substance catches fire and burns, on reaction with oxygen or other gas. The reaction produces heat, light and sound.

   Example   Magnesium burns in oxygen with a dazzling white flame and a fizzing sound:

   magnesium + oxygen ⟶ magnesium oxide
   $2Mg(s) + O_2(g) \longrightarrow 2MgO(s)$

   Because of this vigorous reaction, magnesium is used in fireworks. The magnesium is powdered to make it burn even faster.

Magnesium burning in oxygen.

4 Displacement Here an element swops places with another element, in a compound.

Example When you place copper wire in a solution of silver nitrate, the copper dissolves – and crystals of silver form:

copper + silver nitrate ⟶ copper(II) nitrate + silver
$Cu\,(s) + 2AgNO_3\,(aq) \longrightarrow Cu(NO_3)_2\,(aq) + 2Ag\,(s)$

The reaction occurs because copper is more reactive than silver, so has a stronger drive to form a compound. The solution turns blue as the reaction proceeds. The blue colour is due to copper ions.

5 Precipitation Here solutions react, giving an insoluble product.

Example When solutions of silver nitrate and sodium chloride are mixed, a precipitate of solid silver chloride forms:

silver nitrate + sodium chloride ⟶ silver choride + sodium nitrate
$AgNO_3\,(aq) + NaCl\,(aq) \longrightarrow AgCl\,(s) + NaNO_3\,(aq)$

The silver chloride soon turns dark grey, due to the reaction shown in **2**.

6 Neutralisation This is where an acid reacts with another substance that destroys its acidity, giving water as one product.

Example Hydrochloric acid reacts with sodium hydroxide solution like this:

sodium hydroxide + hydrochloric acid ⟶ sodium chloride + water
$NaOH\,(aq) + HCl\,(aq) \longrightarrow NaCl\,(aq) + H_2O\,(l)$

Acids turn blue litmus paper red. But when the reaction is complete, the products have no affect on blue litmus paper.

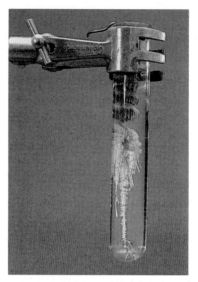

The displacement reaction between copper wire and silver nitrate. Look at the crystals of silver growing. The copper ions that form turn the solution blue.

Three groups of reactions

All the different reactions you meet in chemistry can seem confusing. But in fact most of them fall into these three groups:

- **oxidation and reduction reactions,** or **redox reactions.** In these, electrons are transferred. The examples given in **1–4** are redox.
- **precipitation reactions,** where ions come together to form a solid, but no electrons are transferred. See **5** above.
- **neutralisation reactions** where acids react with substances called bases, and one product is water. But no electrons are transferred. See **6** above.

We look at redox reactions in the rest of this chapter. And you can find out more about precipitation and neutralisation reactions in Chapter 8.

A precipitate of silver chloride forms when you mix solutions of sodium chloride and silver nitrate.

Questions
1. Look at the different types of reaction shown in this unit. Which type is the opposite to combination?
2. Write a balanced equation for the synthesis of water from hydrogen and oxygen.
3. Give an example of a combustion reaction that takes place in the home. Write a word equation if you can.
4. Silver bromide also decomposes in light, to silver and bromine. Write a balanced equation for the reaction.

7.2 Oxidation and reduction

Gaining oxygen during reactions

As you saw on page 102, magnesium burns in oxygen with a dazzling white flame. A white ash is left behind. The reaction is:

magnesium + oxygen ⟶ magnesium oxide
$2Mg\,(s)$ + $O_2\,(g)$ ⟶ $2MgO\,(s)$

The magnesium has gained oxygen. We say it has been **oxidised**.

When a substance gains oxygen, it has been oxidised. Oxidation has taken place.

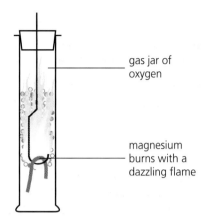

magnesium burns with a dazzling flame

gas jar of oxygen

Losing oxygen during reactions

Now look what happens when hydrogen is passed over heated copper(II) oxide. The black compound turns pink:

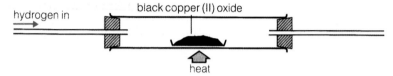

This reaction is taking place:

copper(II) oxide + hydrogen ⟶ copper + water
$CuO\,(s)$ + $H_2\,(g)$ ⟶ $Cu\,(s)$ + $H_2O\,(l)$

This time the heated substance is *losing* oxygen. It is being **reduced**.

When a substance loses oxygen, it has been reduced. Reduction has taken place.

Reductions like this one are important in providing us with metals. Many metals occur in the earth as oxides, or compounds than can be easily changed into oxides. Iron ore is a good example: it occurs as iron(III) oxide. To obtain the metal, we must reduce the oxide.

Oxidation and reduction take place together

Look again at the last reaction above. The hydrogen gains oxygen, to form water. So while copper(II) oxide is reduced, hydrogen is oxidised.

Oxidation and reduction *always* take place together. If one substance is reduced, another is oxidised.

Now look again at the first reaction above. Oxidation and reduction always take place together. The magnesium is being oxidised, which means the oxygen is being reduced. (We will look more closely at this reaction in the next unit.)

Redox reactions

Since oxidation and reduction always take place together, we call the reactions **redox reactions**.

Iron occurs naturally in the earth as iron(III) oxide. This is reduced to iron in the blast furnace. Molten iron runs out from the bottom of the furnace.

When hydrogen is involved

The terms *oxidation* and *reduction* are also used for reactions that involve hydrogen but *no* oxygen. Look at these examples.

1 **Ethanol** is the alcohol in beer and other drinks. It is changed to **ethanal** by a compound called potassium dichromate(VI), in the presence of acid:

$$CH_3CH_2OH\ (l) \xrightarrow[\text{(orange)}]{K_2Cr_2O_7} CH_3CHO\ (l)$$
$$\text{ethanol} \qquad\qquad\qquad \text{ethanal}$$

The ethanol molecule has lost two hydrogen atoms, to become ethanal. We say it has been **oxidised**.
The loss of hydrogen is also called oxidation.

While the ethanol is oxidised, the orange potassium dichromate(VI) is reduced to a green compound. This colour change is the basis of a breathalyser test for alcohol. Look at the photo on the right.

A breathalyser test for drivers. The device contains orange potassium dichromate(VI). Alcohol in the motorist's breath will make it turn green.

2 Soyabean oil and other vegetable oils are runny. But you can 'harden' them by reaction with hydrogen, to give compounds you can spread on bread (instead of butter), or use in baking.

Look at the big molecule on the left below. It has carbon – carbon double bonds. These break and add on hydrogen atoms:

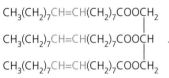 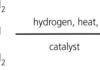

a runny vegetable oil → a fat you can spread with a knife
(hydrogen, heat, catalyst)

The oil molecule has gained hydrogen atoms. We say it has been **reduced**.
The gain of hydrogen is also called reduction.

Oxidation and reduction: a summary so far

- **Oxidation** means gain of oxygen, or loss of hydrogen.
- **Reduction** means loss of oxygen, or gain of hydrogen.

But these are 'old' definitions, used for centuries. Today we have a broader definition for oxidation and reduction, as you will see in the next unit.

Like other vegetable oils, olive oil is runny. It can be 'hardened' by reaction with hydrogen, to give a spread for bread.

Questions

1 $2Mg\ (s) + SO_2\ (g) \rightarrow 2MgO\ (s) + S\ (s)$
In this reaction, which substance is:
  a oxidised? b reduced?
2 Explain where the term *redox* comes from.

2 $N_2\ (g) + 3H_2\ (g) \rightleftharpoons 2NH_3\ (g)$
The two half-arrows show that the reaction is reversible.
  a Write a word equation for the reaction.
  b Is the nitrogen being oxidised, or reduced?

7.3 Redox and ion transfer

Another definition for oxidation and reduction

When magnesium burns in oxygen, magnesium oxide is formed:

$$2Mg\,(s) + O_2\,(g) \longrightarrow 2MgO\,(s)$$

The magnesium has clearly been oxidised. Oxidation and reduction *always* take place together, so the oxygen must have been reduced. But how? Let's see what is happening to the electrons:

magnesium atom — 2,8,2 — two electrons transfer — **oxygen atom** 2,6 — giving — **magnesium ion, Mg^{2+}** $[2,8]^{2+}$ and **oxide ion, O^{2-}** $[2,8]^{2-}$

During the reaction, each magnesium atom loses two electrons and each oxygen atom gains two. This leads us to a new definition:

**If a substance loses electrons during a reaction, it has been oxidised.
If it gains electrons, it has been reduced.**

Writing half-equations to show the electron transfer

You can show the electron transfer for the reaction above very clearly, by using **half-equations**. A half-equation focuses on just one element. Here is how to write half-equations:

1. Write down each reactant, with the electrons it gains or loses.

   magnesium: $\quad Mg \longrightarrow Mg^{2+} + 2e^-$
   oxygen: $\quad O + 2e^- \longrightarrow O^{2-}$

2. Check that each substance is in its correct form (ion, atom or molecule) on each side of the arrow. If it is not, correct it.

   Oxygen is not in its correct form on the left above. It exists as molecules, not single atoms. So you must change O to O_2. And that means you must also double the number of electrons and O^{2-}:

   oxygen: $\quad O_2 + 4e^- \longrightarrow 2O^{2-}$

3. **The number of electrons must be the same in both equations. If it is not, multiply one (or both) equations by a number, to balance them.**

   So we must multiply the magnesium half-equation by 2.

   magnesium: $\quad 2\,Mg \longrightarrow 2\,Mg^{2+} + 4e^-$
   oxygen: $\quad O_2 + 4e^- \longrightarrow 2O^{2-}$

   The equations are now balanced, each with 4 electrons.

Note that if you add the two half-equations, the electrons and ion charges cancel, giving the full equation like the one at the top of the page:

$$2\,Mg + O_2 \longrightarrow 2\,MgO$$

> **Don't forget!**
>
> **O**xidation **I**s **L**oss of electrons.
>
> **R**eduction **I**s **G**ain of electrons.
>
> The term **OIL RIG** will help you remember.

Two other examples of redox reactions

So our definition of redox reactions is now much broader:
Any reaction in which electron transfer takes place is a redox reaction.

Let's look at two more examples:

1 **The reaction between sodium and chlorine.** The equation is:

$$2Na\,(s) + Cl_2\,(g) \longrightarrow 2NaCl\,(s)$$

The sodium atoms give electrons to the chlorine atoms, as shown on the right. So sodium is oxidised, and chlorine is reduced, and the reaction is a redox reaction. Look at the balanced half-equations:

sodium: $2Na \longrightarrow 2Na^+ + 2e^-$ (oxidation)
chorine: $Cl_2 + 2e^- \longrightarrow 2Cl^-$ (reduction)

Now compare these half-equations with the full equation, above.
Are the numbers in front of the symbols the same in both?

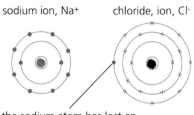

the sodium atom has lost an electron to the chlorine atom

2 **The decomposition of lithium chloride** As you saw on page 102, lithium chloride is broken down by electrolysis to lithium and chlorine:

$$2LiCl\,(l) \xrightarrow{electricity} 2Li\,(l) + Cl_2\,(g)$$

During the reaction, electrons are transferred, as shown on the right. Chloride ions give up electrons, and lithium ions accept them. So this is a redox reaction. Look at the balanced half-equations:

lithium: $2Li^+ + 2e^- \longrightarrow 2Li$ (reduction)
chlorine: $2Cl^- \longrightarrow Cl_2 + 2e^-$ (oxidation)

Once again, compare the half-equations with the full equation.
Are the numbers in front of the symbols the same in both?

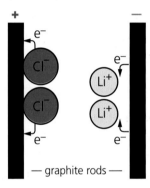

The electron transfer that takes place during the decomposition of lithium chloride by electrolysis.

Redox: a summary

Oxidation is …	Reduction is …
gain of oxygen	loss of oxygen
loss of hydrogen	gain of hydrogen
loss of electrons	gain of electrons

Oxidation and reduction always take place together, so the reaction is called a **redox reaction**.

Questions

1 Give a full definition for:
   a oxidation   b reduction
2 a What does a *half-equation* show?
   b Half-equations are useful. Why?
   c In what ways do the half-equations for a reaction match the full equation?
3 See if you can write the balanced half-equations for:
   a the reaction of potassium with chlorine
   b the decomposition of sodium chloride, by electricity
   c the burning of calcium in oxygen, giving calcium oxide.
4 For each part of 3, add the half-equations, to give the full redox equation for the reaction.

7.4 Redox and changes in oxidation state

What does oxidation state mean?

Oxidation state tells you how many electrons each atom of an element has gained, lost or shared, in forming a compound. For example the oxidation state of chlorine atoms in sodium chloride, NaCl, is -I.

In this unit you will see how oxidation states can help you to identify redox reactions.

The rules for oxidation states

These are the rules for oxidation states:

1 Each atom in a formula has an oxidation state.

2 The oxidation state is often given as a Roman numeral (and *always* when it appears in a compound name). So note these Roman numerals:

number	0	1	2	3	4	5	6	7
Roman numeral	0	I	II	III	IV	V	VI	VII

3 Where an element is not combined with other elements, its atoms are in oxidation state 0.

4 Many elements have the same oxidation state in most or all their compounds. Look at these:

Element	Usual oxidation state in compounds
hydrogen	+I
sodium and the other Group 1 metals	+I
calcium and the other Group 2 metals	+II
aluminium	+III
chlorine and the other Group VII non-metals, in compounds without oxygen	–I
oxygen (except in peroxides)	–II

5 But atoms of transition metals can have variable oxidation states. Look at these:

Element	Oxidation state in compounds
iron	+II and +III
copper	+I and +II
manganese	+VII, +IV, and +II
chromium	+VI and +III

6 Note that in any formula, the oxidation states must add up to zero. Look at the formula of magnesium chloride for example:

   MgCl$_2$
   +II   2 × –I   Total = zero

   So you could use oxidation states to check that formulae are correct.

Copper in its three oxidation states:

                          oxidation state of copper

A copper metal   0
B copper(I) chloride, CuCl   +I
C copper(II) sulphate, CuSO$_4$ +II

Oxidation states change during redox reactions

Look at the equation for the reaction between sodium and chlorine:

$$2Na\ (s)\ +\ Cl_2\ (g)\ \longrightarrow\ 2NaCl\ (s)$$
$$\ \ \ 0\ \ \ \ \ \ \ \ \ \ \ \ \ 0\ \ \ \ \ \ \ \ \ \ \ \ \ \ \ \ \ \ +I\ -I$$

The oxidation states are also shown, from the table on page 108. Notice how they change during the reaction.

Each sodium atom loses an electron during the reaction, to form an Na^+ ion. So sodium is oxidised – and its oxidation state changes from 0 to +I. Each chlorine atom gains an electron, to form a Cl^- ion. So chlorine is reduced. And its oxidation state changes from 0 to -I.

If oxidation states change during a reaction, it is a redox reaction. Oxidation leads to a rise in the oxidation state number. Reduction leads to a fall in the oxidation state number.

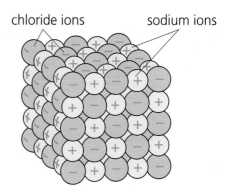

Sodium chloride: made up of sodium ions, Na^+, and chloride ions, Cl^-.

Using oxidation states to identify redox reactions

Example 1 Sodium reacts with oxygen like this:

$$4Na\ (s)\ +\ O_2\ (g)\ \longrightarrow\ 2Na_2O\ (s)$$
$$\ \ \ 0\ \ \ \ \ \ \ \ \ \ \ \ \ 0\ \ \ \ \ \ \ \ \ \ \ \ \ \ \ \ \ \ +I\ -II$$

The oxidation states are shown, from the table on page 108.
There has been a change in oxidation states. So this is a redox reaction.

Example 2 When chlorine is bubbled through a solution of iron(II) chloride, iron (III) choride is formed. The equation and oxidation states are:

$$2FeCl_2\ (aq)\ +\ Cl_2\ (aq)\ \longrightarrow\ 2FeCl_3\ (aq)$$
$$+II\ -I\ \ \ \ \ \ \ \ \ \ \ \ \ \ 0\ \ \ \ \ \ \ \ \ \ \ \ \ \ \ \ +III\ -I$$

There has been a change in oxidation states. So this is a redox reaction.

Example 3 When ammonia and hydrogen chloride gases mix, a white smoke of ammonium chloride forms. The equation and oxidation states are:

$$NH_3\ (g)\ +\ HCl\ (g)\ \longrightarrow\ NH_4Cl\ (s)$$
$$-III\ +I\ \ \ \ \ \ +I\ -I\ \ \ \ \ \ \ \ \ \ \ -III\ +I\ -I$$

There has been no change in oxidation states. So this is *not* a redox reaction.

Questions
1 a Write a word equation for this reaction:
       $2H_2\ (g) + O_2\ (g) \rightarrow 2H_2O\ (l)$
   b Now copy out the chemical equation from **a**. Below each symbol write the oxidation state of the atoms.
   c Is the reaction a redox reaction? Give evidence.
   d Say which substance is oxidised, and which reduced.
2 Repeat the steps in question **1** for each of these equations:
   i $2KBr\ (s) \rightarrow 2K(s) + Br_2\ (l)$
   ii $2KI\ (aq) + Cl_2\ (g) \rightarrow 2KCl\ (aq) + I_2\ (aq)$
3 a Read point 6 on page 108.
   b Using the idea in point 6, work out the oxidation state of the carbon atoms in carbon dioxide, CO_2.
   c Carbon burns in oxygen to form carbon dioxide. Write a chemical equation for the reaction.
   d Now using oxidation states, show that this is a redox reaction, and say which substance is reduced.
4 *Every reaction between two elements is a redox reaction.* Do you agree with this statement? Explain.

7.5 Oxidising and reducing agents

What are oxidising and reducing agents?

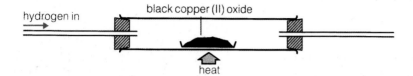

When hydrogen reacts with heated copper(II) oxide, the reaction is:

copper(II) oxide + hydrogen ⟶ copper + water
$CuO\,(s)$ + $H_2\,(g)$ ⟶ $Cu\,(s)$ + $H_2O\,(l)$

The copper(II) oxide is **reduced** to copper by reaction with hydrogen.
So hydrogen acts as a **reducing agent**.
The hydrogen is itself **oxidised** to water, in the reaction. So copper(II) oxide acts as an **oxidising agent**.

An oxidising agent brings about the oxidation of another substance – and is itself reduced.

A reducing agent brings about the reduction of another substance – and is itself oxidised.

Oxidising and reducing agents in the lab

Some substances have a strong drive to gain electrons. So they are strong oxidising agents. They bring about the oxidation of other substances by taking electrons from them.

Other substances are strong reducing agents, readily giving up electrons.

Some of these substances also show a colour change when they react. This makes them useful in the lab, for testing. Let's look at three examples.

1 Potassium manganate(VII): an oxidising agent

Manganese is a transition metal. Like other transition metals, it can exist in different oxidation states. (Look back at point **5** on page 108.)

Potassium manganate(VII) is a purple compound. Its formula is $KMnO_4$. In it, the manganese is in oxidation state (+VII). But it is much more stable in oxidation state (+II). So it is strongly driven to gain electrons, and reduce its oxidation state to (+II).

So potassium manganate(VII) acts as a powerful oxidising agent. It takes electrons from other substances, in the presence of a little acid. It is itself reduced in the reaction – and this causes a colour change:

MnO_4^- $\xrightarrow{\text{reduction}}$ Mn^{2+}
manganate(VII)   manganese(II)
ion (purple)   ion (colourless)

This colour change means that potassium manganate(VII) can be used to test for the presence of a reducing agent. If a reducing agent is present, the purple colour will fade.

Using potassium manganate(VII) to test for the presence of a reducing agent. The purple colour is fading, so the beaker must contain a reducing agent.

2 Potassium dichromate(VI): an oxidising agent

Chromium is also a transition metal, and can exist in different oxidation states. In potassium dichromate(VI) it is in oxidation state (+VI). But oxidation state (+III) is the most stable.

So potassium dichromate(VI) also acts as a strong oxidising agent, in the presence of acid. It reacts in order to gain electrons and reduce the oxidation state to (+III).

Once again there is a colour change on reduction:

$$Cr_2O_7^{2-} \xrightarrow{\text{reduction}} 2Cr^{3+}$$

dichromate(VI)   chromium(III)
ion (orange)   ion (green)

This colour change means that potassium dichromate(VI) can be used to test for the presence of reducing agents.

A contains only a solution of potassium dichromate(VI).
B shows the colour change when the dichromate(VI) ion is reduced to the chromium(III) ion, by a reducing agent.

3 Potassium iodide: a reducing agent

When potassium iodide solution is added to hydrogen peroxide, in the presence of sulphuric acid, this redox reaction takes place:

$$H_2O_2\ (aq) + 2KI\ (aq) + H_2SO_4\ (aq) \longrightarrow I_2\ (aq) + K_2SO_4\ (aq) + 2H_2O\ (l)$$

hydrogen   potassium   iodine   potassium
peroxide   iodide   sulphate

You can see that the hydrogen peroxide loses oxygen: it is reduced. The potassium iodide acts as a reducing agent. At the same time the potassium iodide is itself oxidised to iodine. This also brings a colour change:

$$2I^- \xrightarrow{\text{oxidation}} I_2$$

colourless   red-brown

This colour change means that potassium iodide can be used to test for the presence of an oxidising agent.

Colourless potassium iodide solution has just been added to a colourless solution in this beaker. The red-brown colour that appears shows that the potassium iodide is being oxidised – so there must be an oxidising agent present.

Questions
1. Explain what these are, in terms of electron transfer:
   a an oxidising agent   **b** a reducing agent
2. Explain why potassium manganate(VII) is a powerful oxidising agent.
3. Explain why potassium dichromate(VI) is used in a breathalyser test. (See the photo on page 105.)
4. Potassium iodide is used to test for the presence of oxidising agents. Why is it suitable for this?

Questions on Chapter 7

1. Write balanced equations, using the correct symbols, for these chemical reactions:
   a. combustion of carbon monoxide to form carbon dioxide
   b. decomposition of solid mercury(II) oxide into its elements
   c. precipitation of barium sulphate, on mixing solutions of barium chloride and sodium sulphate
   d. reduction of copper(II) oxide by hydrogen
   e. thermal decomposition of calcium hydroxide to form calcium oxide and water (*thermal* means brought about by heat)
   f. decomposition of silver chloride by light
   g. decomposition of molten aluminium oxide by electricity.

2. If a substance gains oxygen during a reaction, it is being oxidised. If it loses oxygen, it is being reduced.
   Oxidation and reduction always take place together, so if one substance is oxidised, another is reduced.
   a. First, see if you can write a word equation for each redox reaction **A** to **F** below.
   b. Then, using just the ideas above, say which substance is being oxidised, and which is being reduced, in each reaction.
   - **A** $Ca\ (s) + O_2\ (g) \longrightarrow 2\ CaO\ (s)$
   - **B** $2CO\ (g) + O_2\ (g) \longrightarrow 2CO_2\ (g)$
   - **C** $CH_4\ (g) + 2O_2\ (g) \longrightarrow CO_2\ (g) + 2H_2O\ (l)$
   - **D** $2CuO\ (s) + C\ (s) \longrightarrow 2Cu\ (s) + CO_2\ (g)$
   - **E** $2Fe\ (s) + 3O_2\ (g) \longrightarrow 2Fe_2O_3\ (s)$
   - **F** $Fe_2O_3\ (s) + 3CO\ (g) \longrightarrow 2Fe\ (s) + 3CO_2\ (g)$

3. Redox reactions are reactions during which electron transfer takes place. This diagram shows the electron transfer during a reaction.

   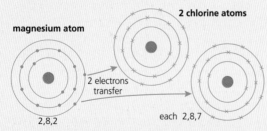

   a. What is the product of this reaction?
   b. Write a balanced equation for the full reaction.
   c. Which element is being oxidised?
   d. Write a half-equation for the oxidation.
   e. i. Which element is being reduced?
      ii. Write a half-equation for the reduction of this element.

4. *Redox* reactions involve electron transfer.
   a. Draw a diagram to show the electron transfer that takes place when fluorine reacts with lithium.
   b. i. Which element is oxidised in the reaction?
      ii. Write a half-equation for this oxidation.
   c. i. Which element is reduced in the reaction?
      ii. Write a half-equation for this reduction.

5. a. Is this a redox reaction? Give your evidence.
      i. $Zn\ (s) + 2HCl\ (aq) \rightarrow ZnCl_2\ (aq) + H_2\ (g)$
      ii. $SiO_2\ (s) + C\ (s) \rightarrow Si\ (s) + CO_2\ (g)$
      iii. $NaOH\ (aq) + HCl\ (aq) \rightarrow NaCl\ (aq) + H_2O\ (l)$
      iv. $Fe\ (s) + CuO\ (s) \rightarrow FeO\ (s) + Cu\ (s)$
      v. $C\ (s) + PbO\ (s) \rightarrow CO\ (g) + Pb\ (s)$
   b. For each of the redox reactions you identify above, name:
      i. the oxidising agent   ii. the reducing agent.

6. Chlorine gas is bubbled into a solution containing sodium bromide. The equation for the reaction is:
   $$Cl_2\ (g) + 2NaBr\ (aq) \rightarrow Br_2\ (aq) + 2NaCl\ (aq)$$
   a. i. Do you agree that chlorine has taken the place of bromine, in the metal compound?
      ii. What is this type of reaction called?
   b. The compounds of Group I metals are white, and give colourless solutions. What would you see as the above reaction proceeds?
   c. i. Write a half-equation for the reaction of the chlorine.
      ii. Is the chlorine oxidised, or reduced, in the reaction? Explain your answer
   d. Write a half-equation for the reaction of the bromide ion.
   e. Why does the above reaction occur?
   f. Which halide ion could be used to convert bromine back to the bromide ion?

7. The oxidation states in a formula add up to zero.
   a. Give the oxidation state of the underlined atom in each formula below:
      i. aluminium oxide, $\underline{Al}_2O_3$
      ii. ammonia, $\underline{N}H_3$
      iii. lead sulphide, $\underline{Pb}S$
      iv. $H_2\underline{C}O_3\ (aq)$, carbonic acid
      v. phosphorus trichloride, $\underline{P}Cl_3$
      vi. copper(I) chloride, $\underline{Cu}Cl$
      vii. copper(II) chloride, $\underline{Cu}Cl_2$
   b. Now comment on the compounds in **vi** and **vii**.

	Group number in Periodic Table							
	1	2	3	4	5	6	7	0
Element	sodium	magnesium	aluminium	silicon	phosphorus	sulphur	chlorine	argon
Ion, where one forms	Na^+	Mg^{2+}	Al^{3+}	none	P^{3-}	S^{2-}	Cl^-	none
Typical compound (not all ionic!)	NaCl	MgO	$AlCl_3$	SiO_2	PH_3	H_2S	HCl	none
Its oxidation state in this compound	+I	+II	+III	+IV	–III	–II	–I	—

8 The table above shows the elements for one period from the Periodic Table.
  a What is the number of the period shown here?
  b Describe the pattern in the oxidation state across the period.
  c i Why are the oxidation states negative, after Group 4?
    ii Why do they decrease after Group 4?
  d Compare the charges on the ions with the oxidation states for the elements. What pattern do you notice?
  e Using point 6 on page 108, predict the formulae of the compounds formed between these pairs of elements:
    i aluminium and sulphur
    ii silicon and chlorine
    iii magnesium and phosphorus
    iv sodium and argon
  f Would you expect similar trends to be found in other periods of the Periodic Table? Explain.

9 Iodine is extracted from seaweed in a redox reaction using acidified hydrogen peroxide. The ionic equation for the reaction is:

$2I^- (aq) + H_2O_2 (aq) + 2H^+ (aq) \longrightarrow I_2 (aq) + 2H_2O (l)$

  a In which oxidation state is the iodine found in seaweed?
  b There is a colour change in this reaction. Why?
  c i Is the iodide ion oxidised or reduced in this reaction?
    ii Write the half equation for this change.
  d In hydrogen peroxide, the oxidation state of the hydrogen is +I.
    i What is the oxidation state of the oxygen in hydrogen peroxide?
    ii How does the oxidation state of oxygen change during the reaction?
    iii Complete the half equation for hydrogen peroxide:
    $H_2O_2 (aq) + 2H^+ (aq) + \longrightarrow 2H_2O (l)$

10 The oxidising agent potassium manganate(VII) can be used to analyse the % of iron(II) present in iron tablets. Below is an **ionic equation**, showing the ions that take part in the reaction:

$MnO_4^- (aq) + 8H^+ (aq) + 5Fe^{2+} (aq) \longrightarrow$
$Mn^{2+} (aq) + 5Fe^{3+} (aq) + 4H_2O (l)$

  a What does the H^+ in the equation tell you about this reaction? (Hint: check page 116.)
  b Describe the colour change.
  c Which is the reducing reagent in this reaction?
  d How could you tell when all the iron(II) had reacted?
  e Write the two half equations for this reaction.

11 Potassium chromate(VI) is yellow. In acid it forms orange potassium dichromate(VI). These are the ions that give those colours:

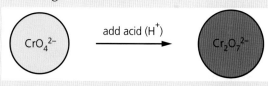

  a What is the oxidation state of chromium in:
    i the yellow compound?
    ii the orange compound?
  b This reaction of chromium ions is *not* a redox reaction. Explain why.

12 When solutions of silver nitrate and potassium chloride are mixed, a white precipitate forms. The ionic equation for the reaction is:
$Ag^+ (aq) + Cl^- (aq) \longrightarrow AgCl (s)$
  a i What is the name of the white precipitate?
    ii Is it a soluble or insoluble compound?
  b Is the precipitation of silver chloride a redox reaction or not? Explain your answer.
  c When left in light, the silver chloride decomposes to form silver and chlorine gas. Write an equation for the reaction and show clearly that this is a redox reaction.

8.1 Acids and alkalis

Acids

One important group of chemicals is called **acids**:

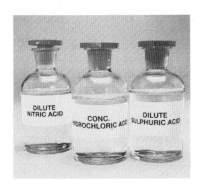
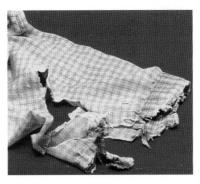

You have probably seen these acids in the lab. They are all solutions of pure compounds in water. They can be dilute, like these, or concentrated.

They must be handled carefully, especially the concentrated solutions, because they are **corrosive**. They can eat away metals, skin, and cloth.

But some acids are not so corrosive, even when concentrated. These are called **weak** acids. Ethanoic acid is one example. It's found in vinegar.

You can tell if something is an acid, by its effect on **litmus**. Litmus is a purple dye. It can be used as a solution, or on paper.
Acids turn litmus red.

> **Remember:**
> acid turns litmus red

Some common acids

The main acids you will meet in chemistry are:

hydrochloric acid	HCl *(aq)*
sulphuric acid	H_2SO_4 *(aq)*
nitric acid	HNO_3 *(aq)*
ethanoic acid	CH_3COOH *(aq)*

But there are many others. For example, lemon and lime juice contain **citric acid**, ant stings contain **methanoic acid**, and fizzy drinks contain **carbonic acid**, formed when carbon dioxide dissolves in water.

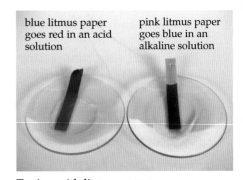

Testing with litmus paper.

Alkalis

There is another group of chemicals that also affect litmus, but in a different way. They are the **alkalis**.
Alkalis turn litmus blue.
Like acids, they must be handled carefully. They too can burn skin.

Some common alkalis

The pure alkalis are solids – except for ammonia, which is a gas. They are used in the lab as aqueous solutions. The main ones you will meet are:

sodium hydroxide	NaOH *(aq)*
potassium hydroxide	KOH *(aq)*
calcium hydroxide	$Ca(OH)_2$ *(aq)*
ammonia	NH_3 *(aq)*

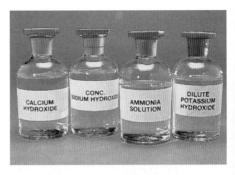

Common laboratory alkalis. The solution of calcium hydroxide is called lime water.

Indicators

Litmus is called an **indicator**, because it indicates whether something is an acid or an alkali. This table shows two others. All show a colour change from acid to alkali. That's why they are used!

Indicator	Colour in acid	Colour in alkali
litmus	red	blue
phenolphthalein	colourless	pink
methyl orange	red	yellow

Neutral substances

Many substances are not acids or alkalis. They are **neutral**. Examples are pure water, and aqueous solutions of sodium chloride and sugar.

The pH scale

You can say how acidic or alkaline a solution is using a scale of numbers called the **pH scale**. The numbers go from 0 to 14:

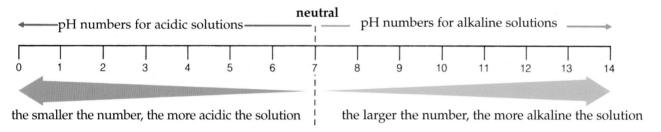

On this scale:
An acidic solution has a pH number less than 7.
An alkaline solution has a pH number greater than 7.
A neutral solution has a pH number of exactly 7.

You can find the pH of any solution by using **universal indicator**. This is a mixture of dyes. Like litmus, it can be used as a solution, or a paper strip. Its colour changes with pH, as shown here:

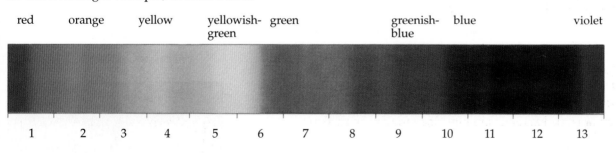

Questions

1. What does *corrosive* mean?
2. How would you test a substance, to see if it is an acid?
3. Write down the formula for:
   sulphuric acid   nitric acid
   calcium hydroxide   ammonia solution
4. Phenolpthalein is an *indicator*. What does that mean?
5. Is a solution acidic, alkaline or neutral, if its pH number is:
   9?   4?   7?   1?   10?   3?
6. What colour is universal indicator, in an aqueous solution of sugar? Why?

8.2 A closer look at acids and alkalis

Acids produce hydrogen ions

Hydrogen chloride is a gas, made of molecules. It dissolves in water to give hydrochloric acid. But this is not molecular. In water, the molecules break up or **dissociate** into ions:

$$HCl\,(aq) \longrightarrow H^+\,(aq) + Cl^-\,(aq)$$

So hydrochloric acid contains hydrogen ions. All other solutions of acids do too. It is the hydrogen ions that give them their 'acidity'. **Solutions of acids contain hydrogen ions.**

Comparing acids

Since solutions of acids contain ions, they conduct electricity. We can measure how well they conduct using a conductivity meter. We can also check their pH using a pH meter.

Samples of acids of the same concentration were tested. This table gives the results. (The unit of conductivity is the siemens, or S.)

Acid	For a 0.1M solution …		
	conductivity (μS/cm)	pH	
hydrochloric acid	25	1.0	strong acids
sulphuric acid	40	0.7	
nitric acid	25	1.0	
methanoic acid	2	2.4	weak acids
ethanoic acid	0.5	2.9	
citric acid	4	2.1	

As you can see, the acids fall into two groups. The first group shows high conductivity, and low pH. These are **strong acids**. The second group does not conduct nearly so well, and has a higher pH. These are **weak acids**.

The difference between strong and weak acids

In a solution of hydrochloric acid, all the molecules of hydrogen chloride have become ions:

$$HCl\,(aq) \xrightarrow{100\%} H^+\,(aq) + Cl^-\,(aq)$$

But in weak acids, only some of the molecules have become ions. For example, for ethanoic acid:

$$CH_3COOH\,(aq) \xrightarrow{\text{much less than }100\%} H^+\,(aq) + CH_3COO^-\,(aq)$$

**In solutions of strong acids, all the molecules become ions.
In solutions of weak acids, only some do.**

So strong acids conduct better because there are more *ions* present. They have a lower pH because there are more *hydrogen ions* present.

The higher the concentration of hydrogen ions, the lower the pH.

Acidic solutions contain the hydrogen ion:

It is what makes them 'acidic'.

A conductivity meter. It measures the current passing through the solution. (The current is carried by the ions.)

Lemon juice contains the weak acid citric acid.

Alkalis produce hydroxide ions

Now let's turn to alkalis, with sodium hydroxide as our example. It is an ionic solid. When it dissolves, all the ions separate:

$$NaOH\ (aq) \longrightarrow Na^+\ (aq) + OH^-\ (aq)$$

So sodium hydroxide solution contains hydroxide ions. The same is true of all alkaline solutions.
Solutions of alkalis contain hydroxide ions.

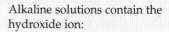

Alkaline solutions contain the hydroxide ion:

It is what makes them alkaline.

Comparing alkalis

We can compare the conductivity and pH of alkalis too. Look at these results:

Alkali	For a 0.1 M solution ...		
	conductivity (µS/cm)	pH	
sodium hydroxide	20	13.0	strong alkalis
potassium hydroxide	15	13.0	
ammonia solution	0.5	11.1	weak alkali

The first two alkalis show high conductivity, and high pH. They are **strong alkalis**. But the ammonia solution shows much lower conductivity, and a lower pH. It is a **weak alkali**.

Why ammonia solution is different

In sodium hydroxide solution, all the sodium hydroxide exists as ions. The same is true for potassium hydroxide.

But ammonia gas is molecular. When it dissolves in water, this is what happens:

$$NH_3\ (aq) + H_2O\ (l) \xrightarrow{\text{much less than 100\%}} NH_4^+\ (aq) + OH^-\ (aq)$$

Only some of the ammonia molecules form ions. So there are fewer hydroxide ions present than in a similar solution of a strong alkali.

So the sodium hydroxide solution is a better conductor than the ammonia solution because it contains more ions. And it has a higher pH because it contains more *hydroxide* ions.

The higher the concentration of hydroxide ions, the higher the pH.

Alkalis react with grease. So the strong alkali sodium hydroxide is used to clear blocked sinks and pipes in homes. What does the drawing tell you?

Questions

1. Write an equation to show what happens when hydrogen chloride dissolves in water.
2. All acids have something in common. What is it?
3. For the table on page 116, explain why ethanoic acid has:
   a lower conductivity   b a higher pH
   than hydrochloric acid.
4. What do all alkaline solutions have in common?
5. Write an equation to show what happens when ammonia gas dissolves in water.
6. For the table above, explain why the ammonia solution has:
   a lower conductivity   b a lower pH
   than the potassium hydroxide solution.

8.3 The reactions of acids and bases

Reactions of acids with metals

Magnesium reacts with dilute sulphuric acid like this:

magnesium + sulphuric acid ⟶ magnesium sulphate + hydrogen
$Mg\,(s) + H_2SO_4\,(aq) \longrightarrow MgSO_4\,(aq) + H_2\,(g)$

The hydrogen bubbles off. When the bubbling stops you can remove any excess magnesium by filtering. You can then evaporate the water from the solution. This leaves a white ionic solid, magnesium sulphate. It is a **salt**.

The metal has displaced the hydrogen from the acid, to form the salt. **When an acid reacts with a metal, hydrogen is displaced, leaving a salt in solution.**

It's a redox reaction Look again at the equation above. The sulphate ions do not change in the reaction. They are the same at each side of the arrow. So we can ignore them, and write an **ionic equation** like this:

$Mg\,(s) + \underset{\text{from the acid}}{2H^+\,(aq)} \longrightarrow Mg^{2+}\,(aq) + H_2\,(g)$

Now it is easier to see that magnesium atoms have lost electrons. The H^+ ions have gained electrons. Electrons have been transferred – so the reaction is a redox reaction.

Magnesium reacting with dilute sulphuric acid.

Naming salts

It depends on the acid you start with:

nitric → **nitrates**
hydrochloric → **chlorides**
sulphuric → **sulphates**

Reactions of acids with bases

Bases are a group of compounds that react with acids, and **neutralise** them, giving a salt *and water*. Bases include alkalis, and insoluble metal oxides, hydroxides and carbonates.

1 With alkalis Alkalis are soluble bases. This shows the reaction between the alkali sodium hydroxide, and hydrochloric acid:

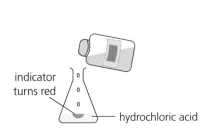

Add a drop of universal indicator to hydrochloric acid in a flask. It turns red, as you'd expect. Then drip in …

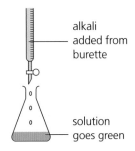

… sodium hydroxide. At a certain point the solution goes green. It is now neutral, like water! The alkali has **neutralised** the acid.

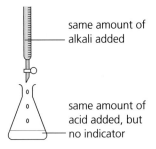

Repeat with the same amounts of acid and alkali, but *no* indicator. Evaporate the solution – and a white solid is left.

The white solid is sodium chloride, a salt. The equation for the reaction is:

hydrochloric acid + sodium hydroxide ⟶ sodium chloride + water
$HCl\,(aq) + NaOH\,(aq) \longrightarrow NaCl\,(aq) + H_2O\,(l)$

Note that water forms too. All acids react with alkalis in the same way:
acid + alkali ⟶ salt + water

2. **With metal oxides and hydroxides** The oxides and hydroxides of many metals are insoluble in water. But if you add the solids to dilute acid, they will neutralise the acid.

For example copper(II) oxide reacts with sulphuric acid like this:

copper(II) oxide + sulphuric acid ⟶ copper(II) sulphate + water
$CuO\ (s)\ +\ H_2SO_4\ (aq) \longrightarrow CuSO_4\ (aq)\ +\ H_2O\ (l)$

The blue copper(II) sulphate that forms is a salt:
acid + metal oxide → salt + water

3. **With carbonates** When calcium carbonate is added to dilute hydrochloric acid, the acid is neutralised. A salt and water are formed. And carbon dioxide is formed too:

calcium + hydrochloric ⟶ calcium + water + carbon
carbonate   acid   chloride   dioxide
$CaCO_3\ (s)\ +\ 2HCl\ (aq) \longrightarrow CaCl_2\ (aq)\ +\ H_2O\ (l)\ +\ CO_2\ (g)$

So with carbonates, the neutralisation reaction is:
acid + metal carbonate ⟶ salt + water + carbon dioxide

In A, black copper(II) oxide is reacting with dilute sulphuric acid. The solution turns blue as copper(II) sulphate forms. B shows how the final solution will look.

A summary of acid reactions

acid + metal ⟶ metal salt + hydrogen
acid + metal oxide or hydroxide ⟶ metal salt + water
acid + metal carbonate ⟶ metal salt + water + carbon dioxide

All three reactions produce a salt. But only the acid/metal reaction is a redox reaction. The acid/base reactions are neutralisations, giving water. As you will see in the next unit, *no electrons are transferred* during a neutralisation, so it is not a redox reaction.

Reactions of bases

1. Bases neutralise acids, as you saw above, giving a salt and water. With carbonates, carbon dioxide is produced too.

2. All the alkalis (except ammonia) will react with ammonium compounds, driving ammonia out. For example:

calcium   + ammonium ⟶ calcium + water + ammonia
hydroxide   chloride   chloride
$Ca(OH)_2\ (s)\ +\ 2NH_4Cl\ (s) \longrightarrow CaCl_2\ (s)\ +\ 2H_2O\ (l)\ +\ 2NH_3\ (g)$

This reaction is used for making ammonia in the laboratory.

Calcium carbonate reacting with dilute hydrochloric acid.

Questions

1. Write a word equation for the reaction of dilute sulphuric acid with:
   a zinc   b sodium hydroxide   c sodium carbonate
2. a Which reaction in question 1 is *not* a neutralisation?
   b Which type of reaction is it? Explain why.
3. Salts are ionic compounds. Name the salt that forms when calcium oxide reacts with hydrochloric acid, and say which ions it contains.
4. Zinc oxide is a base. Suggest a way to make zinc nitrate from it. Write a word equation for the reaction.

8.4 A closer look at neutralisation

The neutralisation of an acid by an alkali (a soluble base)

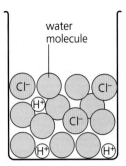

This is a solution of hydrochloric acid. It contains H⁺ ions and Cl⁻ ions. It will turn litmus red.

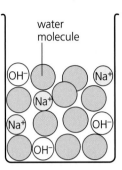

This is a solution of sodium hydroxide. It contains Na⁺ ions and OH⁻ ions. It will turn litmus blue.

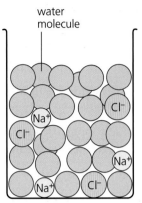

When you mix the two solutions, the OH⁻ ions and H⁺ ions join to form **water molecules**. You end up with a neutral solution of sodium chloride, with no effect on litmus.

The overall equation for this neutralisation reaction is:

$$HCl\ (aq) + NaOH\ (aq) \longrightarrow NaCl\ (aq) + H_2O\ (l)$$

The ionic equation for the reaction

The best way to show what is going on in a neutralisation reaction is to write an **ionic equation** for it. So what is an ionic equation?
An ionic equation shows only the ions that actually take part in a reaction. It leaves out the rest.

This is how to write the ionic equation for the reaction above:

1. **First, write down all the ions present in the equation.**
   The drawings above will help you to do that:
   $$H^+\ (aq) + Cl^-\ (aq) + Na^+\ (aq) + OH^-\ (aq) \longrightarrow Cl^-\ (aq) + Na^+\ (aq) + H_2O\ (l)$$

2. **Now cross out any ions that appear, unchanged, on both sides of the equation.**
   $$H^+\ (aq) + \cancel{Cl^-\ (aq)} + \cancel{Na^+\ (aq)} + OH^-\ (aq) \longrightarrow \cancel{Cl^-\ (aq)} + \cancel{Na^+\ (aq)} + H_2O\ (l)$$

   These crossed-out ions are present in the reaction mixture, but do not actually take part in the reaction. So they are called **spectator ions**.

3. **What's left is the ionic equation for the reaction.**
   $$H^+\ (aq) + OH^-\ (aq) \longrightarrow H_2O\ (l)$$

   It shows that an H⁺ ion combines with an OH⁻ ion to produce a water molecule. That is all that happens in this neutralisation reaction.

   But an H⁺ ion is just a **proton**, as the drawing on the right shows. So, in effect, the acid **donates** (gives) **protons** to the hydroxide ions. The hydroxide ions accept these protons, to form water molecules.

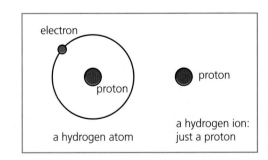

The neutralisation of an acid by an insoluble base

Magnesium oxide is insoluble. It does not produce hydroxide ions. So how does it neutralise an acid? Like this:

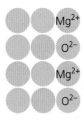 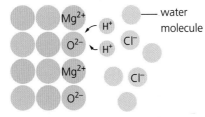 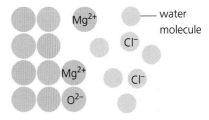

Magnesium oxide consists of a lattice of magnesium and oxygen ions. It is insoluble in water. But when you add it to dilute hydrochloric acid …

… the acid donates protons to the oxide ions. The oxide ions accept them, forming water molecules. So the lattice breaks down.

The magnesium ions join the chloride ions in solution. Evaporating the water gives the salt magnesium chloride.

The equation for this neutralisation reaction is:

$2HCl\ (aq) + MgO\ (s) \longrightarrow MgCl_2\ (aq) + H_2O\ (l)$

The ionic equation for the neutralisation reaction is:

$2H^+\ (aq) + O^{2-}\ (s) \longrightarrow H_2O\ (l)$

Proton donors and acceptors

Now compare the ionic equations for the two neutralisations in this unit:

$H^+\ (aq) + OH^-\ (aq) \longrightarrow H_2O\ (l)$
$2H^+\ (aq) + O^{2-}\ (s) \longrightarrow H_2O\ (l)$

In both:

- the protons are donated by the acids
- ions in the bases accept them, forming water molecules.

So this gives us a new definition for acids and bases:
Acids are proton donors, and bases are proton acceptors.

Questions

1 a What is an *ionic equation*?
   b Hydrochloric acid is neutralised by potassium hydroxide solution. What do you expect the ionic equation for the reaction to be? Write it down.
2 What are *spectator ions*? Explain in your own words.
3 An H^+ ion is just a proton. Explain why. (Do a drawing?)
4 a Acids act as *proton donors*. What does that mean?
   b Bases act as *proton acceptors*. Explain what that means.
5 Neutralisation is *not* a redox reaction. Explain why. Page 119 may help.
6 How to write an ionic equation:
   i Write down all the ions present in the full equation.
   ii Cross out the spectator ions.
   iii Write a new equation showing the remaining ions.
   a Follow those three steps for the reaction between magnesium oxide and hydrochloric acid above.
   b Does your ionic equation match the one shown above? If so, well done!
7 Hydrochloric acid is also neutralised by sodium carbonate solution. Write the ionic equation for this.

8.5 Acids and bases outside the lab

Acids and bases in the home

We often use acids and bases outside the lab – including in the kitchen!

For cooking Lemon juice contains citric acid. And vinegar contains ethanoic acid. Meat is often left to soak in vinegar or lemon juice to make it less tough (and give it more flavour). The acid breaks down the tissues that make the meat tough.

For cleaning Many kitchen cleaners contain sodium hydroxide or ammonia, because alkalis are good at attacking grease. (They feel soapy to the touch because they attack the grease on your skin!)

Some cleaners contain strong acids to dissolve the scale (calcium carbonate) left by hard water, and to remove stains. *Never* mix cleaners, because they can react dangerously together.

For treating insect stings When a bee or ant stings you, it injects an acidic liquid into your skin. You can neutralise it by rubbing on **calamine lotion**, which contains zinc carbonate, or **baking soda**, which is sodium hydrogencarbonate.

Indigestion Your stomach contains some **stomach acid**. (It is in fact hydrochloric acid). You need it for digesting food. But if you have too much acid, you get **indigestion**. You can neutralise the excess acid by taking a drink of baking soda, or an indigestion tablet.

Bee stings are acidic.

Acidity in soil

Most crops grow best when the pH of the soil is near 7. If soil is too acidic, or too alkaline, crops grow badly or not at all. This could be a disaster for farmers and their families.

Usually, the problem is acidity. Soil may be too acidic because it has a lot of rotting vegetation in it. Or because too much fertilizer was used in the past. Or just because the local rock gives acidic soil.

So, to reduce its acidity, the soil is treated with a base. This may be **limestone** (calcium carbonate). You could also use **quicklime** (calcium oxide) or **slaked lime** (calcium hydroxide), which are both made from limestone. It is quite a common rock, so these bases are quite cheap.

Acidic factory waste

Liquid waste from factories often contains acid. If it gets into a river, the acid will kill fish and other river life. So slaked lime is often added to the waste to neutralise it, before it goes down the drain.

Right: slaked lime being spread on soil to neutralise acidity. It is used because it is cheap, and only slightly soluble – so rain won't wash it away quickly.

Attack by acid rain

Oil, coal, and gas are **fossil fuels**. We burn them in factories and homes, and in power stations to make electricity. We burn petrol in car engines. The waste gases from all these reactions include sulphur dioxide, and oxides of nitrogen. They go into the air.

Rainfall is naturally slightly acidic. This is because some carbon dioxide from the air dissolves in it giving **carbonic acid**, a weak acid:

carbon dioxide + water $\longrightarrow$ carbonic acid
CO_2 (g) + H_2O (l) $\longrightarrow$ H_2CO_3 (aq)

But sulphur dioxide reacts with air and water to give sulphuric acid. And the oxides of nitrogen form nitric acid. These strong acids make the rain much more acidic – and it does a lot of damage. For example:

- Many buildings are made of limestone (calcium carbonate). Acid rain eats away their stonework. Concrete and cement also contain calcium carbonate, so they suffer too.
- Metal structures such as bridges, iron railings, and car bodies all come under attack.
- Acid rain kills trees and other plants. And it is carried to rivers, where it kills fish and other river life.

The acidic gases from factories burning coal will result in acid rain.

Tackling the acid rain problem

In many countries, quicklime is added to lakes affected by acid rain. This neutralises the acidity. But the only way to *really* solve the problem is to make sure that power stations, factories and car exhausts do not release acidic gases.

Many modern factories and power stations now spray acidic waste gases with jets of wet slaked lime (calcium hydroxide) to neutralise them, before they leave the chimneys. One product is the salt calcium sulphate, a hard material that is sold for road building.

Many cars also have **catalytic converters** attached to their exhausts, to reduce the acidic oxides of nitrogen to harmless nitrogen. (For more on catalytic converters see page 155.)

But as more countries develop industries, more fossil fuels are being burned. Acidic gases are carried around in the atmosphere. Unless every country takes steps to tackle it, the acid rain problem is likely to grow.

These trees were killed by sulphur dioxide, which dissolved in rain to give acid rain.

Questions

1. Nettle stings contain methanoic acid. Suggest something you could use to neutralise them.
2. Indigestion tablets contain a base. Why?
3. Slaked lime is made from limestone, a common rock. It is not very soluble. Give four reasons to explain why slaked lime is used on soil that is too acidic.
4. a Rain is naturally acidic. Why?
   b Rain in industrial areas may be more acidic. Why?
   c What damage does acid rain do?
   d Industry in one area can cause acid rain to fall in another area, or in another country. Why?
   e Give two examples of actions taken to reduce acid rain.

8.6 Making salts

You can make salts by starting with metals, insoluble bases, or soluble bases (alkalis), and reacting them with acids.

Starting with a metal

Zinc sulphate can be made by reacting dilute sulphuric acid with zinc:

$$Zn\ (s) + H_2SO_4\ (aq) \longrightarrow ZnSO_4\ (aq) + H_2\ (g)$$

These are the steps:

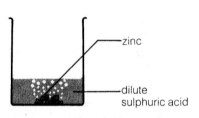 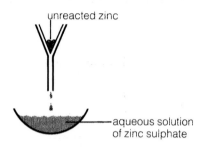

1 Add the zinc to the acid in a beaker. It starts to dissolve, and hydrogen bubbles off. Bubbling stops when all the acid is used up.

2 Some zinc is still left. (The zinc was **in excess**.) Remove it by filtering. This leaves an aqueous solution of zinc sulphate.

3 Heat the solution to evaporate some of the water. Then leave it to cool. Crystals of zinc sulphate soon start to form.

But this method is not suitable for *all* metals, or *all* acids. It is okay for magnesium, aluminium, zinc, and iron. But the reactions of sodium, potassium and calcium with acid are far too violent. Lead reacts too slowly, and copper, silver and gold do not react at all.
(There is more about the reactivity of metals with acids in Unit 12.2.)

Starting with an insoluble base

Copper will not react with dilute sulphuric acid. So to make a copper salt, you must start with a base such as copper(II) oxide, which is insoluble. The blue salt copper(II) sulphate forms:

$$CuO\ (s) + H_2SO_4\ (aq) \longrightarrow CuSO_4\ (aq) + H_2O\ (l)$$

The method is quite like the one above:

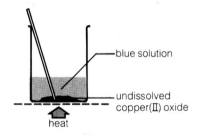 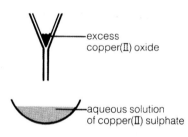 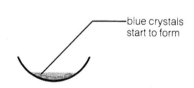

1 Add some copper(II) oxide to dilute sulphuric acid. It dissolves on warming, and the solution turns blue. Add more until no more will dissolve …

2 … which means all the acid has now been used up. Remove the excess solid by filtering. This leaves a blue solution of copper(II) sulphate in water.

3 Heat the solution to evaporate some of the water. (So the blue colour deepens.) Then leave it to cool. Blue crystals of copper(II) sulphate soon start to form.

You could also use copper(II) carbonate as the starting compound here.

Starting with an alkali (soluble base)

It is too dangerous to add sodium to acids. Instead, you can use sodium hydroxide to make sodium salts. You can make sodium chloride like this:

NaOH (*aq*) + HCl (*aq*) ⟶ NaCl (*aq*) + H$_2$O (*l*)

Both reactants are soluble, and no gas bubbles off. So how can you tell when the reaction is complete? Use an indicator! **Phenolphthalein** is the best choice here. It is pink in alkaline solution, but colourless in neutral and acid solutions.

> **Titration**
> - The process shown below is called **titration**.
> - Titration is used to find the amount of acid needed to neutralise a known amount of alkali – or vice versa.

These are the steps in the preparation:

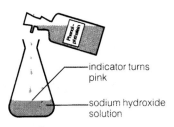 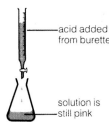 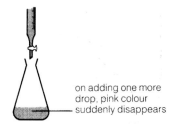

1 Put 25 cm^3 of sodium hydroxide solution into a flask, using a pipette. Add two drops of phenolphthalein. The indicator turns pink.

2 Add the acid from a burette, just a little at a time. Swirl the flask carefully, to help the acid and alkali mix.

3 The indicator suddenly turns colourless. So the alkali has all been used up. The solution is neutral. Add no more acid!

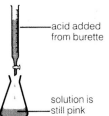 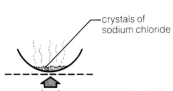

4 You can tell how much acid was added, using the scale on the burette. So this tells you how much acid is needed to neutralise 25 cm^3 of alkali.

5 Carry out the reaction again, but *without* the indicator. Put 25 cm^3 of alkali in the flask, and add the correct amount of acid.

6 Heat the solution from the flask, to evaporate the water. White crystals of sodium chloride will be left behind.

In step 5 the reaction is carried out again, *without* indicator, because the indicator would make the salt impure.

You could use the same method for making potassium salts from potassium hydroxide, and ammonium salts from ammonia solution.

Questions
1. Name the acid and metal you would use for making:
   a zinc chloride   b magnesium sulphate
2. Why would you *not* make potassium chloride from potassium and hydrochloric acid?
3. You would not make lead salts by reacting lead with acids.
   a Why not?   b Suggest a way to make lead nitrate.
4. Look at step 2 at the top of page 124. The zinc was *in excess*. What does that mean? (Check the glossary?)
5. Write instructions for making potassium chloride, starting with solid potassium hydroxide.
6. You are asked to make the salt ammonium nitrate. Which reactants will you choose?

8.7 Making insoluble salts by precipitation

Not all salts are soluble

The salts we looked at so far have all been soluble. You could obtain them as crystals, by evaporating solutions. But not all salts are soluble. This table shows the 'rules' for the solubility of salts:

Soluble		Insoluble
All sodium, potassium, and ammonium salts		
All nitrates		
Chlorides ...	*except*	silver and lead chloride
Sulphates ...	*except*	calcium, barium and lead sulphate
Sodium, potassium, and ammonium carbonates ...		but all other carbonates are insoluble

Insoluble salts can be made by **precipitation**.

Making insoluble salts by precipitation

Barium sulphate is an insoluble salt. You can make it by mixing solutions of barium chloride and magnesium sulphate:

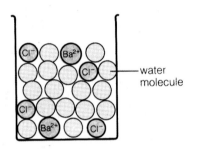 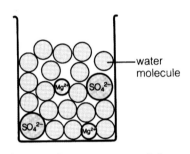 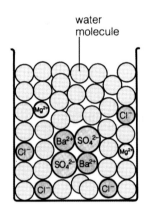

A solution of barium chloride, $BaCl_2$, contains barium ions and chloride ions, as shown here.

A solution of magnesium sulphate, $MgSO_4$, contains magnesium ions and sulphate ions.

When you mix the two solutions, the barium and sulphate ions bond together. Barium sulphate forms as a precipitate.

The equation for the reaction is:

$$BaCl_2\ (aq) + MgSO_4\ (aq) \longrightarrow BaSO_4\ (s) + MgCl_2\ (aq)$$

The ionic equation is:

$$Ba^{2+}\ (aq) + SO_4^{2-}\ (aq) \longrightarrow BaSO_4\ (s)$$

This does not show the magnesium ions and chloride ions, because they are **spectator ions**. They are present, but do not take part in the reaction.

Preparing barium sulphate These are the steps:
1. Make up solutions of barium chloride and magnesium sulphate.
2. Mix them. A white precipitate of barium sulphate forms at once.
3. Filter the mixture. The precipitate is trapped in the filter paper.
4. Rinse the precipitate by running distilled water through it.
5. Then place it in a warm oven to dry.

The precipitation of barium sulphate.

Choosing the starting compounds

Barium sulphate can also be made from barium nitrate and sodium sulphate, since both of these are soluble. As long as barium ions and sulphate ions are present, barium sulphate will precipitate.

To precipitate an insoluble salt, you must mix a solution that contains its positive ions with one that contains its negative ions.

Precipitation and photographic film

Precipitation has some important uses in industry. For example the coloured pigments used in paint are made by precipitation.

It is also used in making film, for photography. For this, solutions of silver nitrate and potassium bromide are mixed with gelatine. A precipitate of tiny crystals of silver bromide forms. The mixture is then coated onto clear film, giving photographic film.

Precipitation is used to make the insoluble pigments that give paint its colour. This pigment has just been dried.

Later, where light strikes the film, the silver bromide breaks down:

$$2AgBr\ (s) \longrightarrow 2Ag\ (s) + Br_2\ (l)$$

This decomposition is a redox reaction. The half-equations are:

silver: $\quad 2Ag^+\ (s) + 2e^- \longrightarrow 2Ag\ (s) \quad$ (reduction)
bromine: $\quad 2Br^-\ (s) \longrightarrow Br_2\ (l) + 2e^- \quad$ (oxidation)

When the film is developed, the remaining silver bromide is converted to a soluble substance and washed away. This leaves a 'negative' image on the film, in silver. It is converted to a 'positive' image by shining a light through it, onto photographic paper.

Silver iodide and silver chloride are also used in making photographic film. Like silver bromide, both break down to silver, on exposure to light.

Putting photographic film in a camera. Most of the film is in the film cartridge at the top, protected from light. When light does strike it, the silver bromide will break down to silver and bromine.

Questions

1. Explain what *precipitation* means, in your own words.
2. **a** Name four salts you could make by precipitation.
   b Now name four salts you could *not* make this way.
3. Choose two starting compounds you could use to make these insoluble salts:
   a calcium sulphate **b** magnesium carbonate
   c zinc carbonate   **d** lead chloride
4. Write a balanced equation for each reaction you chose in 2.
5. **a** What is a spectator ion?
   b Identify the spectator ions for your reactions in 2.
6. **a** Why is precipitation necessary, in making photographic film?
   b What happens to the silver ions, when light strikes the film? Explain why this is a *reduction*.

8.8 Finding concentrations by titration

How to find a concentration by titration

On page 125, the volume of acid needed to neutralise an alkali was found using an indicator, in a process called **titration**. You can find the *concentration* of an acid using the same method.

For this, you titrate the acid against a solution of alkali of known concentration: a **standard solution**.

Example Find the concentration of a solution of hydrochloric acid, using a 1 M solution of sodium carbonate as your standard solution.

First, titrate the acid against your standard solution.
- Measure 25 cm^3 of the sodium carbonate solution into a conical flask, using a pipette. Add a few drops of methyl orange indicator.
- Pour the acid into a 50 cm^3 burette. Record the level.
- Drip the acid slowly into the conical flask. Keep swirling the flask. Stop adding acid when a single drop finally turns the indicator red. Record the new level of acid in the burette.
- Calculate the volume of acid used, as shown on the right.

Now calculate the concentration of the acid.

Step 1 Calculate the number of moles of sodium carbonate used.

1000 cm^3 of 1 M solution contains 1 mole so

25 cm^3 contains $\frac{25}{1000}$ × 1 mole or 0.025 mole.

Step 2 From the equation, find the molar ratio of acid to alkali.

$$2HCl\ (aq) + Na_2CO_3\ (aq) \longrightarrow 2NaCl\ (aq) + H_2O\ (l) + CO_2\ (g)$$
2 moles   1 mole

The ratio is 2 moles of acid to 1 of alkali.

Step 3 Work out the number of moles of acid neutralised.
1 mole of alkali neutralises 2 moles of acid so
0.025 mole of alkali neutralises 2 × 0.025 moles of acid.
0.05 moles of acid were neutralised.

Step 4 Calculate the concentration of the acid.

$$\text{concentration} = \frac{\text{number of moles}}{\text{volume in dm}^3} = \frac{0.05}{0.0278} = 1.8\ \text{mol/dm}^3$$

So the concentration of the hydrochloric acid is **1.8 M**.

Remember
- 1 dm^3 = 1 litre
  = 1000 cm^3
  = 1000 ml
- All these mean the same thing:
  moles per dm^3
  mol/dm^3
  mol dm^{-3}
  moles per litre

How much acid was used?

Final level: 28.8 cm^3
Initial level: 1.0 cm^3
Volume used: 27.8 cm^3

To convert cm^3 to dm^3:

divide by 1000 (or just move the decimal point 3 places to the left). So 27.8 cm^3 = 0.0278 dm^3.

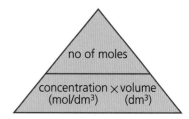

Using the calculation triangle: cover 'concentration' with your finger to see how to calculate it.

Doing a titration: add the solution from the burette a little at a time – and just a drop at a time when the colour change seems near. And keep swirling that flask!

Two more sample calculations

Example 1 25 cm³ of vinegar were neutralised by 20 cm³ of 1 M sodium hydroxide solution. What is the concentration of ethanoic acid in the vinegar?

Step 1 **Calculate the number of moles of sodium hydroxide used.**

1000 cm³ of 1 M solution contains 1 mole so
20 cm³ contains $\frac{20}{1000}$ × 1 mole or 0.02 mole.

Step 2 **From the equation, find the molar ratio of acid to alkali.**

CH_3COOH (aq) + NaOH (aq) ⟶ CH_3COONa (aq) + H_2O (l)
1 mole   1 mole
The ratio is 1 mole of acid to 1 mole of alkali.

Step 3 **Work out the number of moles of acid neutralised.**

1 mole of alkali neutralises 1 mole of acid so
0.02 mole of alkali neutralises 0.02 mole of acid.

Step 4 **Calculate the concentration of the acid.** (25 cm³ = 0.025 dm³)

concentration = $\frac{\text{number of moles}}{\text{volume in dm}^3}$ = $\frac{0.02}{0.025}$ = 0.8 mol/dm³

So the concentration of ethanoic acid in the vinegar is 0.8 M.

Vinegar is a solution of ethanoic acid in water, which accounts for its tang! Ethanoic acid is a weak acid.

Example 2 10 cm³ of a liquid cleaner were neutralised by 12 cm³ of 2 M sulphuric acid. What is the concentration of ammonia in the cleaner?

Step 1 **Calculate the number of moles of acid neutralised.**

1000 cm³ of the sulphuric acid contains 2 moles so
12 cm³ contains $\frac{12}{1000}$ × 2 moles or 0.0024 mole.

Step 2 **From the equation, find the molar ratio of acid to alkali.**

H_2SO_4 (aq) + $2NH_3$ (aq) ⟶ $(NH_4)_2SO_4$ (aq)

so the ratio is 1 mole of acid to 2 moles of alkali.

Step 3 **Work out the number of moles of alkali present.**

1 mole of acid is neutralised by 2 moles of alkali so
0.0024 mole of acid is neutralised by 0.0048 mole of alkali.

Step 4 **Calculate the concentration of the alkali.** (10 cm³ = 0.01 dm³)

concentration = $\frac{\text{number of moles}}{\text{volume in dm}^3}$ = $\frac{0.0048}{0.01}$ = 0.48 mol/dm³

So the concentration of ammonia in the cleaner is **0.48 M**.

This cleaner contains ammonia. A sample of cleaner (in the flask) is being titrated against sulphuric acid (in the burette). The indicator is methyl orange.

Questions

1. What is a *standard* solution?
2. What volume of 2 M hydrochloric acid will neutralise 25 cm³ of 2 M sodium carbonate?
3. 25 cm³ of ammonia solution were neutralised by 20 cm³ of 1 M sulphuric acid. Calculate the concentration of the ammonia solution.

Questions on Chapter 8

1. Rewrite the following, choosing the correct word from each pair in brackets.

   Acids are compounds that dissolve in water giving (hydrogen/hydroxide) ions. Sulphuric acid is an example. It is a (strong/weak) acid, which means that it (partially/fully) dissociates into ions. It can be neutralised by (acids/bases) to form salts called (nitrates/sulphates).

   Many (metals/non-metals) react with acids to give (hydrogen/carbon dioxide). Acids react with (chlorides/carbonates) to give (chlorine/carbon dioxide). Solutions of acids are (good/poor) conductors of electricity. They also affect indicators. Phenolphthalein turns (pink/colourless) in acids, while litmus turns (red/blue). The level of acidity of an acid is shown by its (concentration/pH) number. The (higher/lower) the number, the more acidic the solution.

2. 25 cm³ of potassium hydroxide solution were put in a flask. A few drops of phenolphthalein were added. Dilute hydrochloric acid was added until the indicator changed colour. It was found that 21 cm³ of acid was used.
   a. Draw a labelled diagram of titration apparatus suitable for this neutralisation.
   b. What piece of apparatus should be used to measure 25 cm³ of sodium hydroxide solution accurately?
   c. What colour was the solution in the flask at the start of the titration?
   d. What colour did it turn when the alkali had been neutralised?
   e. Was the acid *more*, or *less*, concentrated than the alkali? Explain your answer.
   f. Name the salt formed in this neutralisation.
   g. Write an equation for the reaction.
   h. How would you obtain pure crystals of the salt?

3. A and B are white powders. A is insoluble in water, but B dissolves. Its solution has a pH of 3. A mixture of A and B bubbles or **effervesces** in water, giving off a gas. A clear solution forms.
   a. One of the white powders is an acid. Which one?
   b. The other powder is a carbonate. Which gas bubbles off in the reaction?
   c. Although A is insoluble in water, a clear solution forms when the mixture of A and B is added to water. Explain why.

4. Magnesium sulphate ($MgSO_4$) is the chemical name for Epsom salts. It can be made in the laboratory by neutralising the base magnesium oxide (MgO).
   a. Which acid should be used to make Epsom salts?
   b. Write a balanced equation for the reaction.
   c. i. The acid is fully dissociated in water. Which term describes this type of acid?
      ii. Which ion, formed in water, causes the 'acidity' of the acid?
   d. i. What is a base?
      ii. Write an ionic equation that shows the oxide ion (O^{2-}) acting as a base.

5. a. Write two lists, dividing the following salts into *soluble in water* and *insoluble in water*.
      (Check the table on page 126?)
         sodium chloride
         calcium carbonate
         potassium chloride
         barium sulphate
         barium carbonate
         silver chloride
         sodium citrate
         zinc chloride
         sodium sulphate
         copper(II) sulphate
         lead sulphate
         lead nitrate
         sodium carbonate
         ammonium carbonate
   b. i. Now write down two starting compounds that could be used to make each *insoluble* salt.
      ii. What is this type of reaction called?
   c. i. Write ionic equations for each reaction that produces an insoluble salt.
      ii. List the spectator ions for each reaction in **i**.

6. 'Aspirin' is 2-ethanoyloxybenzoic acid. This acid is soluble in hot water.
   a. What effect will an aqueous solution of aspirin have on litmus paper?
   b. Do you think aspirin is a strong acid or a weak one? Explain why you think so.
   c. What would you expect to see when baking soda (sodium hydrogen carbonate, $NaHCO_3$) is added to an aqueous solution of aspirin?
   d. Try to write an ionic equation showing how the hydrogen carbonate ion (HCO_3^-) behaves as a base, in reaction with the hydrogen ion (H^+) from the acid.

Method of preparation	Reactants	Salt formed	Other products
a acid + alkali	calcium hydroxide and nitric acid	calcium nitrate	water
b acid + metal	zinc and hydrochloric acid	....	....
c acid + alkali	.... and potassium hydroxide	potassium sulphate	water only
d acid + carbonate	.... and	sodium chloride	water and ....
e acid + metal	.... and	iron(II) sulphate	....
f acid + ...	nitric acid and sodium hydroxide	....	....
g acid + insoluble base	.... and copper(II) oxide	copper (II) sulphate	....
h acid + ...	.... and	copper(II) sulphate	carbon dioxide and ....
i precipitation	silver nitrate and potassium chloride	....	....
j precipitation	lead nitrate and potassium iodide	....	....

7 The table above is about the preparation of salts.
  a Copy it and fill in the missing details.
  b Write balanced equations for the ten reactions.

8 Citric acid has the formula $C_6H_8O_7$.
In a titration, 10 cm³ of 0.05 mol/dm³ citric acid solution were neutralised by 15 cm³ of 0.1 mol/dm³ sodium hydroxide solution.
  a Name a suitable indicator for the titration.
  b What colour change would indicate the end point of the titration?
  c How many *moles* of NaOH were needed, to neutralise the citric acid?
  d How many moles of $C_6H_8O_7$ were present in the flask?
  e How many moles of NaOH would be needed to neutralise 1 mole of $C_6H_8O_7$?
  f i Write a balanced equation for this neutralisation.
    ii Now write the ionic equation for it.

9 **Washing soda** is crystals of hydrated sodium carbonate, $Na_2CO_3.xH_2O$. The value of x can be found by titration.

In the experiment, 2 g of hydrated sodium carbonate neutralised 14 cm³ of a standard 1 M solution of hydrochloric acid.
  a Write a balanced equation for the reaction.
  b How many moles of HCl were neutralised?
  c How many moles of sodium carbonate, Na_2CO_3, were in 2 g of the hydrated salt?
  d What mass of sodium carbonate, Na_2CO_3, is this? (M_r: Ca = 40, C = 12, O = 16)
  e What mass of the hydrated sodium carbonate was water?
  f How many moles of water is this?
  g How many moles of water are there in 1 mole of $Na_2CO_3.xH_2O$?
  h Write the full formula for washing soda.

10 The drawings show the preparation of copper(II) ethanoate, a salt of ethanoic acid.

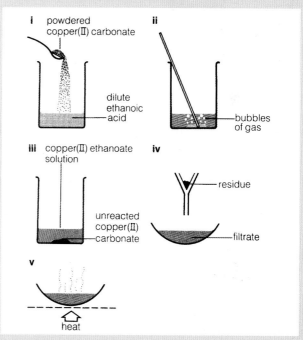

  a Which gas is given off in stage **ii**?
  b Write a word equation for the reaction in **ii**.
  c Write an ionic equation for the reaction in **ii**.
  d How can you tell when the reaction in **ii** is complete?
  e Which of the reactants above is:
    i present in excess? What is your evidence?
    ii completely used up in the reaction?
  f Suggest a reason why copper(II) carbonate powder is used, rather than lumps.
  g The mixture in **ii** is stirred. Suggest a reason.
  h Name the residue in stage **iv**.
  i Write a list of instructions for carrying out this preparation in the laboratory.
  j Suggest another copper compound that could be used instead of copper(II) carbonate, to make copper(II) ethanoate.

9.1 Conductors and insulators

Batteries and electric current

The photograph below shows a battery, a bulb and a rod of graphite (carbon) joined or **connected** to each other by copper wires.

This arrangement is called an **electric circuit**. The bulb is lit: this shows that electricity must be flowing in the circuit.
Electricity is a stream of moving electrons.

> **Reminder: electrons**
>
> An electron is a tiny particle from the atom. It has a negative charge and almost no mass.

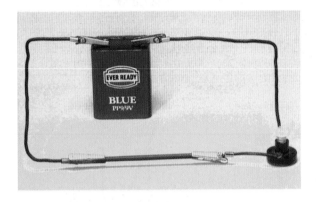
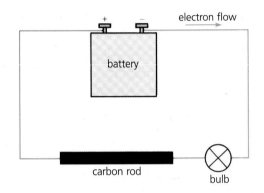

The diagram above shows how the electrons move through the circuit. The battery acts like an electron pump. Electrons leave it through one terminal, called the **negative terminal**. They are pumped through the wires, the bulb and the rod, and enter the battery again through the **positive terminal**.

When the electrons stream through the fine wire in the bulb, they cause it to heat up. It gets white-hot and gives out light.

Conductors

In the circuit above, the light will go out if:

- you disconnect a wire, so that the electricity can't flow. Or
- you connect something into the circuit that won't let electricity flow through it.

Clearly the copper wires and graphite rod allow electricity to flow through them – they conduct **electricity**. Copper and graphite are **conductors**.

Non-conductors

If you connect a plastic rod, or a piece of ceramic tile, in place of the graphite rod above, the bulb will not light. Plastic and ceramics are **non-conductors**, or **insulators**. So plastic is safe to use for the casing in electric plugs. Ceramics are used to support the cables at electricity pylons.

Testing substances to see if they conduct

The circuit above can be used to test any substance to see if it conducts electricity. The substance is simply connected into the circuit. Let's look at some examples.

Copper is the conductor inside this electric drill. But plastic is used for the outer case. Why?

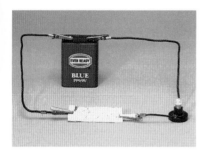

Testing tin to see if it conducts electricity. A strip of tin is connected into the circuit. The bulb lights, so tin must be a conductor.

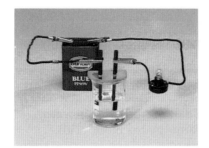

Testing ethanol. The liquid is connected into the circuit by dipping graphite rods into it. The bulb does not light, so ethanol is a non-conductor.

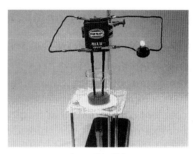

Testing molten lead bromide. A Bunsen is used to melt it. The molten compound conducts, and at the same time gives off a choking brown vapour.

The results   These are the results from a range of tests:

1 **The only solids that conduct are the metals and graphite.**
   These conduct because of their free electrons (pages 51 and 53).
   The electrons get pumped out of one end of the solid by the battery, while more electrons flow in the other end.
   For the same reason, *molten* metals conduct. (It is not possible to test molten graphite, because graphite goes from solid to gas without melting.)

2 **Molecular substances are non-conductors.**
   This is because they contain no free electrons, or other charged particles, that can flow through them.
   Ethanol (above) is a molecular substance. Others are petrol, paraffin, sulphur, sugar and plastic. These never conduct, whether solid or molten.

3 **Ionic substances do not conduct when solid. But they do conduct when melted or dissolved in water – and they decompose at the same time.**
   An ionic substance contains no free electrons. But it does contain **ions**, which are also charged particles. The ions become free to move when the substance is melted or dissolved, and it is they that conduct the electricity.

   The lead bromide above is an example. It is a non-conductor when solid. But it begins to conduct the moment it is melted, and a brown vapour bubbles off at the same time. The vapour is bromine: electricity is causing the lead bromide to **decompose**.

**Decomposition caused by electricity is called electrolysis.
A liquid that conducts current is called an electrolyte.**

Molten lead bromide is therefore an electrolyte. Ethanol is a **non-electrolyte** because it does not conduct.

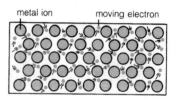

Metals conduct, thanks to their free electrons.

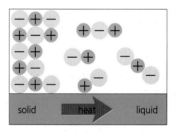

An ionic solid conducts when it melts, because the ions become free to move.

Questions
1 What is a *conductor* of electricity?
2 Draw a circuit to show how you would test whether mercury conducts.
3 Explain why metals are able to conduct electricity.
4 Naphthalene is a molecular substance. Do you think it will conduct electricity when molten? Explain.
5 What is: **a** an electrolyte? **b** a non-electrolyte? Give *three* examples of each.

9.2 The principles of electrolysis

Electrolysis: for ions only!

- Electrolysis is a way to decompose compounds, using electrical energy.
- It works only for :
  - ionic compounds that are melted, or dissolved in water, so that the ions are free to move
  - covalent compounds that form ions when they are dissolved in water. For example hydrogen chloride (which forms hydrochloric acid).
- It works because the ions are free to move.

The electrolysis of molten lead bromide

So let's look more closely at the electrolysis of molten lead bromide, and how the ions move. This diagram shows the apparatus:

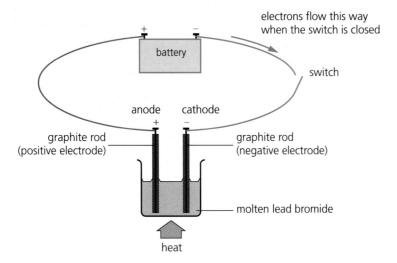

Note that:
- The graphite rods carry current into and out of the molten lead bromide, when the switch is closed. They are called **electrodes**.
- The electrode attached to the positive terminal of the battery is positively charged. It is called the **anode**.
- The other electrode is negatively charged. It is called the **cathode**.
- Molten lead bromide is the **electrolyte** – the liquid for electrolysis.

When the switch is closed, bromine vapour starts to bubble off around the anode. After a time a bead of molten lead forms below the cathode.
So the electrical energy from the battery has caused a **chemical change**.
The lead bromide has **decomposed**:

lead bromide ⟶ lead + bromine
$PbBr_2\ (l)$ ⟶ $Pb\ (l)$ + $Br_2\ (g)$

Note that the carbon electrodes are not changed in the reaction. All they do is carry current. So they are described as **inert**.
Electrodes made of platinum could also be used – they too are inert.
(But later you will see how some electrodes *do* change during electrolysis.)

Which electrode is positive?

Remember **PA**!
Positive **A**node.

What happens to the ions in the lead bromide?

When the lead bromide melts, the ions become free to move. So when the switch is closed, this is what happens:

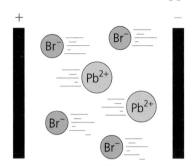 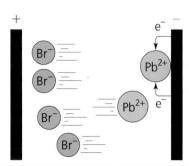 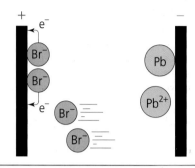

First, the ions move.	At the cathode (−):	At the anode (+):
Opposite charges attract. So the positive lead ions (Pb^{2+}) move to the cathode (−). The negative bromide ions (Br^-) move to the anode (+). The moving ions carry the current.	the lead ions each receive two electrons and become lead atoms. The half-equation is: $Pb^{2+}(l) + 2e^- \rightarrow Pb(l)$ Lead collects on the electrode and eventually drops off it.	the bromide ions each give up an electron, and become atoms. These then pair up to form molecules. The overall half-equation is: $2Br^-(l) \rightarrow Br_2(g) + 2e^-$ The bromine gas bubbles off.
The free ions move.	Ions gain electrons: **reduction**.	Ions lose electrons: **oxidation**.

Remember **OILRIG**:
Oxidation **I**s **L**oss of electrons,
Reduction **I**s **G**ain of electrons.

Overall, electrolysis is a redox reaction.
Reduction takes place at the cathode
and oxidation at the anode.

The electrolysis of other molten ionic compounds

Look at the two other examples in this table. Can you see the pattern?

Electrolyte	The decomposition	At the cathode (−)	At the anode (+)
potassium iodide KI	potassium iodide → potassium + iodine $2KI(l) \rightarrow 2K(l) + I_2(g)$	potassium forms: $2K^+(l) + 2e^- \rightarrow 2K(l)$	iodine forms: $2I^-(l) \rightarrow I_2(g) + 2e^-$
zinc chloride $ZnCl_2$	zinc chloride → zinc + chlorine $ZnCl_2(l) \rightarrow Zn(l) + Cl_2(g)$	zinc forms: $Zn^{2+}(l) + 2e^- \rightarrow Zn(l)$	chlorine forms: $2Cl^-(l) \rightarrow Cl_2(g) + 2e^-$

In each case the compound has been decomposed to its elements.
Electrolysis will decompose a molten ionic compound. A metal is obtained at the cathode, and a non-metal at the anode.

Questions

1. a Which type of compound can be electrolysed? Why?
   b Can it be electrolysed while solid? Explain.
2. For the electrolysis of molten lead bromide, draw diagrams to show:
   a how the ions move when the switch is closed
   b what happens at each electrode
3. What do electrodes do? Why are there two of them?
4. Name the products obtained at the anode and cathode when these molten compounds are electrolysed:
   a sodium bromide   b magnesium chloride
5. Write balanced half-equations for the reactions at the electrodes, for question **4**.

9.3 The electrolysis of solutions

The ions in a solution

When a salt such as sodium chloride dissolves in water, its ions become free to move. So the solution can be electrolysed.

But the products may be different from when you electrolyse the *molten* salt, *because water itself also produces ions.* Although water is molecular, a tiny % of its molecules is split up into ions:

some water molecules → hydrogen ions + hydroxide ions
H_2O (l) → H^+ (aq) + OH^- (aq)

The rules for the electrolysis of a solution

In a solution, H^+ and OH^- ions from water compete with the other ions, to receive or give up electrons. So which ions win? These are the rules.

At the cathode (−), either a metal or hydrogen is formed.
1 The more reactive an element, the more it 'likes' to exist as ions. So if a metal is more reactive than hydrogen, its ions stay in solution. The H^+ ions accept electrons, and hydrogen bubbles off.
2 But if the metal is less reactive than hydrogen, its ions will accept the electrons. The metal forms, leaving the H^+ ions in solution.

At the anode (+), a non-metal other than hydrogen is formed.
1 If a halide ion (Cl^-, Br^- or I^-) is present in sufficient concentration, it will give up electrons more readily than the OH^- ion does. Molecules of chlorine, bromine or iodine are formed.
2 But if no halide ion is present, *or* if the halide solution is dilute, the OH^- ions will give up electrons, and oxygen will be formed.

So ions are discharged **selectively**. Let's look at some examples.

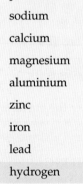

Order of reactivity

This shows the order of reactivity of some metals compared to hydrogen:

potassium
sodium
calcium
magnesium
aluminium
zinc
iron
lead
hydrogen
copper
silver

increasing reactivity

A concentrated solution of sodium chloride

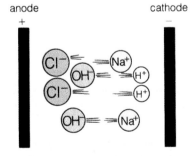

The solution contains Na^+ ions and Cl^- ions from the salt, and H^+ and OH^- ions from water. The positive ions go to the cathode, and negative ions to the anode.

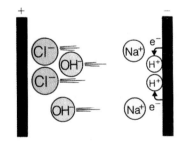

At the cathode, the H^+ ions accept electrons, since hydrogen is less reactive than sodium:

$2H^+$ (aq) + $2e^-$ → H_2 (g)

This is a **reduction**.
The hydrogen gas bubbles off.

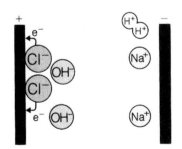

At the anode, the Cl^- ions give up electrons more readily than the OH^- ions do.

$2Cl^-$ (aq) → Cl_2 (aq) + $2e^-$

This is an **oxidation**.
The chlorine gas bubbles off.

When hydrogen and chlorine gases bubble off, Na^+ and OH^- ions are left behind – so a solution of sodium hydroxide is formed.

Two more examples of concentrated solutions

Concentrated solution	Obtained at the cathode …	Obtained at the anode …
potassium bromide KBr (aq)	hydrogen $2H^+ (aq) + 2e^- \longrightarrow H_2 (g)$	bromine $2Br^- (aq) \longrightarrow Br_2 (g) + 2e^-$
hydrochloric acid HCl (aq)	hydrogen $2H^+ (aq) + 2e^- \longrightarrow H_2 (g)$	chlorine $2Cl^- (aq) \longrightarrow Cl_2 (g) + 2e^-$

Check the two examples above. Do they obey the rules on page 136? Note how the concentrated hydrochloric acid has been decomposed!

Dilute solutions

What happens if the sodium chloride solution is dilute? Let's see.

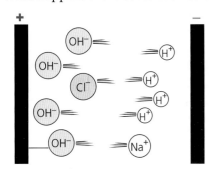
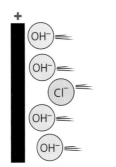
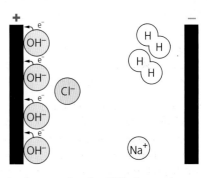

The same ions are present as before. But now the proportion of Na^+ and Cl^- ions is lower, since this is a dilute solution. So this time the electrolysis will give a different result.

At the cathode, hydrogen 'wins' as before, and bubbles off:
$4H^+ (aq) + 4e^- \longrightarrow 2H_2 (g)$
(The 4 electrons are to balance the half-equation at the anode.)

At the anode, the OH^- ions give up electrons, since not many Cl^- ions are present. Oxygen bubbles off:
$4OH^- (aq) \longrightarrow$
$\qquad O_2 (g) + 2H_2O (l) + 4e^-$

When the hydrogen and oxygen bubble off, the Na^+ and Cl^- ions are left behind. So we *still* have a solution of sodium chloride!
The overall result is that water has been decomposed.

Now check the results in the table below. Do they obey the rules?

Dilute solution	Obtained at the cathode …	Obtained at the anode …
copper(II) chloride $CuCl_2$ (aq)	copper $2Cu^{2+} (aq) + 4e^- \longrightarrow 2Cu (s)$	oxygen $4OH^- (aq) \longrightarrow 2H_2O (l) + O_2 (g) + 4e^-$
hydrochloric acid HCl (aq)	hydrogen $4H^+ (aq) + 4e^- \longrightarrow 2H_2 (g)$	oxygen $4OH^- (aq) \longrightarrow 2H_2O (l) + O_2 (g) + 4e^-$

Questions

1. Water conducts electricity very slightly. Why?
2. List the products, for the electrolysis of:
   a. a concentrated solution of hydrochloric acid
   b. a dilute solution of hydrochloric acid
3. Write *balanced* half-equations for the reaction at each electrode, for the electrolysis of:
   a. a dilute solution of sodium nitrate, $NaNO_3$
   b. a concentrated solution of copper(II) chloride, $CuCl_2$

9.4 The electrolysis of brine

What is brine?

Brine is a concentrated solution of sodium chloride, or common salt. It is the starting point for a whole range of chemicals.

The brine can be obtained by pumping water into salt mines to dissolve the salt, or by evaporating sea water.

During electrolysis, the overall reaction is:

$$\text{brine} \xrightarrow{\text{electrolysis}} \text{sodium hydroxide} + \text{chlorine} + \text{hydrogen}$$
$$2NaCl\ (aq) + 2H_2O\ (l) \longrightarrow 2NaOH\ (aq) + Cl_2\ (g) + H_2\ (g)$$

The electrolysis

The diagram below shows a cell for this electrolysis: the **diaphragm cell.**

The anode is made of titanium, and the cathode of steel. Now look at the diaphragm down the middle of the cell. It is made of asbestos. Its function is to let ions through, but keep the gases apart.

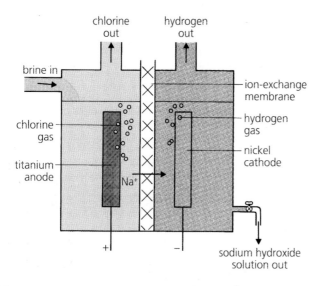

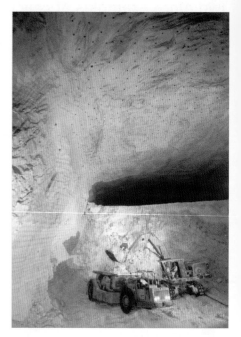

Inside a salt mine. Many countries have underground salt beds. They were deposited millions of years ago, when the sea drained away from the land.

The ions present are Na⁺ and Cl⁻ from the salt, and H⁺ and OH⁻ from the water. The reactions at the electrodes are exactly as shown at the bottom of page 136. (Look back at them!)

At the cathode Hydrogen is discharged in preference to sodium:
$$2H^+\ (aq) + 2e^- \longrightarrow H_2\ (g)$$

As usual at the cathode, this is a reduction.

At the anode Chlorine is discharged in preference to oxygen:
$$2Cl^-\ (aq) \longrightarrow Cl_2\ (g) + 2e^-$$

As usual at the anode, this is an oxidation.

The two gases bubble off. Na⁺ and OH⁻ ions are left behind, giving a solution of sodium hydroxide. Some of the solution is evaporated to give a more concentrated solution, and some is evaporated to dryness, giving solid sodium hydroxide.

Chlorine has many uses. One is to kill germs in water. Behind the scenes at a swimming pool, this man makes sure there is chlorine in the water.

What the products are used for

The electrolysis of brine is an important process, because the products are so useful. Look at these:

Chlorine, a poisonous yellow-green gas
Used for making ...
- the plastic PVC (nearly 1/3 of it used for this)
- solvents for degreasing and drycleaning
- medical drugs (85% of these involve chlorine)
- weedkillers and pesticides (95% of these involve chlorine)
- paints and dyestuffs
- bleaches
- hydrogen chloride and hydrochloric acid

It is also used as a sterilising agent, to kill bacteria in water supplies and swimming pools.

Sodium hydroxide solution, alkaline and corrosive
Used in making ...
- soaps
- detergents
- viscose (rayon) and other textiles
- paper (like the paper in this book)
- ceramics (tiles, furnace bricks, and so on)
- dyes
- medical drugs

Hydrogen, a colourless flammable gas
Used ...
- in making nylon
- to make hydrogen peroxide
- to 'harden' vegetable oils to make margarine
- as a fuel in hydrogen fuel cells

Of the three chemicals, chlorine is the most widely used. Around 50 million tonnes of it are produced each year around the world.

Tanks of chlorine at a waterworks. What's it doing here?

All three products from the electrolysis of brine must be transported with care. Why?

Questions
1. What is brine? Where is it obtained from?
2. Write a word equation for the electrolysis of brine.
3. Draw a sketch of the diaphragm cell. Mark in where the oxidation and reduction reactions take place in it, and write the half-equations for them.
4. What is the diaphragm for, in the diaphragm cell?
5. The electrolysis of brine is a very important process.
   a. Explain why.
   b. Give three uses for each of the products.
6. Your job is to keep a brine electrolysis plant running safely and smoothly. Try to think of four safety precautions you might need to take.

9.5 Two more uses of electrolysis

First, the effect of changing electrodes

Copper(II) sulphate solution contains contains blue Cu^{2+} ions, and SO_4^{2-} ions, from copper(II) sulphate, and H^+ and OH^- ions from water. Now compare these two electrolyses.

1 Copper(II) sulphate solution with carbon electrodes

At the cathode Copper ions are discharged:

$2Cu^{2+} (aq) + 4e^- \longrightarrow 2Cu (s)$ (copper ions reduced)

The copper coats the electrode.

At the anode Oxygen bubbles off:
$4OH^- (aq) \longrightarrow 2H_2O (l) + O_2 (g) + 4e^-$ (hydroxide ions oxidised)

So copper and oxygen are produced. This fits the rules on page 136. And when the copper is scraped off it, the cathode is the same as before. So the electrodes have not changed. They are **inert**.

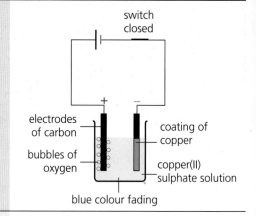

2 Copper(II) sulphate solution with copper electrodes

At the cathode Again, copper is formed, and coats the electrode:

$Cu^{2+} (aq) + 2e^- \longrightarrow Cu (s)$ (copper ions reduced)

At the anode The anode dissolves:

$Cu (s) \longrightarrow Cu^{2+} (aq) + 2e^-$ (copper oxidised to ions)

So copper and copper ions are formed at the same time. Here the electrodes are *not* inert. The anode dissolves, forming ions. These move to the cathode and deposit as atoms. And this leads to two important uses of electrolysis, as you will see below.

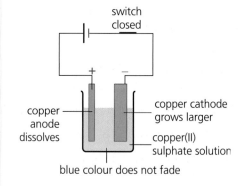

1 Refining (purifying) copper

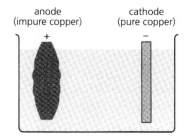

The anode is made of impure copper. The cathode is pure copper. The electrolyte is dilute copper(II) sulphate solution.

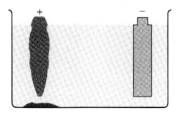

The copper in the anode dissolves. But the impurities do not dissolve. They just drop to the floor of the cell as a sludge.

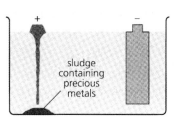

A layer of pure copper builds up on the cathode. When the anode is almost gone, the anode and cathode are replaced.

The copper deposited on the cathode is over 99.9% pure. The sludge may contain valuable metals such as platinum, gold, silver, and selenium. These are recovered and sold.

The purer it is, the better copper conducts electricity. Highly refined copper is used for the electrics in cars. A car like this will have hundreds of copper wires, well over 1 km long in total.

A chromium-plated bathroom tap. Chromium does not stick well to steel. So the steel is first electroplated with copper or nickel, and then chromium.

2 Electroplating

Electroplating is where one metal is coated with another, to make it look better, or to prevent corrosion. For example, steel car bumpers are coated with chromium. Steel cans are coated with tin to make tins for food. And cheap metal jewellery is often coated with silver.

The drawing on the right shows how to electroplate a steel jug with silver. The jug is used as the cathode of an electrolytic cell. The anode is made of silver. The electrolyte is a solution of a silver compound, such as silver nitrate.

At the anode The silver dissolves, forming ions in solution:

$$Ag\,(s) \longrightarrow Ag^+\,(aq) + e^- \qquad \text{(silver is oxidised)}$$

At the cathode The silver ions in the solution are attracted to the cathode. There they receive electrons, forming a coat of silver on the jug:

$$Ag^+\,(aq) + e^- \longrightarrow Ag\,(s) \qquad \text{(silver ions are reduced)}$$

When the layer of silver is thick enough, the jug is removed.

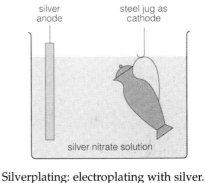

Silverplating: electroplating with silver. When the elecrtrodes are connected to a power source, electroplating begins.

> In general, to electroplate any object with metal X, the set-up is:
> **cathode** – object to be electroplated
> **anode** – metal X
> **electrolyte** – solution of a soluble compound of X.

Questions

1. Copper(II) ions are blue. When copper(II) sulphate solution is electrolysed, the blue solution:
   a loses its colour when carbon electrodes are used
   b keeps its colour when copper electrodes are used.
   Explain each of these observations.
2. If you want to purify a metal by electrolysis, will you make it the anode or the cathode? Why?
3. Describe the process of refining copper.
4. What does *electroplating* mean?
5. Steel cutlery is often electroplated with nickel. Why?
6. You plan to electroplate steel cutlery with nickel.
   a What will you use as the anode?
   b What will you use as the cathode?
   c Suggest a suitable *electrolyte*. (Check the glossary?)

Questions on Chapter 9

1 In which of these would the bulb light?

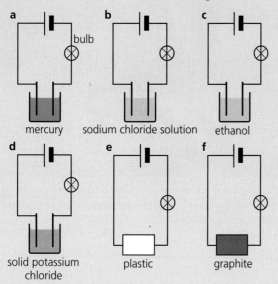

2 a Which of the substances in question 1 are:
   i conductors?   ii non-conductors?
   iii electrolytes?   iv non-electrolytes?
  b What is the difference between a conductor and an electrolyte?
  c For which substances above would you expect to see changes taking place at the electrodes? Explain your choice.

3 The electrolysis of lead bromide can be investigated using this apparatus.

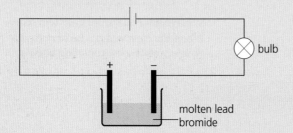

  a The bulb will not light until the lead bromide is melted. Why not?
  b What would be *seen* at the anode during the experiment?
  c Name the substance in **b**.
  d What is formed at the cathode?
  e Write a half-equation for the reaction at each electrode.
  f i At which electrode do positive ions gain electrons?
    ii What is this process called?
    iii Name the process at the other electrode.

4 Six substances A to F were dissolved in water, and connected in turn into this circuit. Ⓐ represents an ammeter, used to measure the current.

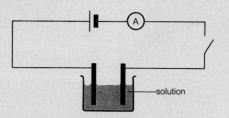

This table shows the results:

Substance	Current (amperes)	Cathode (−)	Anode (+)
A	0.8	copper	chlorine
B	1.0	hydrogen	chlorine
C	0.0	—	—
D	0.8	copper	oxygen
E	1.2	hydrogen	oxygen
F	0.7	silver	oxygen

  a Which solution conducts best?
  b Which solution is a non-electrolyte?
  c Which solution could be:
   i silver nitrate?   ii copper(II) sulphate?
   iii copper(II) chloride?   iv sodium hydroxide?
   v potassium chloride?   vi sugar?
  d Explain how the current is carried:
   i *within* the electrolytes
   ii in the rest of the circuit

5 a List the ions that are present in concentrated solutions of:
   i sodium chloride
   ii copper(II) chloride
  b Explain why and how the ions move, when each solution is electrolysed using platinum electrodes.
  c Write the half-equation for the reaction at:
   i the anode   ii the cathode
   during the electrolysis of each solution.
  d Explain why the anode reactions for both solutions are the same.
  e i The anode reactions will be different if the solutions are made very dilute. Explain why.
   ii Write the equation for the new reactions.
  f Explain why copper is obtained at the cathode, but sodium is not.

6 Molten lithium chloride contains lithium ions (Li⁺) and chloride ions (Cl⁻).
  a Copy the following diagram and use arrows to show which way:
    i the ions move when the switch is closed
    ii the electrons flow in the wires

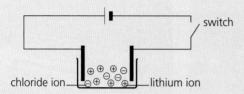

  b i Write equations for the reaction at each electrode, and the overall reaction.
    ii Describe each of the reactions using the terms *reduction*, *oxidation* and *redox*.

7 This question is about the electrolysis of a dilute aqueous solution of lithium chloride.
  a Give the names and symbols of the ions present in the solution.
  b Say what will be formed, and write a half-equation for the reaction:
    i at the anode   ii at the cathode
  c Name another compound that will give the same products at the electrodes.

8 This diagram shows apparatus that can be used to purify impure copper by electrolysis:

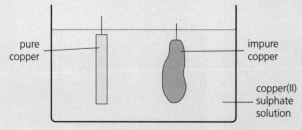

  a i Copy and complete the diagram, adding wires and a battery to complete the circuit.
    ii Mark + and − on the correct electrodes.
    iii Mark in where the impurities will be found after the purification.
  b What you would *see* during the electrolysis?
  c What name is given to the copper(II) sulphate solution, in this type of apparatus?
  d Write a half-equation for the reaction at each electrode.
  e The above electrolysis is a very important industrial process. Explain why.

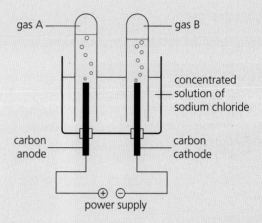

9 The electrolysis above produces gases A and B.
  a Why does the solution conduct electricity?
  b Identify each gas, and describe a test you could carry out, to confirm its identity.
  c Name one product manufactured from:
    i gas A   ii gas B
  d i Write balanced half-equations to show how the two gases are produced.
    ii The overall reaction is a *redox* reaction. Explain why.
  e The solution remaining after the electrolysis will turn litmus paper blue.
    i What is the name of this solution?
    ii Give one chemical property for it.

10 An experiment is needed, to see if an iron object can be electroplated with chromium.
  a Suggest a solution to use as the electrolyte.
  b i Draw a labelled diagram of the apparatus that could be used for the electroplating.
    ii Show how the *electrons* will travel from one electrode to the other.
  c Write half-equations for the reactions at each electrode.
  d At which electrode does oxidation take place?
  e The concentration of the solution does not change. Why not?

11 Nickel(II) sulphate ($NiSO_4$) is green. A solution of this salt is electrolysed using nickel electrodes.
  a Write a half-equation for the reaction at each electrode.
  b What happens to the size of the anode?
  c The colour of the solution does not change, during the electrolysis. Explain why.
  d Give one industrial use for this electrolysis.

10.1 Rates of reaction

Fast and slow

Some reactions are **fast** and some are **slow**. Look at these examples:

Silver chloride precipitating, when solutions of silver nitrate and sodium chloride are mixed. This is a very fast reaction.

Concrete setting. This reaction is quite slow. It will take a couple of days for the concrete to fully harden.

Rust forming on an old car. This is usually a very slow reaction. It will take years for the car to rust completely away.

But it is not always enough to know just that a reaction is fast or slow. In factories where they make products from chemicals, they need to know *exactly* how fast a reaction is going, and how long it will take to complete. In other words, they need to know the **rate** of the reaction.

What is rate?

Rate is a measure of how fast or slow something is. Here are some examples.

This plane has just flown 2000 kilometres in 1 hour. It flew at a **rate** of 2000 kilometres per hour.

This petrol pump can pump out petrol at a **rate** of 50 litres per minute.

This machine can print newspapers at a **rate** of 10 copies per second.

From these examples you can see that:
Rate is a measure of the change that happens in a single unit of time.
Any suitable unit of time can be used – a second, a minute, an hour, even a day.

Rate of a chemical reaction

When zinc is added to dilute sulphuric acid, they react together. The zinc disappears slowly, and a gas bubbles off.

After a time, the gas bubbles off more and more slowly. This is a sign that the reaction is slowing down.

Finally, no more bubbles appear. The reaction is over, because all the acid has been used up. Some zinc remains behind.

The gas that bubbles off is hydrogen. The equation for the reaction is:

zinc + sulphuric acid ⟶ zinc sulphate + hydrogen
$Zn\ (s)$ + $H_2SO_4\ (aq)$ ⟶ $ZnSO_4\ (aq)$ + $H_2\ (g)$

Both zinc and sulphuric acid get used up in the reaction. At the same time, zinc sulphate and hydrogen form.

You could measure the rate of the reaction, by measuring:
- the amount of zinc used up per minute *or*
- the amount of sulphuric acid used up per minute *or*
- the amount of zinc sulphate produced per minute *or*
- the amount of hydrogen produced per minute.

For this reaction, it is easiest to measure the amount of hydrogen produced per minute. The hydrogen can be collected as it bubbles off, and its volume can then be measured.

In general, to find the rate of a reaction, you should measure:
the amount of a reactant used up per unit of time *or*
the amount of a product produced per unit of time.

Questions

1. Here are some reactions that take place in the home. Put them in order of decreasing rate (the fastest one first).
   a raw egg changing to hard-boiled egg
   b fruit going rotten
   c cooking gas burning
   d bread baking
   e a metal tin rusting

2. Which of these rates of travel is slowest?
   5 kilometres per second
   20 kilometres per minute
   60 kilometres per hour

3. Suppose you had to measure the rate at which zinc is used up in the reaction above. Which of these units would be suitable? Explain your choice.
   a litres per minute
   b grams per minute
   c centimetres per minute

4. Iron reacts with sulphuric acid like this:
   $Fe\ (s)$ + $H_2SO_4\ (aq)$ ⟶ $FeSO_4\ (aq)$ + $H_2\ (g)$
   a Write a word equation for this reaction.
   b Write down four different ways in which the rate of the reaction could be measured.

10.2 Measuring the rate of a reaction

A reaction that produces a gas

The rate of a reaction is found by measuring the amount of a **reactant** used up per unit of time, or the amount of a **product** produced per unit of time. Look at this reaction:

magnesium + hydrochloric acid ⟶ magnesium chloride + hydrogen
$Mg(s)$ + $2HCl(aq)$ ⟶ $MgCl_2(aq)$ + $H_2(g)$

Here hydrogen is the easiest substance to measure, because it is the only gas in the reaction. It bubbles off and can be collected in a **gas syringe**, where its volume is measured.

Testing a substance as an explosive for industry. The rate of a fast reaction like this, giving a mixture of gases, is not easy to measure.

The method

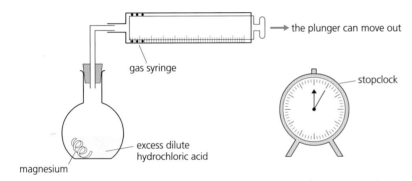

The magnesium is cleaned with sandpaper. Dilute hydrochloric acid is put into the flask. The magnesium is dropped in and the stopper and syringe are inserted immediately. The clock is started at the same time.

Hydrogen begins to bubble off. It rises up the flask and into the gas syringe, pushing the plunger out:

At the start, no gas has yet been produced or collected. So the plunger is all the way in.

Now the plunger has been pushed out to the 20 cm³ mark. 20 cm³ of gas have been collected.

The volume of gas in the syringe is noted at intervals – for example every half a minute. How will you know when the reaction is complete?

Typical results

Time/minutes	0	$\frac{1}{2}$	1	$1\frac{1}{2}$	2	$2\frac{1}{2}$	3	$3\frac{1}{2}$	4	$4\frac{1}{2}$	5	$5\frac{1}{2}$	6	$6\frac{1}{2}$
Volume of hydrogen/cm³	0	8	14	20	25	29	33	36	38	39	40	40	40	40

This table shows some typical results for the experiment.

You can tell quite a lot from this table. For example, you can see that the reaction lasted about five minutes. But a graph of the results is even more helpful. The graph is shown on the next page.

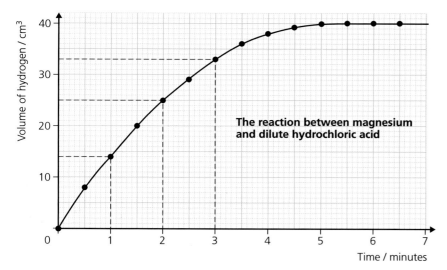

Notice these things about the results:

1. In the first minute, 14 cm³ of hydrogen are produced.
   So the rate for the first minute is 14 cm³ of hydrogen per minute.
   In the second minute, only 11 cm³ are produced. (25 − 14 = 11)
   So the rate for the second minute is 11 cm³ of hydrogen per minute.
   The rate for the third minute is 8 cm³ of hydrogen per minute.
   So you can see that the rate decreases as time goes on.
   The rate changes all through the reaction. It is greatest at the start, but gets less as the reaction proceeds.

2. The reaction is fastest in the first minute, and the curve is steepest then. It gets less steep as the reaction gets slower.
   The faster the reaction, the steeper the curve.

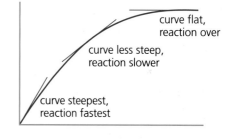

3. After 5 minutes, no more hydrogen is produced, so the volume no longer changes. The reaction is over, and the curve goes flat.
   When the reaction is over, the curve goes flat.

4. Altogether, 40 cm³ of hydrogen are produced in 5 minutes.

   The average rate for the reaction $= \dfrac{\text{total volume of hydrogen}}{\text{total time for the reaction}}$

   $= \dfrac{40 \text{ cm}^3}{5 \text{ minutes}}$

   $= $ **8 cm³ of hydrogen per minute.**

Note that this method can be used for *any* reaction where one product is a gas.

Questions

1. For the experiment in this unit, explain why:
   a. the magnesium ribbon is first cleaned
   b. the clock is started the moment the reactants are mixed
   c. the stopper is replaced immediately
2. From the graph at the top of the page, how can you tell when the reaction is over?
3. Look again at the graph at the top of the page.
   a. How much hydrogen is produced in the first:
      i. 2.5 minutes?   ii. 4.5 minutes?
   b. How long did it take to get 20 cm³ of hydrogen?
   c. What is the rate of the reaction during:
      i. the fourth minute?   ii. the sixth minute?

10.3 Changing the rate of a reaction (I)

The effect of concentration

A reaction can be made to go faster or slower by changing the **concentration** of a reactant.

The experiment with magnesium and excess hydrochloric acid is repeated twice (A and B below). Everything is kept the same each time, *except* the concentration of the acid:

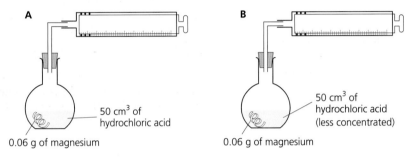

The acid in A is *twice as concentrated* as the acid in B.

Here are both sets of results, shown on the same graph.

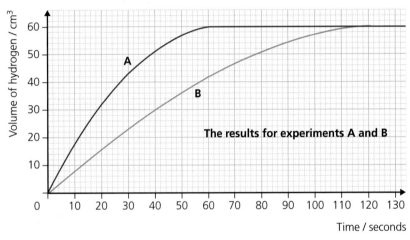

The results for experiments A and B

Notice these things about the results:

1. Curve A is steeper than curve B. From this you can tell straight away that the reaction was faster in A than in B.
2. In A, the reaction lasts for 60 seconds. In B it lasts for 120 seconds.
3. Both reactions produced 60 cm³ of hydrogen. Do you agree?
   In A it was produced in 60 seconds, so the average rate was 1 cm³ of hydrogen per second. In B it was produced in 120 seconds, so the average rate was 0.5 cm³ of hydrogen per second.
   The average rate in A was twice the average rate in B.

These results show that:
A reaction goes faster when the concentration of a reactant is increased.

For the reaction between magnesium and hydrochloric acid, the rate doubles when the concentration of acid is doubled. Does this surprise you? Can you think of a way to make the reaction go even faster?

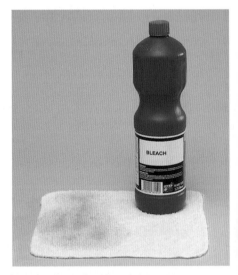

This stain could be removed with a solution of bleach. The more concentrated the solution, the faster the stain will disappear.

The effect of temperature

A reaction can also be made to go faster or slower by changing the **temperature** of the reactants.

This time, a different reaction is used. When dilute hydrochloric acid is mixed with sodium thiosulphate solution, a fine yellow precipitate of sulphur forms. The rate can be followed like this:

1 Mark a cross on a piece of paper.
2 Place a beaker containing sodium thiosulphate solution on top of the paper, so that you can see the cross through it, from above.
3 Quickly add hydrochloric acid, and start a clock at the same time. The cross grows fainter as the precipitate forms.
4 Stop the clock the moment you can no longer see the cross.

The low temperature in the fridge slows down the reactions that make food rot.

View from above the beaker:

    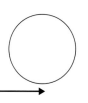

The cross grows fainter with time

Now repeat the experiment several times, changing *only* the temperature. This table shows some typical results:

Temperature/°C	20	30	40	50	60
Time for cross to disappear/seconds	200	125	50	33	24

The higher the temperature, the faster the cross disappears

The cross disappears when enough sulphur has formed to hide it. You can see that this took 200 seconds at 20°C, but only 50 seconds at 40°C. So the reaction is *four times faster* at 40°C than at 20°C.

A reaction goes faster when the temperature is raised. When the temperature increases by 10°C, the rate generally doubles.

That it why food cooks much faster in pressure cookers than in ordinary saucepans. (The temperature in a pressure cooker can reach 125 °C.)

Comparing rates for this reaction

$$\text{average rate} = \frac{\text{mass of sulphur (g)}}{\text{time to produce it (s)}}$$

The mass of sulphur needed to hide the cross is the same each time. Suppose it is 1 gram. Then:

At 20°C av. rate $= \frac{1}{200}$ g/s
$= 0.005$ g/s

At 40°C av. rate $= \frac{1}{50}$ g/s
$= 0.02$ g/s

So the reaction is 4 times faster at 40°C than at 20°C.
($4 \times 0.005 = 0.02$)

Questions

1 Look at the graph on the opposite page.
   a How much hydrogen was obtained after 2 minutes in:
      i experiment A? ii experiment B?
   b How can you tell which reaction was faster, from the shape of the curves?
2 Explain why experiments A and B both gave the same amount of hydrogen.
3 Copy and complete: A reaction goes ...... when the concentration of a ...... is increased. It also goes ...... when the ...... is raised.
4 Raising the temperature speeds up a reaction. Give two (new) examples of how this is used in everyday life.
5 What happens to the rate of a reaction when the temperature is *lowered*? How do we make use of this?

149

10.4 Changing the rate of a reaction (II)

The effect of surface area

In many reactions, one reactant is a solid. The reaction between hydrochloric acid and calcium carbonate (marble chips) is an example. Carbon dioxide gas is produced:

$$CaCO_3\,(s) + 2HCl\,(aq) \longrightarrow CaCl_2\,(aq) + H_2O\,(l) + CO_2\,(g)$$

The rate can be measured using the apparatus on the right.

The method The marble chips are placed in the flask and the acid is added. The flask is quickly plugged with cotton wool to stop any liquid splashing out. Then it is quickly weighed, and the clock is started. The mass is noted regularly until the reaction is complete.

Since carbon dioxide can escape through the cotton wool, the flask gets lighter as the reaction proceeds. (It is a heavy gas, so the loss of mass is appreciable.) So by weighing the flask at regular intervals, you can follow the rate of reaction.

The experiment is repeated twice. Everything is kept exactly the same each time, except the *surface area* of the marble chips.

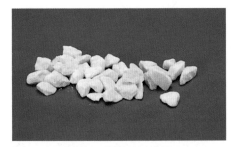

For experiment 1, large chips are used. Their surface area is the total area of exposed surface.

For experiment 2, the same *mass* of marble is used – but the chips are small so the surface area is greater.

The results The results of the two experiments are plotted here:

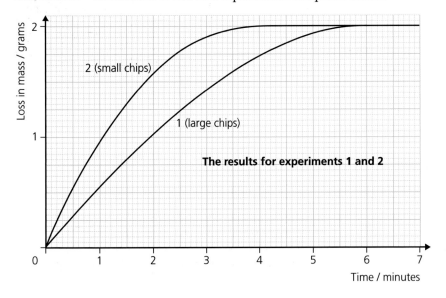

The results for experiments 1 and 2

How to draw the graph

First you have to find the *loss in mass* at different times:

loss in mass at a given time = mass at start – mass at that time

Then you plot the values for loss in mass against the times.

You should notice these things about the results:

1 Curve 2 is steeper than curve 1. This shows immediately that the reaction is faster for the small chips.
2 In both experiments, the final loss in mass is 2.0 grams. In other words, 2.0 grams of carbon dioxide are produced each time.
3 For the small chips, the reaction is complete in 4 minutes. For the large chips, it lasts for 6 minutes.

These results show that:
The rate of a reaction increases when the surface area of a solid reactant is increased.

When a large surface area can mean danger

A large surface area can make a reaction dangerously fast. For example flour dust has a large surface area, and it is **combustible**: it will burn in air. There can be a lot of flour dust lying around in a flour mill. So a spark from a machine, or a lit match, can cause an explosion.

There is the same risk in wood mills, with fine wood dust. And in factories making things like custard powder, instant coffee, sugar, and dried milk.

Seven people died when fine wheat dust exploded in this grain storage complex, in Kansas, USA, in 1998. (Flour is made from wheat.)
(Photo courtesy of Battalion Chief Michael Steele and the Wichita Fire Department, Kansas, USA.)

The effect of light

Some chemical reactions obtain the energy they need from light. They are called **photochemical reactions**. For example:

1 Silver bromide is pale yellow, but darkens on exposure to light because the light causes it to decompose to silver:

$$2AgBr \xrightarrow{light} 2Ag + Br_2$$

(This is also a redox reaction. The silver ions, Ag^+, are reduced to silver.)

2 Plants use carbon dioxide from the air to make a sugar called **glucose**, in a reaction called **photosynthesis**. This uses the energy in sunlight. The green substance – **chlorophyll** – in leaves speeds up the reaction:

$$\underset{\text{carbon dioxide}}{6CO_2 (g)} + \underset{\text{water}}{6H_2O (l)} \xrightarrow[\text{chlorophyll}]{light} \underset{\text{glucose}}{C_6H_{12}O_6 (s)} + \underset{\text{oxygen}}{6O_2 (g)}$$

In both these reactions, the stronger the light, the more energy it provides – so the faster the reaction goes.

The plant on the right is unhealthy because it did not get enough light – so it made glucose too slowly.

Questions

1 This question is about the graph on the opposite page. For each experiment find:
   a the loss in mass in the first minute
   b the mass of carbon dioxide gas produced in the first minute
   c the average rate of production of the gas, for the complete reaction.

2 a Which has the largest surface area: 1 g of large marble chips, or 1 g of small marble chips?
   b Which 1g sample will disappear first when reacted with excess hydrochloric acid? Why?
3 Fine wood dust can be a danger, in wood mills. Why?
4 a What is a *photochemical reaction*? Give an example.
   b How could you speed up a photochemical reaction?

10.5 Explaining rates

A closer look at a reaction

Magnesium and dilute hydrochloric acid react together:

magnesium + hydrochloric acid ⟶ magnesium chloride + hydrogen
$Mg\,(s) + 2HCl\,(aq) \longrightarrow MgCl_2\,(aq) + H_2\,(g)$

In order for the magnesium and acid particles to react together:
- **the particles must collide with each other**, and
- **the collision must have enough energy to be successful.**

This is called the **collision theory**. It is shown by the drawings below.

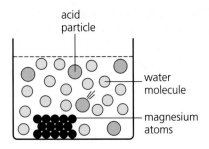
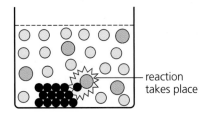
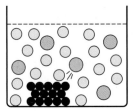

The particles in the liquid move around non-stop. To react, an acid particle must collide with a magnesium atom.

This collision has enough energy, so it is successful: the particles react. Magnesium chloride and hydrogen are formed.

But this collision did not have enough energy. So it was not successful. The acid particle has just bounced away again.

If there are lots of successful collisions in a given minute, then a lot of hydrogen is produced in that minute. In other words, the rate of reaction is high. If there are not many, the rate of reaction is low.

The rate of a reaction depends on how many successful collisions there are in a given unit of time.

Changing the rate of a reaction

Why rate increases with concentration If the concentration of the acid is increased, the reaction goes faster. It is easy to see why:

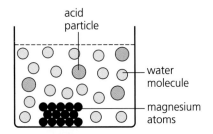
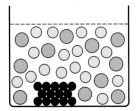

In dilute acid, there are not so many acid particles. So there is less chance of an acid particle hitting a magnesium atom.

Here the acid is more concentrated – there are more acid particles. So there is now more chance of a successful collision.

The more successful collisions there are, the faster the reaction.

Reactions between gases
- When you increase the pressure on two reacting gases, it means you squeeze more gas molecules into a given space.
- So there is a greater chance of successful collisions.
- So if pressure ↑ then rate ↑ for a gaseous reaction.

The same idea also explains why the reaction between magnesium and hydrochloric acid slows down as time goes on:

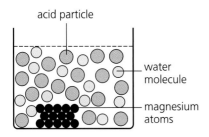

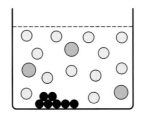

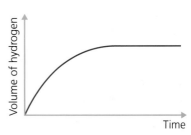

At the start, there are plenty of magnesium atoms and acid particles. But they get used up in successful collisions.

After a time, there are fewer magnesium atoms, and the acid is less concentrated. So the reaction slows down.

This means that the slope of the reaction curve decreases with time, as shown above. It goes flat when the reaction is over.

Why rate increases with temperature On heating, *all* the particles take in heat energy.

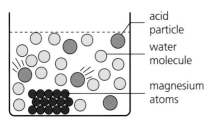

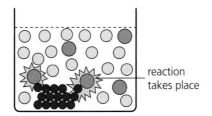

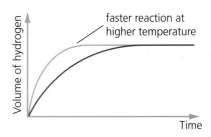

This makes the acid particles move faster – so they collide more often with magnesium particles.

The extra energy also means that more collisions are successful. So the reaction rate increases.

In fact, as you saw earlier, the rate generally doubles for an increase in temperature of 10° C.

Why rate increases with surface area The reaction between the magnesium and acid is much faster when the metal is powdered:

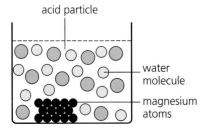

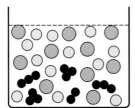

The acid particles can collide only with those magnesium atoms in the outer layer of the metal ribbon.

In the powdered metal, many more atoms are exposed. So the chance of a collision increases.

Questions

1. Copy and complete: Two particles can react together only if they …… and the …… has enough …… to be …….
2. What is meant by:
    a a successful collision?
    b an unsuccessful collision?
3. Reaction between magnesium and acid speeds up when:
    a the concentration of the acid is doubled. Why?
    b the temperature is raised. Why?
    c the acid is stirred. Why?
    d the metal is ground to a powder. Why?

10.6 Catalysts

What is a catalyst?

Look at this amazing experiment:

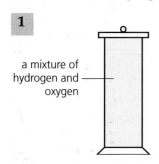

Fill a gas jar with a mixture of hydrogen and oxygen, and cover it. Even if you leave it for hours, no reaction takes place.

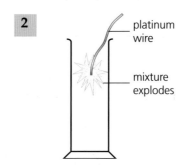

Now dip a platinum wire into the mouth of the jar. The gas mixture explodes immediately with a pop, producing water.

The reaction in **2** above is:

$$2H_2\,(g) + O_2\,(g) \longrightarrow 2H_2O\,(l)$$

The platinum wire has helped the reaction to take place, but has not itself been changed. It is a **catalyst** for the reaction.

A catalyst is a substance that speeds up a chemical reaction, but remains chemically unchanged itself.

How does a catalyst work?

For hydrogen and oxygen to react together, their molecules must collide. Then H–H bonds and O=O bonds must get broken. This needs energy!

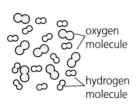

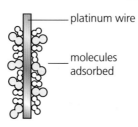

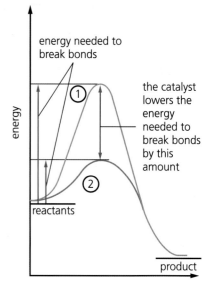

At room temperature, the molecules in **1** do not have enough energy for successful collisions to take place. So no bonds get broken, and no reaction takes place.

But in **2**, the molecules are **adsorbed** onto the platinum wire. This way, less energy is needed for the bonds to break. So now there is more chance of the molecules reacting.

In other words, by adsorbing the molecules, platinum lowers the energy needed to break the bonds to start the reaction. So reaction can begin at room temperature.

This energy needed to break the bonds is called the **activation energy**.

A catalyst works by lowering the activation energy for the reaction.

Once the bonds in the molecules are broken, new bonds form between hydrogen and oxygen atoms, giving water molecules. This *gives out* a lot of energy – which is then taken in by hydrogen and oxygen molecules. The reaction rate now increases rapidly, and you get an explosion.

Catalysts in industry

In industry, many reactions need heat. The heat may come from burning oil, gas, or coal, or from electricity. All these are expensive!

With a catalyst, a reaction goes faster *at a given temperature*. So you get the product faster, saving time. Even better, it may go fast enough *at a lower temperature* – which means a lower fuel bill.

So catalysts are very important in industry. They are usually **transition metals** or their **oxides.** Two examples are:

- **iron** used in the manufacture of ammonia
- **vanadium pentoxide** used in the manufacture of sulphuric acid.

Some different catalysts. Catalysts are made with a large surface area. Why?

Catalysts in car exhausts

Petrol is a mixture of **hydrocarbons** – compounds that contain only hydrogen and carbon. When it burns in a car engine, many harmful gases flow out the exhaust pipe. They include:

- oxides of nitrogen (NO and NO_2), which form when nitrogen and oxygen from the air react in the hot engine. These cause acid rain.
- carbon monoxide, CO, which forms when carbon compounds burn in insufficient oxygen. It is poisonous.
- unburnt hydrocarbons from the petrol. These can cause cancer.

To tackle this problem, modern car exhausts contain a **catalytic converter**. In this, the harmful gases are adsorbed onto catalysts, and react to produce harmless gases. The converter usually has two catalyst compartments, marked A and B below.

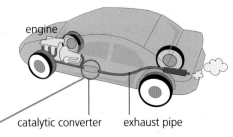

In **A**, harmful compounds are **reduced**. For example:

$$2NO\,(g) \longrightarrow N_2\,(g) + O_2\,(g)$$

The nitrogen and oxygen from this reaction then flow into **B**.

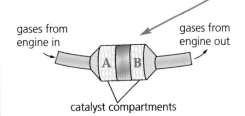

catalyst compartments

In **B**, harmful compounds are **oxidised**, using the oxygen from **A**. For example:

$$2CO\,(g) + O_2\,(g) \longrightarrow 2CO_2\,(g)$$

The harmless products then flow out the exhaust pipe.

The usual catalysts are the transition metals platinum, palladium, and rhodium. They are coated onto a ceramic support. This is in the form of beads, or a honeycomb, to give a large surface area for adsorbing the gases. The harmless products flow out the exhaust pipe.

Questions

1. Which does a catalyst *not* change?
   a. the speed of a reaction   b. the products formed
   c. the total amount of each product formed
2. a. Why is platinum able to speed up the reaction between hydrogen and oxygen? (Use the term *adsorbed*.)
   b. Do you think powdered platinum will work too?
3. What is *activation energy*?
4. Catalysts are very important in industry. Why?
5. Explain why a catalytic converter is needed in car exhausts, and describe one.
6. The same catalyst will *not* work for all reactions. See if you can think of a reason to explain this.

10.7 Enzymes: biological catalysts

What are enzymes?

Enzymes are **proteins** that act as catalysts. They are different from other catalysts in one big way: *they are made by living cells*. So they are often called **biological catalysts**.

Hundreds of different reactions go on inside living things – including you. For example, digestion involves many different reactions. But your body temperature is quite low. To make reactions go fast enough to keep you alive, your cells use enzymes as catalysts.

Enzymes have large complex molecules. This is a model of an enzyme molecule.

Studying an enzyme in the lab

Hydrogen peroxide is a colourless liquid that decomposes like this:

hydrogen peroxide $\longrightarrow$ water + oxygen
$2H_2O_2\ (l) \longrightarrow 2H_2O\ (l) + O_2\ (g)$

Some hydrogen peroxide is poured into three measuring cylinders. One is used as a control. Manganese(IV) oxide is added to the second, and raw liver to the third.

A glowing wooden splint is used to test the three cylinders for oxygen. The splint will burst into flame if there is enough oxygen present. Look at these results:

The test for oxygen

1. Light a wooden splint, and blow it out again.
2. The splint will continue to glow in air.
3. It will burst into flame in oxygen.

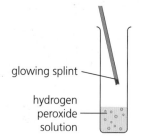
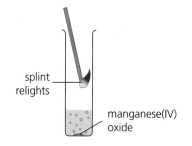
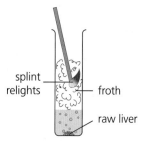

Hydrogen peroxide decomposes *very* slowly. There is not enough oxygen to relight the splint.

Manganese(IV) oxide makes the reaction go thousands of times faster. The splint bursts into flame.

Raw liver also speeds it up. The liquid froths as the oxygen bubbles off – and the splint relights.

So manganese(IV) oxide acts a catalyst for the reaction. So does something in the raw liver. That 'something' is **catalase**, an enzyme made by liver cells.

How enzymes work

This simple model shows how an enzyme catalyses the breakdown of a molecule (such as a hydrogen peroxide molecule):

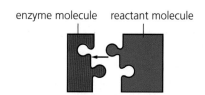

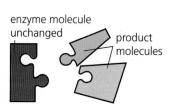

First, the two must fit together like jigsaw pieces. So the reactant molecule has to be the right shape.

The 'complex' that forms makes it easier for the reactant molecule to break down.

The product molecules then move away. Another molecule of reactant will take their place.

Some points to note about enzymes

- An enzyme works best in conditions that match those in the living cells it came from.
- This means most enzymes work best in the temperature range 25–45 °C. (The temperature in your cells is 36 °C, for example.)
- If the temperature is too high, an enzyme loses its shape – it becomes **denatured**. So it stops working. If the temperature is too low, an enzyme becomes inactive.
- An enzyme also works best in a particular pH range. You can denature it by adding acid or alkali.

Making use of enzymes

Enzymes have many uses. For example:

1. **In making ethanol** The alcohol ethanol is now being used as a fuel for cars. It is made by the action of yeast on a sugar called glucose. The glucose is obtained from crops such as sugar cane and corn.

   Yeast consists of millions of tiny living cells. These get their energy by breaking down the glucose into ethanol and carbon dioxide. They use enzymes to catalyse the reaction, which is called **fermentation**:

   $$C_6H_{12}O_6 \ (aq) \xrightarrow[\text{yeast}]{\text{enzymes in}} 2C_2H_5OH \ (l) + 2CO_2 \ (g) + \text{energy}$$
   glucose ethanol carbon dioxide

   The ethanol is removed from the fermenting mixture by fractional distillation. You can find out more on page 250.

2. **In making bread** The reaction above is also used in making bread. But this time carbon dioxide is the important product.

   Bread dough contains yeast and sugar. When it is left sitting in a warm place, the yeast feeds on the sugar, giving carbon dioxide gas. This makes the dough rise. Later, in the hot oven, the gas expands so the dough rises even further. The heat kills off the yeast.

3. **In biological detergents** Some detergents contain enzymes that can break down the compounds in grease, sweat, and other stains.

In **1** and **2** above, the cells that make the enzymes are present (as yeast). But in **3**, the enzymes are made by bacteria living in tanks, in a liquid containing food for them. The enzymes are then separated from the liquid, and taken away to use in detergents.

Enzymes helped! Yeast was used in the dough for the pizza base.

A biological detergent. It contains enzymes that break down compounds in sweat and stains. The enzymes were made by bacteria living in tanks.

Questions

1. What is: **a** an enzyme? **b** a biological catalyst?
2. What is the difference between an enzyme such as catalase, and a catalyst such as manganese dioxide?
3. **a** Which gas is produced when hydrogen peroxide decomposes?
   b How would you test for this gas?
4. Explain how the enzyme catalase works, in catalysing the decomposition of hydrogen peroxide.
5. What is meant by *denatured*?
6. Cooked liver does *not* catalyse the decomposition of hydrogen peroxide. Suggest a reason.
7. Give two examples of the use of enzymes in industry.

Questions on Chapter 10

1 The rate of the reaction between magnesium and dilute hydrochloric acid can be measured using this apparatus:

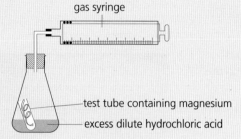

   a What is the purpose of:
     i the test tube?   ii the gas syringe?
   b How would you get the reaction to start?

2 Some magnesium and an *excess* of dilute hydrochloric acid were reacted together.
The volume of hydrogen produced was recorded every minute, as shown in the table:

Time/min	0	1	2	3	4	5	6	7
Volume of hydrogen/cm^3	0	14	23	31	38	40	40	40

   a What does an *excess* of acid mean?
   b Plot a graph of the results.
   c What is the *rate of reaction* (in cm^3 of hydrogen per minute) during:
     i the first minute?
     ii the second minute?
     iii the third minute?
   d Why does the rate change during the reaction?
   e How much hydrogen was produced in total?
   f How long does the reaction last?
   g What is the *average rate* of the reaction?
   h How could you make the reaction go *slower*, while keeping the reactants unchanged?

3 You will need your graph from question **2**.
The experiment with magnesium and an excess of dilute hydrochloric acid was repeated.
This time a different concentration of hydrochloric acid was used. The results were:

Time/min	0	1	2	3	4	5	6
Volume of hydrogen/cm^3	0	22	34	39	40	40	40

   a Plot these results on your graph for question 2.
   b Which reaction was faster? How can you tell?
   c In which experiment was the acid more concentrated? Give a reason for your answer.
   d The same volume of hydrogen was produced in each experiment. What does that tell you about the mass of magnesium used?

4 Suggest a reason for each observation below.
   a Hydrogen peroxide decomposes much faster in the presence of the enzyme catalase.
   b The reaction between manganese carbonate and dilute hydrochloric acid speeds up when some concentrated hydrochloric acid is added.
   c In fireworks, powdered magnesium is used rather than magnesium ribbon.
   d In most countries, dead animals decay quite quickly. But in Siberia, bodies of mammoths that died 30 000 years ago have been found fully preserved in ice.
   e Zinc and dilute sulphuric acid react much more quickly when a few drops of copper(II) sulphate solution are added.

5 When sodium thiosulphate and hydrochloric acid react together, a precipitate forms.
The effect of thiosulphate concentration on the rate of the reaction was investigated. The time it took for the precipitate to hide the cross was recorded, for different concentrations of thiosulphate:

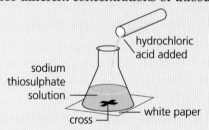

These are the results:

Experiment	A	B	C	D
Time taken/s	42	71	124	63

   a How is the time measured?
   b Name the precipitate that forms.
   c What would be *observed* during the experiment?
   d In which experiment was the reaction:
     i fastest? ii slowest?
   e In which experiment was the sodium thiosulphate solution the most concentrated? How can you tell?
   f Suggest two other ways to speed up the reaction.

6 Copper(II) oxide catalyses the decomposition of hydrogen peroxide. 0.5 g of the oxide was added to a flask containing 100 cm³ of hydrogen peroxide solution. A gas was released. It was collected, and its volume noted every 10 seconds.
This table shows the results:

Time/s	0	10	20	30	40	50	60	70	80	90
Volume/cm³	0	18	30	40	48	53	57	58	58	58

a What is a catalyst?
b Draw a diagram of suitable apparatus for this experiment.
c Name the gas that is formed.
d Write a balanced equation for the decomposition of hydrogen peroxide.
e Plot a graph of the volume of gas (vertical axis) against time (horizontal axis).
f Describe how rate changes during the reaction.
g What happens to the concentration of hydrogen peroxide as the reaction proceeds?
h What chemicals are present in the flask after 90 seconds?
i What mass of copper(II) oxide would be left in the flask at the end of the reaction?
j Sketch on your graph the curve that might be obtained for 1.0 g of copper(II) oxide.
k Name one other substance that catalyses this decomposition.

7 In two separate experiments, two metals A and B were reacted with an excess of dilute hydrochloric acid. The volume of hydrogen was measured every 10 seconds. These graphs show the results:

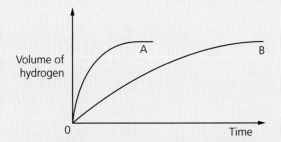

a i Which piece of apparatus can be used to measure the volume of hydrogen produced?
ii What other measuring equipment is needed?
b Which metal, A or B, reacts faster with hydrochloric acid? Explain your choice clearly.
c Sketch and label the curves that will be obtained for metal B if:
i more concentrated acid is used (curve X)
ii the reaction is carried out at a lower temperature (curve Y)

8 Marble chips (lumps of calcium carbonate) react with hydrochloric acid as follows:

$$CaCO_3\ (s) + 2HCl\ (aq) \longrightarrow CaCl_2\ (aq) + CO_2\ (g) + H_2O\ (l)$$

a What gas is released during this reaction?
b Describe a laboratory method that could be used to investigate the rate of the reaction.
c How will this affect the rate of the reaction?
i increasing the temperature
ii adding water to the acid
d Explain each of the effects in c in terms of collisions between reacting particles.
e If the lumps of marble are crushed first, will the reaction rate change? Explain your answer.

9 Zinc and iodine solution react like this:

$$Zn\ (s)\ +\ I_2\ (aq) \longrightarrow ZnI_2\ (aq)$$

The rate of reaction can be followed by measuring the mass of zinc metal at regular intervals, until all the iodine has been used up.
a What will happen to the mass of the zinc, as the reaction proceeds?
b Which reactant is in excess? Explain your choice.
c The reaction rate slows down with time. Why?
d Sketch a graph showing the mass of zinc on the y axis, and time on the x axis.
e How will the graph change if the temperature of the iodine solution is increased by 10 °C?
f Explain your answer to e using the idea of collisions between particles.

10 Ethanol can be made by the fermentation of sugar, in this apparatus:

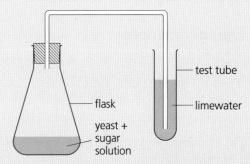

a What is present in the yeast, that causes the sugar to ferment?
b Which temperature is best for the reaction?
i 0 °C   ii 10 °C   iii 25 °C   iv 75 °C
Explain your answer.
c i Which gas is released during fermentation?
ii How could you prove this?
d Write the equation for the fermentation reaction.
e Name two industries that use this reaction.

11.1 Energy changes in reactions

Energy changes in reactions

During a chemical reaction, an energy change takes place.
Energy is given out or taken in, often in the form of heat.
So reactions can be divided into two groups, **exothermic** and **endothermic**.
(*therm* is from a Greek word for *heat*.)

Exothermic reactions

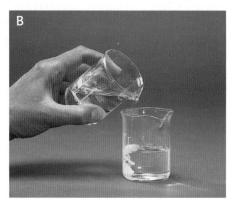
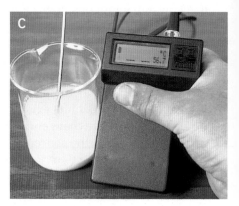

Once iron starts reacting with sulphur, it's easy to see that energy is given out. The mixture glows – with the Bunsen turned off.

Mixing silver nitrate and sodium chloride solutions gives a white precipitate of silver chloride – and a temperature rise.

When you add water to quicklime (calcium oxide) heat is given out, so the temperature rises. Here the rise is being measured.

The three reactions above are **exothermic.** They give out heat energy.
(*Exo means out.*) This can be summarized as:

   reactants ⟶ **products + heat energy**

The heat energy is given out to the surroundings, as shown on the right.
So the products have *lower energy* than the reactants. We can show this on an **energy level diagram** like the one below.

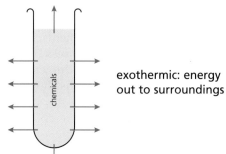

exothermic: energy out to surroundings

The change in energy

The energy change during a reaction can be measured.
The unit of energy is the kilojoule or kJ.
For reaction **A** above, the equation and energy change are:

   Fe (s) + S (s) ⟶ FeS (s)   energy change = –100 kJ/mole

So 100 kJ of energy is given out when 1 mole of iron atoms reacts with 1 mole of sulphur atoms.

The minus sign shows that energy is given out.

Other examples of exothermic reactions

- The neutralisation of an acid by an alkali.
- The combustion of fuels. We burn fuels to obtain heat energy.
- Respiration in your body cells. It provides the energy to keep your heart and lungs working, and for warmth and movement.

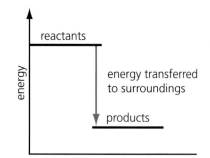

An energy level diagram for an exothermic reaction. The products have lower energy than the reactants.

Endothermic reactions

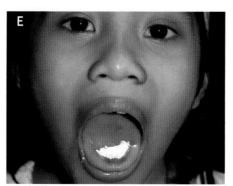

When barium hydroxide reacts with ammonium chloride, the temperature falls so sharply it makes water freeze.

Sherbet is citric acid plus the base sodium hydrogencarbonate. This neutralisation cools your tongue.

The crucible contains calcium carbonate. Strong heating will decompose it to calcium oxide.

This time the reactions *take in* heat energy. They are **endothermic**. (*Endo* means *in*.) The reactions can be summarized as:

reactants + heat energy ⟶ products

The heat energy is transferred from the surroundings. For example in **D**, it comes from the air and the wet wood. In **F**, from the bunsen burner.

Since energy is taken in, it means the products have *higher energy* than the reactants. You can show this on an energy level diagram like the one on the right below. The reactants must climb the energy gap for the reaction to go ahead – so they take energy from the surroundings.

Look again at reaction **F** above. The equation and energy change are:

$CaCO_3 (s) \longrightarrow CaO (s) + CO_2 (g)$ energy change = +178 kJ/mole

So 178 kJ of energy is needed to make 1 mole of $CaCO_3$ decompose. **The plus sign shows that energy is taken in.**

Other examples of endothermic reactions

Reactions **D** and **E** above are spontaneous, at room and body temperature. But many endothermic reactions are like **F**, where energy must be put in to start, and keep, the reaction going. For example:

- the reactions that take place when you cook food.
- photosynthesis. It takes in energy from sunlight.

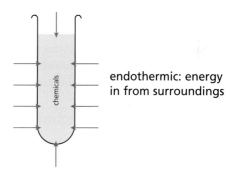

endothermic: energy in from surroundings

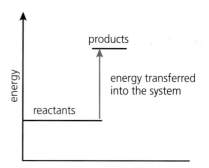

An energy level diagram for an endothermic reaction. The products have higher energy than the reactants.

Questions

1. What is: **a** an endothermic reaction?
   b an exothermic reaction?
2. Is it exothermic or endothermic?
   a the burning of a candle
   b the reaction between sodium and water
   c the change from raw egg to fried egg
3. Which unit is used to measure energy changes?
4. $2Na (s) + Cl_2 (g) \longrightarrow 2NaCl (s)$
   The energy change for this reaction is – 822.4 kJ/mole. What can you conclude about this reaction?
5. Write a balanced equation for reaction B on page 160, and say what sign the energy change will have.
6. Draw an energy level diagram for:
   a an endothermic reaction   **b** an exothermic reaction

11.2 Explaining energy changes

An exothermic reaction

A mixture of hydrogen and chlorine will explode in sunshine:

   hydrogen + chlorine ⟶ hydrogen chloride

The explosion is a sign that a lot of energy is given out. But where does it come from? Let's look more closely:

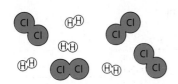

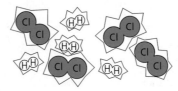

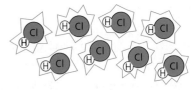

1 Molecules of hydrogen and chlorine are mixed together. The atoms in each molecule are held together by covalent bonds.

2 These bonds must be broken before reaction can take place. This step takes in energy. It is endothermic.

3 Now new bonds form between the hydrogen and chlorine atoms. This step releases energy. It is exothermic.

The energy given out in step **3** is greater than the energy taken in in step **2**. And that is why the reaction is exothermic overall.
If the energy released in bond making is greater than the energy needed for bond breaking, then the reaction is exothermic.

An endothermic reaction

If you heat ammonia strongly it breaks down. This time we show the molecules with a line for each bond. (Nitrogen has a triple bond.)

```
   H                H
   |                 \
H—N—H          H—N—H          (step 2 diagrams)
   |                 /
   H                H
```

1 In ammonia gas, the molecules are made of hydrogen atoms and nitrogen atoms, held together by covalent bonds.

2 For reaction to take place, these bonds must break. This step is **endothermic**. You supply the energy for it by heating.

3 Now the hydrogen atoms bond together. So do the nitrogen atoms. This step releases energy. It is **exothermic**.

But this time the energy released in step **3** is less than the energy supplied for step **2**. So *overall*, the reaction is endothermic.
If the energy released in bond making is less than the energy needed for bond breaking, then the reaction is endothermic.

Bond energies

Chemists have worked out how much energy is needed to break different types of bonds. This energy is called the **bond energy.**

Look at the list on the right. 242 kJ must be supplied to break the bonds in a mole of chlorine molecules, to give chlorine atoms. If the atoms join again to form molecules, 242 kJ of energy are given out.

The bond energy is the energy needed to break bonds, or released when these bonds form. It is given in kJ /mole.

Bond energy (kJ / mole)	
H—H	436
Cl—Cl	242
H—Cl	431
C—C	346
C=C	612
C—O	358
C—H	413
O=O	498
O—H	464
N≡N	946
N—H	391

Calculating the energy changes in reactions

For the reaction between hydrogen and chlorine
H–H + Cl–Cl ⟶ 2 H–Cl

Energy in to break each mole of bonds:
1 × H—H	436 kJ
1 × Cl—Cl	242 kJ
Total energy in	678 kJ

Energy out from the two moles of bonds forming:
2 × H—Cl   2 × 431 = 862 kJ

Energy in – energy out = 678 kJ – 862 kJ = – 184 kJ

So the reaction gives out **184 kJ** of energy, overall.

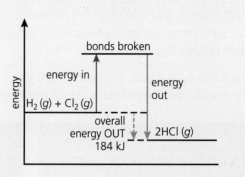

For the decomposition of ammonia

2 N(–H)(–H)–H ⟶ N≡N + 3 H—H

Energy in to break the two moles of bonds:
6 × N—H   6 × 391 = 2346 kJ

Energy out from the four moles of bonds forming:
1 × N≡N		946 kJ
3 × H—H	3 × 436 =	1308 kJ
Total energy out		2254 kJ

Energy in – energy out = 2346 kJ – 2254 kJ = 92 kJ

So the reaction takes in **92 kJ** of energy, overall.

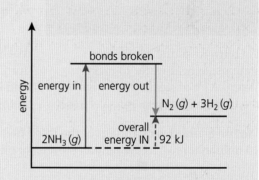

Activation energy

In the first stage of a reaction, bonds must be broken. The energy needed to break them is called the **activation energy**. Look at the diagram on the right.

You usually have to provide the activation energy, to start bonds breaking. For example, for reaction A on page 160, you add heat from a Bunsen flame. But once an exothermic reaction begins, the energy given out when new bonds form is used to break more bonds in the reactants. So you can turn the Bunsen burner off again!

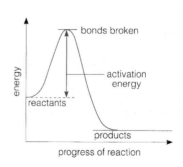

Questions

1. **a** When chlorine and hydrogen react, which part of the reaction is: **i** endothermic? **ii** exothermic?
   b Why does the reaction *give out* energy overall?
2. Draw the equation for the reaction between hydrogen and oxygen, with lines to show bonds, as above.
3. Now calculate the energy change for the reaction in **2**.
4. What does *activation energy* mean? (Check the glossary?)
5. Draw an energy diagram for the combustion of natural gas (methane) in a gas cooker. How would you provide the activation energy for this reaction?

11.3 Energy from fuels

What is a fuel?

A fuel is any substance we use to obtain energy, for cooking, heating, lighting, driving cars, and making electricity. Most fuels have to be burned to release their energy – but not all, as you'll see.

The fossil fuels

The fossil fuels – coal, oil, and natural gas – are the main fuels used around the world. They are all burned to release heat:

We burn fossil fuels in power stations, to heat water to make steam. A jet of steam then drives the turbines that generate electricity.

We burn them in factories to heat furnaces, and in homes for cooking and heating. (And some homes burn kerosene, from oil, for light.)

We burn petrol and diesel (from oil) in engines, to give the hot gas that makes the pistons move. The pistons then make the wheels turn.

We use enormous quantities of the fossil fuels. For example, around 10 million tonnes (or 80 million barrels) of oil *every day*!

So what makes a good fuel?

These are the main questions to ask about a fuel:

- **How much heat does it give out?** We want as much heat as possible, per mole or gram or tonne of the fuel.
- **Does it cause pollution?** If it causes a lot of pollution, we may be better off without it.
- **Is it easily available?** Since we depend so much on fuel, we need a steady and reliable source.
- **Is it easy and safe to store and transport?** Since most fuels catch fire quite easily, safety is always an issue.
- **How much does it cost?** The cheaper the better.

So, how do the fossil fuels rate? They do give out a lot of heat. But they all cause pollution, with coal the worst culprit, and oil next. The pollutants include carbon dioxide, which is linked to global warming, and acidic gases that cause acid rain. (There is more about these pollutants in Unit 14.3.)

What about availability? We are using up the fossil fuels fast. Some experts say that we may run out of oil and gas within 50 years. But there is probably enough coal to last a couple of hundred years.

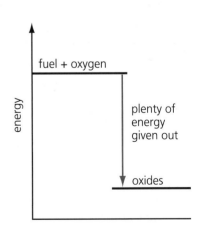

The burning of fuel is an exothermic reaction. The more heat given out the better – as long as the fuel is safe to use.

Two fuels growing in importance

Because of global warming, and the threat of oil and gas running out, there is a push to develop new fuels. Like these two:

Ethanol This is an alcohol, with the formula C_2H_5OH. It can be made from any plant material. For example, in Brazil it is made from sugar cane. It is used in car engines, on its own or mixed with petrol. You can find out more about ethanol in Units 17.6 and 17.7.

Hydrogen This gas burns explosively in oxygen, giving out a lot of energy – so it is used to fuel space rockets. But in a **fuel cell** it combines with oxygen without burning. The current from the cell can be used to power homes and cars. You can read more about the fuel cell on page 213.

Filling up with hydrogen!

Now compare those fuels with natural gas (methane).

Fuel	Equation for burning in oxygen	Heat given out/mole (kJ)	Heat given out/g of fuel (kJ)
methane	$CH_4\ (g) + 2O_2\ (g) \longrightarrow CO_2\ (g) + 2H_2O\ (l)$	– 882	– 55
ethanol	$C_2H_5OH\ (l) + 3O_2\ (g) \longrightarrow 2CO_2\ (g) + 3H_2O\ (l)$	– 1371	– 86
hydrogen	$2H_2\ (g) + O_2\ (g) \longrightarrow 2H_2O\ (g)$	– 286	– 143

Which of the three fuels produces no pollutant when it burns? Which one gives out most energy per gram?

Nuclear fuels

Nuclear fuels are *not* burned. They contain unstable atoms or **radioisotopes**. Over time, these break down naturally. (See page 30.) But you can also *force* them to break down, by shooting neutrons at them. They break into new atoms, giving out radiation and a lot of energy. In a nuclear power station, this energy is used to heat water, to make the steam to drive turbines.

Uranium-235 is one radioisotope used in nuclear fuel. The new atoms that form are very unstable, and immediately break down further.

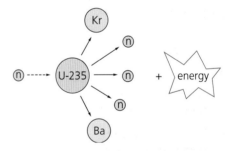

When hit by a neutron, a U-235 atom breaks down to other atoms, giving out a huge amount of energy. Power stations use the energy to provide electricity.

Nuclear fuel has two big advantages:
- It gives out huge amounts of energy. A pellet of nuclear fuel the size of a pea can give as much energy as a tonne of coal.
- No carbon dioxide or other greenhouse gases are produced.

But there is danger. An explosion in a nuclear power station could pollute a large area with radiation. The waste material is also radioactive, and may be so for hundreds of years. Finding a place to store it safely is a big problem.

Questions

1. **a** Sketch an energy level diagram that you think would show:
   i a good fuel
   ii a very poor fuel
   b What else would you think about, to decide whether something would be a good fuel?

2. Look at the table above. From the information given, which of the three fuels do you think is best? Explain.

3. The fuel butane (C_4H_{10}) gives the same products as methane when it burns. Write a balanced equation for its combustion.

11.4 Reversible reactions

When you heat copper(II) sulphate crystals ...

When you heat blue crystals of copper(II) sulphate, they break down into *anhydrous* copper(II) sulphate, a white powder:

$CuSO_4 \cdot 5H_2O$ (s) $\longrightarrow$
$\qquad CuSO_4$ (s) + $5H_2O$ (l)

The reaction is easy to reverse: just add water! The white powder turns blue again. In fact this is used as a test for water:

$CuSO_4$ (s) + $5H_2O$ (l) $\longrightarrow$
$\qquad CuSO_4 \cdot 5H_2O$ (s)

Water of crystallization
- The water in blue copper(II) sulphate crystals is called **water of crystallization**.
- The blue compound is **hydrated**.
- The white compound is **anhydrous**.

The above reaction is **reversible**: it can go in either direction.

Many chemical reactions are reversible. We use the symbol ⇌ to show that a reaction is reversible.

So the equation for the reaction above is written like this:

$CuSO_4 \cdot 5H_2O$ (s) ⇌ $CuSO_4$ (s) + $5H_2O$ (l)

Reaction 1 above is called the **forward** reaction.
Reaction 2 is the **back** reaction.

Two tests for water
- It turns white anhydrous copper(II) sulphate blue.
- It turns blue cobalt chloride paper pink.

Both changes are reversible.

What about the energy change?

In **1** above, you must heat the blue crystals to obtain the white powder – so the reaction is **endothermic**. In **2**, the white powder gets hot and spits when you drip water on – so this reaction is **exothermic**. In fact it gives out *the same amount of heat* as reaction **1** took in.

If a reversible reaction is endothermic in one direction, it is exothermic in the other. The same amount of heat is transferred each time.

This is easy to see from the energy level diagram on the right.

It's always the same energy gap between white anhydrous copper(II) sulphate and the blue crystals.

Another example of a reversible reaction

Thermal decomposition means breaking a substance down into simpler substances, by heating it.

The thermal decomposition of ammonium chloride is another example of a reversible reaction:

NH_4Cl (s) ⇌ NH_3 (g) + HCl (g)

Look what happens when it is heated in a test tube.

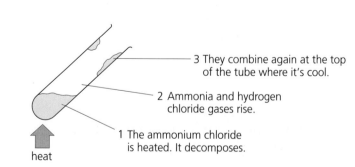

3 They combine again at the top of the tube where it's cool.
2 Ammonia and hydrogen chloride gases rise.
1 The ammonium chloride is heated. It decomposes.
heat

Reversible reactions and dynamic equilibrium

1

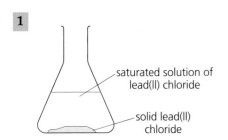

This is a saturated solution of lead(II) chloride in water. Some dissolved:

$$PbCl_2\ (s) \longrightarrow Pb^{2+}\ (aq) + 2Cl^-\ (aq)$$

The rest has sunk to the bottom.

2

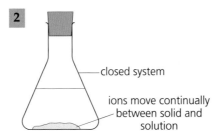

With the stopper on, the flask is a closed system. You might think nothing is happening. But inside, ions move non-stop between the solid and the solution.

3

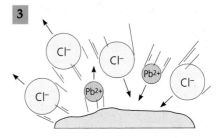

It happens *at the same rate in both directions*, so no extra lead(II) chloride dissolves. We can show it like this:

$$PbCl_2\ (s) \rightleftharpoons Pb^{2+}\ (aq) + 2Cl^-\ (aq)$$

Inside the flask, the solid and its saturated solution have reached a state of **dynamic equilibrium**. (It is often just called **equilibrium**.)

Equilibrium means there is no *overall* change. But *dynamic* means there is continual change.

You can prove this by adding lead(II) chloride containing radioactive lead ions to the flask in **2** above. It falls to the bottom. After 20 minutes the mass of solid in the flask is still the same – but there are radioactive lead ions in the solution.

Dissolving is a physical change. But exactly the same thing happens with a reversible chemical reaction.

In a closed system, a reversible reaction reaches a state of dynamic equilibrium. The forward and back reactions then take place at the same rate, so there is no overall change.

Dynamic equilibrium and industry

Many important reactions in industry are reversible. For example the reaction between nitrogen and hydrogen to make ammonia:

$$N_2\ (g) + 3H_2\ (g) \rightleftharpoons 2NH_3\ (g)$$

The reaction reaches equilibrium. At that point, every time a molecule of ammonia forms, another breaks down. So the reaction *is never complete*.

This is a problem for companies that make ammonia. They want the yield of ammonia to be as high as possible. So what can they do to increase it? You will find out in the next unit.

Nitrogen, hydrogen, and ammonia in equilibrium. The mixture is in balance.

Questions

1. What is a *reversible* reaction?
2. Write an equation for the reversible reaction between solid copper(II) sulphate and water.
3. What does *dynamic equilibrium* mean?
4. Explain how you could establish a state of dynamic equilibrium between table salt and water.

11.5 Shifting the equilibrium

The manufacture of ammonia

Imagine you run a factory that makes ammonia from nitrogen and hydrogen. The reaction is **reversible**:

$$N_2 (g) + 3H_2 (g) \rightleftharpoons 2NH_3 (g)$$

So let's see what happens during the reaction:

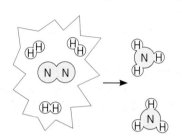
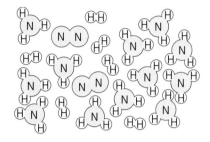
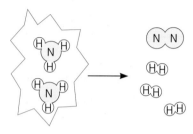

Three molecules of hydrogen react with one of nitrogen to form two of ammonia. So if you mix the right amounts of nitrogen and hydrogen …

… will it all turn into ammonia? No! Once a certain amount of ammonia is formed, the system reaches a state of dynamic equilibrium. From then on …

… every time two ammonia molecules form, another two break down into nitrogen and hydrogen. So the level of ammonia remains unchanged.

But the more ammonia that forms, the higher your profits. So how can you increase the yield?

This idea, called Le Chatelier's principle, will help you:
When a reversible reaction is in equilibrium and you make a change, it will do what it can to oppose that change.

You can't make a reversible reaction go to completion. It *always* ends up in a state of equilibrium. But by changing the conditions you can *shift equilibrium to the right* and obtain more product. So let's see what changes you can make, to obtain more ammonia.

Shifting the equilibrium to the right will give you more ammonia.

1 Increasing the temperature

$$N_2 + 3H_3 \xrightarrow{\text{heat out}} 2NH_3$$

$$2NH_3 \xrightarrow{\text{heat in}} N_2 + 3H_3$$

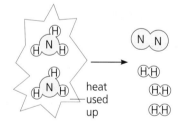

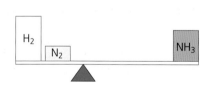

The forward reaction is exothermic – it gives out heat. The back reaction is endothermic – it takes it in. Heating speeds up *both* reactions …

… but if you heat the equilibrium mixture, it acts to oppose the change. More ammonia breaks down in order to use up the heat you've added.

The result is that the reaction reaches equilibrium faster, which is good, but the level of ammonia *decreases*. So you're worse off than before.

You could improve the yield by carrying out the reaction at a very *low* temperature. That will encourage lots of ammonia to form. But then the reaction will take too long to reach equilibrium! In industry, time is money. A slow reaction is not a good idea.

2 Increasing the pressure

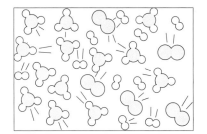
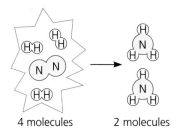
4 molecules   2 molecules
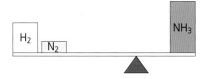

Pressure is caused by the gas molecules colliding with the walls of the container. So the more molecules present, the higher the pressure.

Push the gas into a smaller space: pressure rises. The mixture acts to oppose this. More ammonia forms, to reduce the number of molecules.

The result is that the level of ammonia in the mixture *increases*. So equilibrium shifts to the right. Well done. You're on the right track.

3 Removing the ammonia

The equilibrium mixture is a balance between nitrogen, hydrogen, and ammonia. Suppose you cool the mixture. Ammonia condenses first, so you can run it off as a liquid. Then warm the remaining nitrogen and hydrogen again – and more ammonia forms.

4 Adding a catalyst

Iron is a catalyst for this reaction. A catalyst speeds up the forward and back reactions *equally*. So the reaction reaches equilibrium faster, but the amount of ammonia does not change. Still, the catalyst saves time – so is worth using.

Choosing the optimum conditions

So to get the best yield of ammonia, it is best to:
- use high pressure, and remove ammonia, to shift equilibrium to the right
- use a moderate temperature, as a compromise
- use a catalyst to reach equilibrium quickly, at that temperature.

Page 216 shows how these conditions are applied in an ammonia factory.

> **Making ammonia: a summary**
>
> $$N_2 + 3H_2 \underset{\text{endothermic}}{\overset{\text{exothermic}}{\rightleftharpoons}} 2NH_3$$
> 4 molecules   2 molecules
>
> **To improve the yield:**
> - high pressure
> - low temperature
> - remove ammonia
>
> **To get a decent reaction rate:**
> - raise temperature (compromise!)
> - use a catalyst

Remember, for reversible reactions between gases …

If the forward reaction is:
- exothermic, temperature ↑ means yield ↓
- endothermic, temperature ↑ means yield ↑

If there are fewer molecules of product than reactant, in the equation:
pressure ↑ means yield ↑

Questions

1 a Explain what goes on in an equilibrium mixture of nitrogen, hydrogen, and ammonia.
   b Why does this cause a problem in industry?
2 What is Le Chatelier's principle? Write it down.
3 In manufacturing ammonia, explain why:
   a high pressure is used   b ammonia is removed
4 Sulphur dioxide (SO_2) and oxygen react to form sulphur trioxide (SO_3). The reaction is exothermic and reversible.
   a Write a balanced equation for the reaction.
   b What happens to the yield of sulphur trioxide if you:
      i increase the pressure?   ii raise the temperature?
      iii use a catalyst?

Questions on Chapter 11

1 Sodium hydroxide and hydrochloric acid react as shown in this equation:

   NaOH *(aq)* + HCl *(aq)* ⟶ NaCl *(aq)* + H₂O *(l)*

   a Which type of chemical reaction is this?
   b The reaction is *exothermic*. Explain what that means.
   c So what happens to the temperature of the solution, as the chemicals react?
   d Draw an energy diagram for the reaction.

2 Water at 25 °C was used to dissolve two compounds. The temperature of each solution was measured immediately afterwards.

Compound	Temperature of solution/°C
ammonium nitrate	21
calcium chloride	45

   a Calculate the temperature change on dissolving each compound.
   b i Which one dissolved exothermically?
     ii What can you say about the bonds that are formed with water in this solution?
   c i Which one dissolved endothermically?
     ii What can you say about the bonds that are formed with water this time?
   d For each solution, estimate the temperature of the solution if:
     i the amount of water is halved, but the same mass of compound is used
     ii the mass of the compound is halved, but the volume of water is unchanged
     iii both the mass of the compound, and the volume of water, are halved.

3 In this reaction of methane (CH₄), 1664 kJ of energy are *taken in*, per mole of methane:

   H–C(H)(H)(H)–H *(g)* ⟶ C *(g)* + 4H *(g)*

   (Note: no bonds are *formed* in this reaction.)
   a Why is this reaction *endothermic*?
   b How many bonds are broken?
   c Find the bond energy of the C–H bond.
   d i In a similar reaction for ethane (C₂H₆), 2826 kJ of energy are taken in, per mole of ethane. Write an equation, displayed like the one above, to show this reaction.
     ii List the bonds broken during the reaction.
     iii Work out the bond energy of the C–C bond.

4 Hydrogen peroxide decomposes very slowly to water and oxygen, in this reaction:

   2H₂O₂ *(aq)* ⟶ 2H₂O *(l)* + O₂ *(g)*

   a This is the energy level diagram for the reaction:

   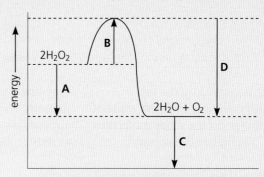

   i Which represents the *overall* energy change for the reaction: **A**, **B**, **C** or **D**?
   ii Overall, is the reaction exothermic or endothermic? Explain your answer.
   b Which energy change, **A**, **B**, **C**, or **D**, represents the activation energy?
   c i When bonds are *formed*, is energy released, or is it taken in?
     ii Which energy change, A, B, C, or D, is due to bonds being formed, in this reaction?
   d This shows the decomposition of hydrogen peroxide, using displayed formulae:

   2 (H–O–O–H) ⟶ 2 (H–O–H) + O=O

   These are the bond energies:

Bond	Bond energy/kJ per mol
O=O	498
O–O	146
H–O	464

   i Calculate the energy needed to break all the bonds in the reactant.
   ii Calculate the energy given out when new bonds are formed to give the products.
   iii Calculate the energy change for this reaction.
   iv Is the reaction exothermic or endothermic? Explain your answer, and compare it with your answer to **a ii** above.
   v A thermometer would not be suitable for measuring the energy change. Why not?
   e i Manganese(IV) oxide acts a catalyst for this reaction. What is a catalyst?
     ii How will the energy level diagram change when a catalyst is used? (Check page 154!)

5 The fuel *natural gas* is mostly methane. Here is the equation for the combustion of methane in oxygen. The reaction is exothermic:

   $CH_4 (g) + 2O_2 (g) \longrightarrow CO_2 (g) + 2H_2O (l)$

   a Explain, in terms of bond breaking and bond making, why this reaction is exothermic.
   b i Copy and complete this energy diagram for the reaction, indicating:
         the overall energy change
         the energy needed to break bonds
         the energy given out when new bonds form.

   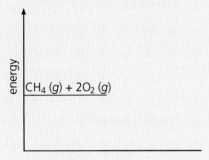

      ii Methane will not burn in air until a spark or flame is applied. Why not?
   When 1 mole of methane burns in oxygen, the energy change is − 890 kJ.
   c i What does the − sign tell you?
      ii Which word describes a reaction with this type of energy change?
   d How much energy is given out when 1 gram of methane burns?
      (The A_r values are: C = 12, H = 1.)

6 The gas hydrazine, N_2H_4, burns in oxygen like this:

   H₂N−NH₂ (g) + O=O (g) ⟶ N≡N (g) + 2O(H)₂ (g)

   a Count and list the bonds broken in this reaction.
   b Count and list the new bonds formed.
   c Calculate the total energy:
      i   required to break the bonds
      ii released when the new bonds form.
      (The bond energies in kJ/mole are:
      N−H 391; N−N 158; N≡N 945; O−H 464;
      O=O 498.)
   d Calculate the energy change in the reaction.
   e Is the reaction exothermic, or endothermic?
   f Where is energy transferred from, and to?
   g Do you think hydrazine would be a suitable fuel? Explain your answer.

7 When chlorine gas is passed over crystals of iodine, the following reactions take place.
   Reaction 1: $I_2 (s) + Cl_2 (g) \longrightarrow 2 ICl (l)$
                    black   brown
   Reaction 2: $ICl (l) + Cl_2 (g) \rightleftharpoons ICl_3 (s)$
                    brown   yellow

   a What is the first change you would see as the chlorine is passed over the iodine?
   b As more chlorine is passed, what further change would you see?
   c Which of the reactions can be *reversed*?
   d What change would you see as the chlorine gas supply is turned off? Explain your answer.

8 Hydrogen and bromine react reversibly:

   $H_2 (g) + Br_2 (g) \rightleftharpoons 2HBr (g)$

   a Which of these will favour the formation of more hydrogen bromide?
      i   adding more hydrogen
      ii   removing bromine
      iii removing the hydrogen bromide as it forms.
   b Explain why increasing the pressure will have no effect on the amount of product formed.
   c However, the pressure *is* likely to be increased, when the above reaction is carried out in industry. Suggest a reason for this.

9 Ammonia is made from nitrogen and hydrogen. The energy change in the reaction is −92kJ/mole. The reaction is reversible, and reaches equilibrium.
   a Write the equation for the reaction.
   b Is the forward reaction endothermic, or exothermic? How can you tell?
   c Explain why the *yield* of ammonia:
      i   rises if you increase the pressure
      ii falls if you increase the temperature
   d What effect does increasing:
      i the pressure   ii the temperature
      have on the *rate* at which ammonia is made?
   e Why is the reaction carried out at 450 °C rather than at a lower temperature?

10 The dichromate and chromate ions, $Cr_2O_7^{2-}$ and CrO_4^{2-}, exist in equilibrium like this:

   $Cr_2O_7^{2-} (aq) + H_2O (l) \rightleftharpoons 2CrO_4^{2-} (aq) + 2H^+ (aq)$
   orange   yellow

   a What would you *see* if you added dilute acid to a solution containing chromate ions?
   b How would you reverse the change?
   c Use Le Chatelier's principle to explain why adding hydroxide ions shifts the equilibrium.

12.1 Metals and non-metals

Most elements are metals

The Periodic Table on page 58 shows 105 elements. Of these, 84 are **metals,** and 21 are **non-metals.** So over three-quarters are metals. But you are likely to meet only about 15 of them in the lab and in everyday life.

The properties of metals

Metals *usually* have these properties:

1 They are **strong**. If you press on them, or drop them, or try to tear them, they won't break – and it is hard to cut them.
2 They are **malleable**. That means they can be hammered and bent into shape without breaking.
3 They are **ductile**: they can be drawn out into wires.
4 They are **sonorous**: they make a ringing noise when you strike them.
5 They are shiny when polished.
6 They are good conductors of electricity and heat.
7 They have high melting and boiling points. (They are all solid at room temperature, except mercury.)
8 They have high **density** – they feel 'heavy'.
9 They react with oxygen to form **oxides**. For example, magnesium burns in air to form magnesium oxide. Metal oxides are **bases**, which means they neutralise acids, forming salts and water.
10 When they react, metals form positive ions. For example, magnesium forms magnesium ions (Mg^{2+}) when it reacts with oxygen, as shown on page 39.

The last two properties above are called **chemical properties,** because they are about chemical change. The others are **physical properties.**

Some of the metals

aluminium, Al
calcium, Ca
copper, Cu
gold, Au
iron, Fe
lead, Pb
magnesium, Mg
potassium, K
silver, Ag
sodium, Na
tin, Sn
zinc, Zn

What is density?

The density of a substance is a measure of how 'heavy' it is:

$$\text{density} = \frac{\text{mass (in grams)}}{\text{volume (in cm}^3\text{)}}$$

Compare these:

1 cm³ of iron
mass = 7.86 g
density =
　7.86 g/cm³

1 cm³ of lead
mass = 11.34 g
density =
　11.34 g/cm³

Think of two reasons why metals are used to make drums . . .

. . . and three reasons why they are used for saucepans.

All metals are different

The properties on the opposite page are typical of metals. But not all metals have *all* of these properties. For example:

Iron is a typical metal. It is used for gates like these because it is both malleable and strong. It is used for anchors because of its high density. It melts at 1530°C. But unlike most other metals, it is **magnetic**.

But sodium is so soft that you can cut it with a knife, and it melts at only 98°C. It is so light that it floats on water, but it reacts immediately with the water, forming a solution. No good for gates then!

Gold melts at 1064°C. Unlike most other metals it does not form an oxide – it is very unreactive. But it is malleable and ductile, and looks attractive. So it is used for jewellery.

In fact no two metals have exactly the same properties. You will find out more about differences between them in the rest of this chapter.

Comparing metals with non-metals

Non-metals are different from metals. They *usually* have these properties.

1. They are not strong, or malleable, or ductile, or sonorous. In fact, when solid non-metals are hammered, they break up – they are **brittle**.
2. They have lower melting and boiling points than metals. (One of them is a liquid and eleven are gases, at room temperature.)
3. They are poor conductors of electricity. Graphite (carbon) is the only exception. They are also poor conductors of heat.
4. They have low densities.
5. Like metals, most of them react with oxygen to form oxides:

   sulphur + oxygen ⟶ sulphur dioxide

   But unlike metal oxides, these oxides are not bases. Many of them dissolve in water to give *acidic* solutions.
6. When the non-metals form ions, the ions are negative. For example oxygen forms the ion O^{2-}. But hydrogen is an exception – it forms the ion H^+.

Some of the non-metals

bromine, Br
carbon, C
chlorine, Cl
helium, He
hydrogen, H
iodine, I
nitrogen, N
oxygen, O
sulphur, S

Questions

1. Make two lists, one showing 20 metals, and the other 15 non-metals. Give their symbols too.
2. Not all metals share the typical metal properties. See if you can name three metals which are NOT:
   a hard and strong   b malleable at room temperature.
3. Suggest reasons for this use of a metal:
   a silver for jewellery   b copper for electrical wiring.
4. For some uses, a highly *sonorous* metal is needed. See if you can give three examples.
5. Try to think of *two* reasons why:
   a mercury is used in thermometers
   b aluminium is used for beer cans.
6. Look at the properties of the non-metals, above. Which are *physical* properties? Which are *chemical*?

12.2 Comparing metals for reactivity (I)

What does *reactive* mean?

A *reactive* element has a strong drive to become a compound. So it reacts readily with other elements and compounds. Here we compare some metals to see how reactive they are, and if there is a pattern.

1 The reaction of metals with water

Compare these:

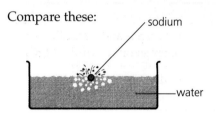
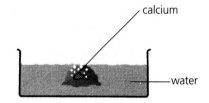
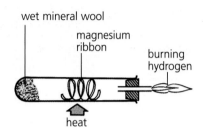

Sodium reacts violently with cold water, whizzing over the surface. Hydrogen gas and a clear solution of sodium hydroxide are formed.

The reaction between calcium and cold water is slower. Hydrogen bubbles off, and a cloudy solution of calcium hydroxide forms.

Magnesium reacts very slowly with cold water, but vigorously on heating in steam: it glows brightly. Hydrogen and solid magnesium oxide are formed.

You can tell from the violence of its reaction that sodium is the most reactive of the three metals, and magnesium the least.

Compare the equations for the three reactions, given below. What pattern do you notice?

$$2Na\,(s) + 2H_2O\,(l) \longrightarrow 2NaOH\,(aq) + H_2\,(g)$$
$$Ca\,(s) + 2H_2O\,(l) \longrightarrow Ca(OH)_2\,(aq) + H_2\,(g)$$
$$Mg\,(s) + H_2O\,(g) \longrightarrow MgO\,(s) + H_2\,(g)$$

Now compare the reactions of those metals with the others in this table:

Metal	Reaction	Order of reactivity	Products
potassium	very violent with cold water; catches fire	most reactive	hydrogen and a solution of potassium hydroxide, KOH
sodium	violent with cold water		hydrogen and a solution of sodium hydroxide, NaOH
calcium	less violent with cold water		hydrogen and calcium hydroxide, $Ca(OH)_2$, which is only slightly soluble
magnesium	very slow with cold water, but vigorous with steam		hydrogen and solid magnesium oxide, MgO
zinc	quite slow with steam		hydrogen and solid zinc oxide, ZnO
iron	slow with steam		hydrogen and solid iron oxide, Fe_3O_4
copper silver gold	no reaction	least reactive	

Note that only the first three metals in the list produce hydroxides. The others produce insoluble oxides, if they react at all.

2 The reaction of metals with hydrochloric acid

It is not safe to add sodium or potassium to acid in the lab, because the reactions are explosively fast. But other metals can be tested safely. Compare these reactions with hydrochloric acid:

Metal	Reaction with hydrochloric acid	Order of reactivity	Products
magnesium	vigorous	most reactive	hydrogen and a solution of magnesium chloride, $MgCl_2$
zinc	quite slow		hydrogen and a solution of zinc chloride, $ZnCl_2$
iron	slow		hydrogen and a solution of iron(II) chloride, $FeCl_2$
lead	slow, and only if the acid is concentrated		hydrogen and a solution of lead(II) chloride, $PbCl_2$
copper silver gold	no reaction, even with concentrated acid	least reactive	

The equation for the reaction with magnesium this time is:

$$Mg\,(s) + 2HCl\,(aq) \longrightarrow MgCl_2\,(aq) + H_2\,(g)$$

Now compare the order of the metals in the two tables, and the equations for the reactions. What patterns can you see?

Hydrogen is displaced

When a metal *does* react with water or hydrochloric acid, it displaces hydrogen. This shows that the metal is *more reactive* than hydrogen. It has a stronger drive to be a compound. But copper and silver don't react with water or acid. So they are *less reactive* than hydrogen.

Note that the displacement reactions that do take place are **redox reactions**. When magnesium reacts with hydrochloric acid, its atoms lose electrons, so it is oxidised. The hydrogen ions from the acid gain electrons, so they are reduced. The half-equations are:

magnesium: $\quad Mg\,(s) \longrightarrow Mg^{2+}\,(aq) + 2e^-$ (oxidation)

hydrogen ions: $\quad 2H^+\,(aq) + 2e^- \longrightarrow H_2\,(g)$ (reduction)

> **Redox again!**
>
> Remember OIL RIG:
> Oxidation Is Loss of electrons
> Reduction Is Gain of electrons.

> **Questions**
> 1 Write a balanced equation for the reaction of potassium with water.
> 2 Which is more reactive? And what is your evidence?
>   a potassium or sodium?   b copper or zinc?
> 3 Which gas is always produced if a metal reacts with water, or dilute acid?
> 4 Explain why the reaction of iron with hydrochloric acid is a redox reaction.

12.3 Comparing metals for reactivity (II)

Metals in competition

You saw in the last unit how metals can be put in order of reactivity, using their reactions with water and hydrochloric acid.

We can also test what happens when they compete with each other, and with carbon and hydrogen, to form a compound. The more reactive element will always win.

1 Competing for oxygen

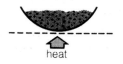

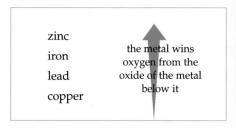

Some powdered iron is heated with copper(II) oxide, CuO. Can the iron grab the oxygen from the copper(II) oxide?

The reaction gives out heat, once it gets going. The mixture glows. Iron(II) oxide and copper are formed. The iron has won.

Other metals were compared in the same way. This shows their order of reactivity – the same as in the table on page 175.

The tests confirm that iron, zinc and lead are all more reactive than copper. The equation for the reaction with iron is:

$$Fe\,(s) + CuO\,(s) \longrightarrow FeO\,(s) + Cu\,(s)$$
$$\text{iron} + \text{copper(II) oxide} \longrightarrow \text{iron(II) oxide} + \text{copper}$$

The iron is acting as a **reducing agent,** removing oxygen.

When a metal is heated with the oxide of a less reactive metal, it acts as a reducing agent. The reaction always gives out heat – it is exothermic.

2 Competing to be the compound in solution

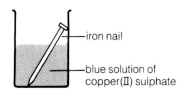
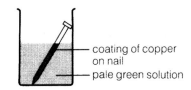

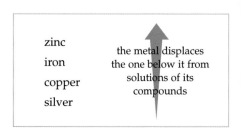

This time, an iron nail is placed in copper(II) sulphate solution. Will there be a reaction?

Yes! Copper soon coats the nail. The iron has displaced copper from the solution, which turns green.

Other metals and solutions were tested too, with these results. What do you notice?

Once again, iron wins against copper. It displaces it from the copper(II) sulphate solution, just as it drove it from its oxide:

$$Fe\,(s) + CuSO_4\,(aq) \longrightarrow FeSO_4\,(aq) + Cu\,(s)$$
$$\text{iron} + \text{copper(II) sulphate} \longrightarrow \text{iron(II) sulphate} + \text{copper}$$
$$\qquad\quad\text{(blue)} \qquad\qquad\quad \text{(green)}$$

Zinc and other metals that were tested displaced the less reactive metals in the same way.

A metal will always displace a less reactive metal from solutions of its compounds.

3 Competing with carbon, for oxygen

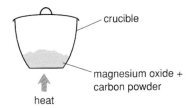

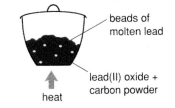

calcium	more reactive than carbon
aluminium	
carbon	
zinc	
iron	less reactive than carbon
copper	

Magnesium oxide is mixed with powdered carbon and heated. No reaction! So magnesium must be more reactive than carbon.

But when lead(II) oxide is used instead, it is reduced to molten lead. This shows that carbon is more reactive than lead.

The metal oxides above were also tested. Two were found to be more reactive than carbon. The others were less reactive than than carbon.

The equation for the reduction of the lead(II) oxide is:

$$2PbO\,(s) + C\,(s) \rightarrow 2Pb\,(s) + CO_2\,(g)$$
lead(II) oxide + carbon → lead + carbon dioxide

4 Competing with hydrogen, for oxygen

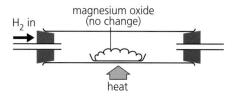

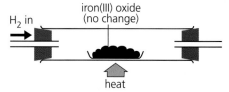

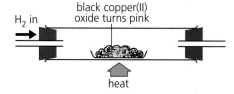

Hydrogen does not react with heated magnesium oxide …

… or with iron(II) oxide, lead oxide or zinc oxide. But when it passes …

… over heated copper(II) oxide, the black powder turns to copper.

Hydrogen reduces the copper(II) oxide, since it is more reactive than copper:

$$CuO\,(s) + H_2\,(g) \rightarrow Cu\,(s) + H_2O\,(l)$$

But it can't reduce the other oxides because it is less reactive than those metals.

They are all redox reactions

In all the reactions in this unit, electrons are transferred – so they are all redox reactions. For the reaction between copper(II) sulphate and iron in **2**, for example, the half-equations are:

iron: $\quad Fe(s) \rightarrow Fe^{2+}\,(aq) + 2e^-$ (oxidation)
copper: $\quad Cu^{2+}\,(aq) + 2e^- \rightarrow Cu\,(s)$ (reduction)

REMEMBER!

Oxidation is …
gain of oxygen
loss of hydrogen
loss of electrons

Reduction is …
loss of oxygen
gain of hydrogen
gain of electrons

Questions

1 Write a balanced equation for each reaction, and then show that it is a redox reaction:
   a the reaction between iron and silver oxide, Ag_2O
   b the reaction between carbon and iron(II) oxide, FeO

2 When copper wire is put in a solution of silver nitrate, $AgNO_3$, crystals of silver form on it. The solution goes blue.
   a Explain why the solution goes blue.
   b Show that a redox reaction is taking place.

12.4 The reactivity series

Pulling it all together: the reactivity series

We can use the results of the experiments in the last two units to put the metals in final order, with the most reactive one first. The list is called **the reactivity series**. Here it is.

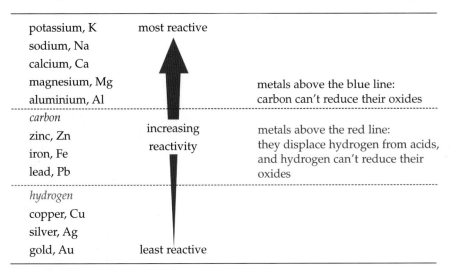

potassium, K	most reactive
sodium, Na	
calcium, Ca	
magnesium, Mg	
aluminium, Al	metals above the blue line: carbon can't reduce their oxides
carbon	
zinc, Zn	increasing reactivity
iron, Fe	metals above the red line: they displace hydrogen from acids, and hydrogen can't reduce their oxides
lead, Pb	
hydrogen	
copper, Cu	
silver, Ag	
gold, Au	least reactive

The non-metals carbon and hydrogen are included because they are useful for reference. The list is not 'complete', of course. You could test many other metals, and add them in the right place.

Things to remember about the reactivity series

1. The reactivity series is really a list of the metals in order of their drive to form positive ions. The more reactive the metal, the more easily it gives up electrons to form positive ions.

2. So a metal will react with a compound of a less reactive metal (for example an oxide, or a salt in solution) by pushing the less reactive metal out of the compound and taking its place, as ions.

3. The more reactive the metal, the more **stable** its compounds are. They do not break down easily.

4. The more reactive the metal, the more difficult it is to extract from its ores, since those compounds are stable. For the most reactive metals you need the toughest method of extraction: electrolysis.

5. The less reactive the metal, the less it likes to form compounds. That is why copper, silver and gold are found as elements in the Earth's crust. The other metals are *always* found as compounds.

Because they are easy to obtain from their ores, the less reactive metals have been known and used for thousands of years. Copper has been in wide use for 6000 years, and iron for 3500 years.

But the more reactive metals had to wait for the invention of electrolysis, in 1800, for their discovery. Sodium and potassium were discovered in 1807, and calcium in 1808.

A metal's position in the reactivity series will give you clues about its uses. Only unreactive metals are used to make coins.

The stability of some metal compounds

Many compounds break down easily on heating. In other words, they undergo **thermal decomposition.**

But the more reactive a metal is, the more stable its compounds will be. They will not break down so easily on heating. So let's see how compounds of the metals in the reactivity series behave, when you heat them.

1 Carbonates

Most decompose to the **oxide** and **carbon dioxide**, on heating.
- But the carbonates of potassium and sodium do not decompose.
- Strong heating is needed to break down calcium carbonate – and the reaction is reversible.
- The further down the series, the more easily the other carbonates break down. Copper(II) carbonate breaks down very easily, like this:

    $CuCO_3$ (s) $\longrightarrow$ CuO (s) + CO_2 (g)
    green   black

2 Hydroxides

Most decompose to the **oxide** and **water** on heating, like this:

$Zn(OH)_2$ (s) $\longrightarrow$ ZnO (s) + H_2O (l)

- But the hydroxides of potassium and sodium do not decompose.
- The further down the series, the more easily the others break down.

3 Nitrates

All decompose on heating – but not all to the same products.
- Potassium and sodium nitrates break down to **nitrites**, releasing only **oxygen**, like this:

    $2NaNO_3$ (s) $\longrightarrow$ $2NaNO_2$ (s) + O_2 (g)

- But the nitrates of the other metals break down further to **oxides**, releasing the brown gas **nitrogen dioxide** as well as **oxygen**:

    $2Pb(NO_3)_2$ (s) $\longrightarrow$ $2PbO$ (s) + $4NO_2$ (g) + O_2 (g)

- The further down the series, the more easily they break down.

A simple lime kiln, in which crushed limestone (calcium carbonate) is broken down by heat to quicklime (calcium oxide). Quicklime is used in making steel, and to neutralise acidity in soil.

Questions

1. Explain what *the reactivity series* is, in your own words.
2. Magnesium is never found as the element, in nature. Why not?
3. Gold has been known and used for thousands of years longer than aluminium. Explain why. (Page 190 may help!)
4. What will you see if you drop some magnesium ribbon into a blue solution of copper(II) sulphate?
5. When you heat aluminium with chromium oxide, you get chromium metal and aluminium oxide. What can you say about the reactivity of chromium?
6. Which will break down more easily on heating, magnesium nitrate or silver nitrate? Why?
7. Write a balanced equation for the thermal decomposition of lead(II) carbonate.

12.5 Making use of the reactivity series

Those differences in reactivity are useful!
The reactivity of metals is a big factor in deciding how we use them. For example gold is used for jewellery because it looks good *and* is unreactive. But we can also make good use of the *differences* in reactivity of the metals. We look at four examples in this unit.

1 The thermite process
This is used to repair railway lines. Powdered aluminium and iron(III) oxide are put in a container over the damaged rail. When the mixture is lit, the aluminium reduces the iron(II) oxide to molten iron, in a very vigorous reaction. The iron runs into the cracks and gaps in the rail, and hardens:

$$Fe_2O_3(s) + 2Al(s) \longrightarrow 2Fe(s) + Al_2O_3(s)$$

A similar thermite reaction is used to extract chromium from its ore. First the ore is converted to chromium(III) oxide. Then this is heated with aluminium, to obtain chromium metal:

$$Cr_2O_3(s) + 2Al(s) \longrightarrow Al_2O_3(s) + 2Cr(s)$$

2 The sacrificial protection of iron
Iron is widely used in structures such as oil rigs, ships, underground pipes, and car bodies. But it has one huge drawback: it reacts with oxygen and water, forming iron(III) oxide or **rust**.

This problem can be solved by teaming it up with a more reactive metal such as zinc or magnesium. For example a block of zinc can be welded to the side of a ship. Zinc is more reactive than iron – it gives up electrons to form positive ions more readily. So the zinc dissolves:

$$2Zn(s) \longrightarrow 2Zn^{2+}(aq) + 4e^- \qquad \text{(oxidation)}$$

The electrons flow to the iron, which passes them on, in this reaction:

$$O_2(g) + 2H_2O(l) + 4e^- \longrightarrow 4OH^-(aq) \qquad \text{(reduction)}$$

Adding those half-equations gives the full equation for the reaction:

$$2Zn(s) + O_2(g) + 2H_2O(l) \longrightarrow 2Zn(OH)_2(aq)$$

So the zinc is oxidised instead of the iron. This is called **sacrificial protection** because the zinc is sacrificed to protect the iron. The zinc block must be replaced before it all dissolves away.

3 Galvanising
The same idea is used to protect iron roofs, and the steel in car bodies.
- The iron or steel is plated with zinc, by electrolysis. Metal that has been coated like this is called **galvanised**.
- The zinc coating keeps air and moisture away.
- But if the coating gets broken, the zinc will still protect the iron by sacrificial protection.

The thermite process at work.

Galvanised iron, used for this modern home.

4 Making cells (batteries)

If you stand strips of magnesium and copper in an electrolyte, and connect them to a light bulb, the bulb will light! So electrons must be flowing.

The electrons flow because magnesium is more reactive than copper: it has a stronger drive to give up electrons. So its atoms give up electrons and go into solution as ions. The electrons flow along the wire to the copper strip.

The set-up is called a **cell**. The magnesium strip is called the **negative pole** since it is the source of the electrons. The copper strip is the **positive pole**.

A cell consists of two metals and an electrolyte. The more reactive metal is the negative pole. Electrons flow from it.

You can measure the 'push' or **voltage** that makes the electrons flow, using a voltmeter. For this cell it is 2.7 volts.

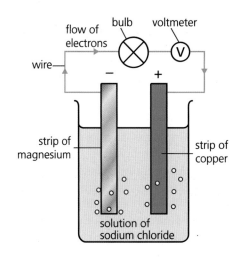

Using other metals for the cell

- As long as the strips are made of different metals, electrons will flow – from the more reactive metal to the less reactive one.
- The voltage depends on the metals you choose. Look at the table on the right. Copper and magnesium are furthest apart in the reactivity series, so they give the highest voltage, and the bulb light will be brightest.

Metal strips	Volts
copper and magnesium	2.70
copper and iron	0.78
lead and zinc	0.64
lead and iron	0.32

Batteries

Batteries are just portable cells. This diagram shows a torch battery.

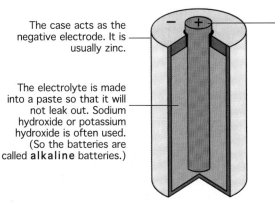

The case acts as the negative electrode. It is usually zinc.

The electrolyte is made into a paste so that it will not leak out. Sodium hydroxide or potassium hydroxide is often used. (So the batteries are called **alkaline** batteries.)

The positive electrode is down the middle.

Manganese(IV) oxide is often used, packed around a carbon rod. It is suitable because the Mn^{4+} ions can accept electrons to become Mn^{3+} ions.

The battery 'dies' when the reactions at the electrodes stop.

Battery companies are always trying out different electrodes and electrolytes, to produce better batteries that last longer.

Light, thanks to chemical reactions in a torch battery.

Questions

1. Write a word equation for the thermite reaction used to:
   a. repair railway lines
   b. extract chromium from its oxide
2. a. Steel for cars is *galvanised*. What does that mean?
   b. Explain how this protects the steel.
3. a. What is a *cell*?
   b. Explain why cells are able to make bulbs light up.
   c. Give examples of how *you* use portable cells.
4. Which pair of metals will give the highest voltage? iron, zinc; copper, iron; silver, magnesium. Why?

Questions on Chapter 12

1. Read the following passage about metals.
   Elements are divided into metals and non-metals. All metals are <u>electrical conductors</u>. Many of them have a high <u>density</u> and they are usually <u>ductile</u> and <u>malleable</u>. All these properties influence the way the metals are used. Some metals are <u>sonorous</u> and this leads to special uses for them.
   a Explain the underlined terms.
   b Copper is ductile. How is this property useful in everyday life?
   c Aluminium is hammered and bent to make large structures for use in ships and planes. What property allows it to be shaped like this?
   d Name one metal that has a *low* density.
   e Some metals are cast into bells. What property must the chosen metals have?
   f Add the correct word: *Metals are good conductors of ………. and electricity.*
   g Choose another physical property of metals, and give two examples of how it is useful.

2.
Metal	Density (g/cm³)
aluminium	2.7
calcium	1.6
copper	8.9
gold	19.3
iron	7.9
lead	11.4
magnesium	1.7
sodium	0.97

   a List the metals given in the table above in order of increasing density.
   b i   What is meant by *density*?
      ii Which of the metals is the least dense?
      iii Which one is the most dense?
      iv A block of metal has a volume of 20 cm³ and a mass of 158 g. Which metal is it?
   c Now list the metals in order of reactivity.
   d i   The most reactive metal in the list has a density of …..?
      ii The least reactive one has a density of …..?
      iii Does there appear to be a link between density and reactivity? If yes, what?
   e Using low-density metals for vehicles saves money on fuel, and on road and rail repairs. Explain why.
   f Which of the low-density metals above is the most suitable for vehicles? Why? Give three reasons.

3. This shows metals in order of reactivity:
   sodium   (most reactive)
   calcium
   magnesium
   zinc
   iron
   lead
   copper
   silver   (least reactive)

   a Which element is stored in oil?
   b Which elements will react with cold water?
   c Choose one metal that will react with steam but *not* with cold water. Draw a diagram of suitable apparatus for this reaction. (You must show how the steam is generated.)
   d Name the gas given off in **b** and **c**.
   e Name another reagent that reacts with many metals to give the same gas.
   f Write balanced chemical equations for the reactions in **b**, **c** and **e**.

4. Look again at the list of metals in **3**. Because zinc is *more reactive* than iron, it will remove the oxygen from iron(III) oxide, on heating.
   a Write a word equation for the reaction.
   b Will these react together, when heated?
      i   magnesium + lead(II) oxide
      ii   copper + lead(II) oxide
      iii magnesium + copper(II) oxide
      iv   iron + magnesium oxide.
   c For those pairs that will react:
      i   describe what you would see
      ii   write a word equation
      iii write a balanced equation.
   d What name is given to reactions of this type?

5. When magnesium and copper(II) oxide are heated together, this redox reaction occurs:
   $Mg\,(s) + CuO\,(s) \longrightarrow MgO\,(s) + Cu\,(s)$
   a What does the word *redox* stand for?
   b For the above reaction, name:
      i the reducing agent   ii the oxidising agent
   c Describe how electrons are transferred in the reaction.
   d Explain as fully as you can why the *reverse* reaction does not occur.
   e i   Name one metal that would remove the oxygen from magnesium oxide.
      ii Does this metal *gain* electrons, or *lose* them, more easily than magnesium does?

6 For each description below, choose one metal that fits the description. Name the metal. Then write a balanced equation for the reaction that takes place.
   a A metal that burns in oxygen.
   b A metal that displaces copper from copper(II) sulphate solution.
   c A metal that reacts gently with dilute hydrochloric acid.
   d A metal that floats on water and reacts vigorously with it.
   e A metal that reacts quickly with steam but very slowly with cold water.

7 When magnesium powder is added to copper(II) sulphate solution, a displacement reaction occurs and solid copper forms.

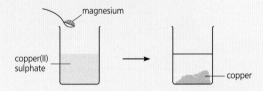

   a Write a word equation for the reaction.
   b Why does the displacement reaction occur?
   c i Write a half-equation to show what happens to the magnesium atoms.
     ii Which type of reaction is this?
   d i Write a half-equation to show what happens to the copper ions.
     ii Which type of reaction is this?
   e i Write the ionic equation for the displacement reaction. (See page 120 for a reminder!)
     ii Which type of reaction is this?
   f Use the reactivity series of metals to decide whether these will react together:
     i iron + copper(II) sulphate solution
     ii silver + calcium nitrate solution
     iii zinc + lead(II) nitrate solution
   g For those that react:
     i describe what you would see
     ii write the ionic equations for the reactions.

8 Nickel is below iron, but above lead, in the reactivity series. Use the series to predict the reaction of nickel:
   a with cold water
   b with concentrated hydrochloric acid
   c on heating with copper(II) oxide
   d when a strip of nickel is placed in a solution of lead(II) nitrate.

9 Use the reactivity series to explain why:
   a calcium is not used for household articles
   b copper is used for water pipes
   c powdered magnesium is used in fireworks
   d gold is an excellent material for filling teeth

10 Strips of copper foil and magnesium ribbon were cleaned with sandpaper and then connected as shown below. The bulb lit up.

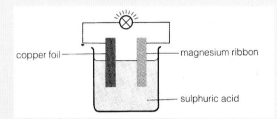

   a Why were the metals cleaned?
   b Name the electrolyte used.
   c Explain why the bulb lit up.
   d Which of the two metals is more reactive?
   e Which one releases electrons into the circuit?
   f In this arrangement, energy is being changed from one form to another. Explain.
   g What is this type of arrangement called?
   h Give reasons why the set-up shown above would not be used as a torch battery.

11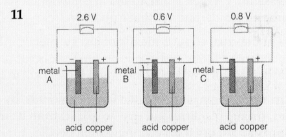

   Look at the three cells above.
   a How can you tell that the three unknown metals are all more reactive than copper?
   b Place the metals in order, most reactive first.
   c What voltage will be obtained if the cell is constructed using: i A and B? ii B and C?
   d For each of the cells in c state which metal is the negative terminal.

12 Write balanced equations for these reactions:
   a magnesium reacts with steam to form an oxide
   b zinc reacts with dilute hydrochloric acid
   c copper wire is placed in silver nitrate solution
   d potassium is added to water.

13.1 Metals in the Earth's crust

The composition of the Earth's crust

We get some metals from the sea, but most from the Earth's **crust** – the Earth's outer layer.

The crust is made up of many different **compounds**. It also contains **elements** such as sulphur, copper, silver, platinum, and gold. These occur uncombined, or **native**, because they are unreactive.

If you could dig up the Earth's crust and break down all the compounds to elements, you would find it is almost half oxygen! This pie chart shows its composition:

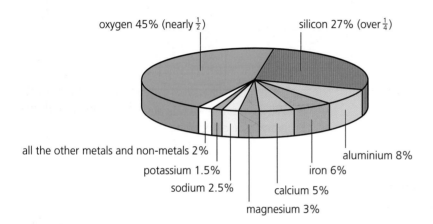

We use about nine times more iron than all the other metals put together.

Nearly three-quarters of the crust is made of just two non-metals, oxygen and silicon. These occur together in compounds such as silicon dioxide (**silica** or **sand**). Oxygen is also found in compounds such as aluminium oxide, iron(III) oxide, and calcium carbonate.

The rest of the crust consists mainly of just six metals. Aluminium is the most abundant of these, and iron is next. All six occur only as compounds, because they are all reactive metals – and these 'like' to form compounds.

Of all the metals, iron is used in the largest amounts, by far. That's because it is hard, strong, plentiful, and easy to extract from its main compound, iron(III) oxide. We use it for everthing from needles to ships. But as you will see later, it is not perfect. It rusts too easily!

Some metals are scarce

Look again at the pie chart. It shows that there are just six plentiful metals. All the other metals *together* make up less than one-fiftieth of the Earth's crust, so they are not plentiful.

If a metal makes up less than one-thousandth of the Earth's crust, it is considered **scarce**. Here are some of the scarce metals:

```
copper    zinc     lead    tin
mercury   silver   gold    platinum
```

We are using these up quite quickly, so some people are worried that they may soon run out.

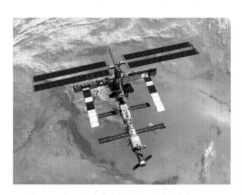

Light strong metals such as aluminium and titanium are used in the International Space Station, which orbits the Earth once every 92 minutes, 360 km above us.

Metal ores

The rocks in the Earth's crust are a mixture of substances. Some contain a large amount of one metal compound, or one metal, so it may be worth digging them up to extract the metal. Rocks from which metals are obtained are called **ores**. For example:

This is a chunk of **rock salt**, the main ore of sodium. It is mostly **sodium chloride**.

This is a piece of **bauxite**, the main ore of aluminium. It is mostly **aluminium oxide**.

Since gold is unreactive, it is found as a free element. This sample is almost pure gold.

To mine or not to mine?

Before starting to mine an ore, the mining company must decide whether it is economical. It must find answers to questions like these:

1. How much ore is there?
2. How much metal will we get from it?
3. Are there any special problems about getting the ore out?
4. How much will it cost to mine the ore and extract the metal from it? (The cost will include roads, buildings, mining equipment, the extraction plant, transport, fuel, chemicals, and wages.)
5. How much will we be able to sell the metal for?
6. So will we make a profit if we go ahead?

The answers to these questions will change from year to year. For example if the selling price of a metal rises, even a low-quality or **low-grade** ore may become worth mining.

The local people may worry that the area will be spoiled, and the air and rivers polluted. So they may object to plans for a new mine. On the positive side, they may welcome the new jobs that mining will bring.

A landscape spoiled by mining. These days, mining companies are usually forced to clear and restore the land when a mine is exhausted.

Questions

1. Which is the main *element* in the Earth's crust?
2. Which is the most common *metal* in the Earth's crust? Which is the second?
3. Gold occurs *native* in the Earth's crust. Explain.
4. Is it true that the most reactive metals are quite plentiful in the Earth's crust?
5. What is a *scarce* metal? Name four.
6. One metal is used more than all the others put together. Which one? Why is it used so much?
7. What is a *metal ore*?
8. Name the main ore of: **a** sodium **b** aluminium What is the main compound in each ore?

13.2 Extracting metals from their ores

Extraction

An ore is the rock from which a metal is obtained. After mining the ore, the next step is to remove or **extract** the metal from it. How you do this depends on the metal's **reactivity**.

- The most unreactive metals – such as silver and gold – occur in their ores as elements. All you need to do is separate the metal from sand and other impurities. This is like removing stones from soil, and does not involve chemical reactions.
- The ores of all the other metals contain the metals as compounds. These have to be reduced, to give the metal:

    metal compound $\xrightarrow{\text{reduction}}$ metal

- The compounds of the more reactive metals are very stable, and need electrolysis to reduce them. This is a powerful method, but it uses a lot of electricity so it is expensive.
- The compounds of the less reactive metals are not so stable, and can be reduced using a suitable **reducing agent**.

Reduction of metal ores

Remember, you can define reduction as:

- loss of oxygen
  $Fe_2O_3 \longrightarrow Fe$
- or gain of electrons
  $Fe^{3+} + 3e^- \longrightarrow Fe$

Either way, the ore is reduced to the metal.

Extraction and the reactivity series

So the method of extraction is strongly linked to the **reactivity series**, as shown below. Carbon is included for reference.

As you saw on page 177, carbon can act as a reducing agent. It will react with oxides of metals that are less reactive than itself, reducing them to the metals. Luckily, many ores are oxides, or compounds that can easily be changed to oxides.

Carbon is made from coal and is quite cheap. The gas carbon monoxide (CO) also works well as a reducing agent. So both can be used to reduce the ores of zinc, iron and lead. But they cannnot reduce the ores of the metals above carbon. Electrolysis is needed for these.

Metal			Method of extraction from ore		
potassium sodium calcium magnesium aluminium	metals more reactive ↑	ores more difficult to decompose ↑	electrolysis	method of extraction more powerful ↑	method of extraction more expensive ↑
carbon					
zinc iron lead			heating with a reducing agent – carbon or carbon monoxide		
silver gold			occur naturally as elements so no chemical reaction needed		

Three examples of ore extraction

1 **Iron ore** This is mainly iron(III) oxide. It is reduced like this:

iron(III) oxide + carbon monoxide ⟶ iron + carbon dioxide
Fe_2O_3 (s) + 3CO (g) ⟶ 2Fe (l) + $3CO_2$ (g)

We will look more closely at this extraction in Unit 13.3.

2 **Aluminium ore** This is mainly aluminium oxide. Aluminium is more reactive than carbon, so electrolysis is needed for this reduction:

aluminium oxide ⟶ aluminium + oxygen
$2Al_2O_3$ (l) ⟶ 4Al (l) + $3O_2$ (g)

We will look more closely at this extraction in Unit 13.4.

3 **Zinc blende** This is mainly zinc sulphide, ZnS. First it is roasted in air, giving zinc oxide:

zinc sulphide + oxygen ⟶ zinc oxide + sulphur dioxide
2ZnS (s) + $3O_2$ (g) ⟶ 2ZnO (s) + $2SO_2$ (g)

Then the oxide is reduced in one of the two ways below.

i Using carbon monoxide This is carried out in a furnace:

zinc oxide + carbon monoxide ⟶ zinc + carbon dioxide
ZnO (s) + CO (g) ⟶ Zn (s) + CO_2 (g)

The final mixture contains zinc and a slag of impurities. The zinc is separated from it by fractional distillation. (It boils at 907 °C.)

ii Using electrolysis For this, a compound must be melted, or in solution. But zinc oxide has a very high melting point (1975°C), and is insoluble in water!

important

ZnO has a very high melting point and its insoluble in H_2O

So instead, the zinc oxide is dissolved in dilute sulphuric acid (made from sulphur dioxide from the first reaction above). Zinc oxide is a base, so neutralisation takes place, giving a solution of zinc sulphate. This undergoes electrolysis, and zinc is deposited at the cathode:

Zn^{2+} (aq) + $2e^-$ ⟶ Zn (s)   (reduction)

The zinc is scraped off the cathode, and melted into bars to sell.

In fact most zinc is extracted by electrolysis, because this gives zinc of very high purity. Cadmium and lead occur as impurities in the zinc blende, and these metals are recovered and sold too.

After extraction, some of the aluminium is made into flat sheets, then rolls like these, ready for sale.

Zinc metal is used ...

- to **galvanise** iron – coat it to stop it rusting (page 196)
- in the **sacrificial protection** of ships, oil rigs and other iron structures (page 180)
- to make **alloys** such as brass and bronze (page 195)
- to make batteries (page 181)

Questions
1 Why is no chemical reaction needed to get gold?
2 Lead is extracted by heating its oxide with carbon:
    lead oxide + carbon → lead + carbon monoxide
    a Why can carbon be used for this reaction?
    b One substance is *reduced*. Which one?
    c Which substance is the reducing agent?
3 The reaction in question **2** is a *redox* reaction. Why?
4 Sodium is extracted from rock salt (sodium chloride).
    a Electrolysis is needed for this. Explain why.
    b Write a balanced equation for the reaction.
    c Say which ions are oxidised, and which reduced.
5 Describe the extraction of zinc by electrolysis.

13.3 Extracting iron

The blast furnace

This diagram shows the **blast furnace** used for extracting iron from its ore. It is an oven, shaped like a chimney, at least 30 metres tall.

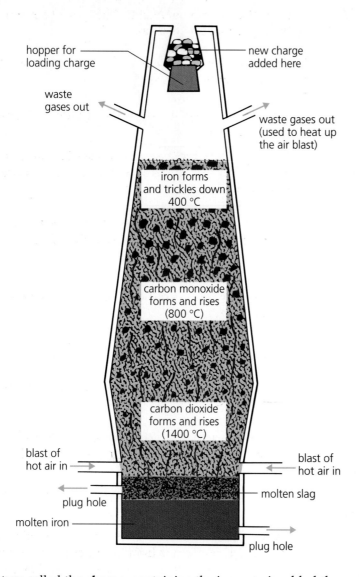

Blast furnaces run non-stop 24 hours a day.

A mixture called the **charge**, containing the iron ore, is added through the top of the furnace. Hot air is blasted in through the bottom. After a series of reactions, liquid iron collects at the bottom of the furnace.

What's in the charge?

The charge contains three things:

1 **Iron ore.** The chief ore of iron is called **haematite**. It is mainly iron(III) oxide, Fe_2O_3, mixed with sand.

2 **Limestone.** This common rock is mainly calcium carbonate, $CaCO_3$.

3 **Coke.** This is made from coal, and is almost pure carbon.

A stockpile of iron ore.

The reactions in the blast furnace

Reactions, products, and waste gases	Comments
Stage 1: The coke burns, giving off heat The blast of hot air starts the coke burning. It reacts with the oxygen in the air, giving carbon dioxide: carbon + oxygen → carbon dioxide $C(s) + O_2(g) \rightarrow CO_2(g)$	This, like all combustion, is a **redox reaction**. The carbon is **oxidised** to carbon dioxide. The blast of air provides the oxygen for the reaction. The reaction is **exothermic** – it gives off heat, which helps to heat the furnace.
Stage 2: Carbon monoxide is made The carbon dioxide reacts with more coke, giving carbon monoxide: carbon + carbon dioxide → carbon monoxide $C(s) + CO_2(g) \rightarrow 2CO(g)$	In this redox reaction, the carbon dioxide loses oxygen. It is **reduced**. The reaction is **endothermic** – it takes in heat from the furnace. That's good, because stage 3 needs a lower temperature.
Stage 3: The iron(III) oxide is reduced This is where the actual extraction occurs. Carbon monoxide reacts with the iron ore, giving liquid iron: iron(III) oxide + carbon monoxide → iron + carbon dioxide $Fe_2O_3(s) + 3CO(g) \rightarrow 2Fe(l) + 3CO_2(g)$ The iron trickles to the bottom of the furnace.	In this redox reaction, carbon monoxide acts as the **reducing agent**. It reduces the iron(III) oxide to the metal. At the same time the carbon monoxide is **oxidised** to carbon dioxide.
What is the limestone for? The limestone reacts with sand (silica) in the ore, to form **calcium silicate** or **slag**. limestone + silica → calcium silicate + carbon dioxide $CaCO_3(s) + SiO_2(s) \rightarrow CaSiO_3(s) + CO_2(g)$ The slag runs down the furnace and floats on the iron.	The purpose of this reaction is to remove impurities from the molten iron. Silica is an acidic oxide. Its reaction with limestone is a **neutralisation**, giving calcium silicate, a salt. The molten slag is drained off. When it solidifies it is sold, mostly for road building.
The waste gases These are **carbon dioxide** and **nitrogen**. They come out at the top of the furnace.	The carbon dioxide is from the reduction reaction in stage **3**. The nitrogen is from the air blast. It has not taken part in the reactions so has not been changed.

The molten iron is tapped from the bottom of the furnace at intervals. It is impure, with carbon as the main impurity. Some is run into moulds to give **cast iron**. This is hard but brittle – so is used only for things like iron railings. But most of the iron is turned into **steel**s.

Questions

1. Name the raw materials for extracting iron.
2. Write the equation for the reaction that gives iron.
3. The calcium carbonate in the blast furnace helps to purify the iron. Explain how, with an equation.
4. What is the 'blast' of the blast furnace?
5. Name the waste gases from the blast furnace.
6. The slag and waste gases are both useful. How?
7. What is: **a** cast iron? **b** steel?

13.4 Extracting aluminium

From rocks to rockets

Aluminium is the most abundant metal in the Earth's crust. Its main ore is **bauxite**, which is aluminium oxide mixed with impurities such as sand and iron oxide. The impurities make it reddish brown.

These are the steps in obtaining aluminium:

1 First, geologists test rocks and analyse the results, to find out how much bauxite there is. If the tests are satisfactory, mining begins.

2 Bauxite is usually near the surface, so is easy to dig up. This is a bauxite mine in Jamaica. Everything gets coated with red-brown bauxite dust.

3 From the mine, the ore is taken to a bauxite plant, where it is treated to remove the impurities. The result is white **aluminium oxide**, or **alumina**.

4 The alumina is taken to another plant for electrolysis. Much of the Jamaican alumina is shipped to plants like this one, in Canada or the USA . . .

5 . . . where electricity is cheaper. There it is electrolysed to give aluminium. The metal is made into sheets and blocks, and sold to other industries.

6 It is used to make drinks cans, food cartons, cooking foil, bikes, TV aerials, electricity cables, ships, planes, trains, trams, space rockets, and in buildings.

A closer look at the electrolysis

Aluminium is obtained from alumina by electrolysis. This is carried out in a huge steel tank. (See next page.) It is lined with carbon, which acts as the cathode (−). Huge blocks of carbon hang in the middle of the tank, and act as the anode (+).

Alumina melts at 2045°C. It would be impossible to keep the tank that hot. Instead, the alumina is dissolved in molten **cryolite**, or sodium aluminium fluoride, which has a much lower melting point.

This is the tank for the electrolysis:

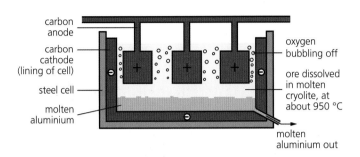

Aluminium being dissolved in molten cryolite, ready for electrolysis.

Once the alumina dissolves, the aluminium ions and oxide ions are free to move. They move to the electrode of opposite charge.

At the cathode The aluminium ions gain electrons:

$$4Al^{3+} (l) + 12e^- \longrightarrow 4Al (l) \qquad \text{(reduction)}$$

The aluminium drops to the bottom of the cell as molten metal. This is run off at intervals, and taken to a holding furnace. Some of it is mixed with other metals to make alloys. Some is run into molds, to harden into blocks.

At the anode The oxygen ions lose electrons:

$$6O^{2-} (l) \longrightarrow 3O_2 (g) + 12e^- \qquad \text{(oxidation)}$$

The oxygen gas bubbles off, and reacts with the anode:

$$C (s) + O_2 (g) \longrightarrow CO_2 (g) \qquad \text{(oxidation of carbon)}$$

So the carbon blocks get eaten away, and need to be replaced.

The overall reaction The overall reaction is the decomposition of alumina:

aluminium oxide $\longrightarrow$ aluminium + oxygen
$2Al_2O_3 (l) \longrightarrow 4Al (l) + 3O_2 (g)$

Some properties of aluminium

1. Aluminium is a bluish-silver, shiny metal.
2. Unlike most metals, it has a low density – it is 'light'.
3. It is a good conductor of heat and electricity.
4. It is malleable and ductile.
5. It is non-toxic, and resists corrosion.
6. It is not very strong when pure, but it can be made stronger by mixing it with other metals to form **alloys**. (See page 195.)

Aluminium is used for electricity cables because it is light, does not corrode, is a good conductor, and cheaper than copper. The cables have a steel core, for strength.

Questions

1. Which compounds are used in the extraction of aluminium? Say what role each plays.
2. a Sketch the electrolysis cell for extracting aluminium.
   b Why do the aluminium ions move to the cathode?
   c What happens at the cathode? Give an equation.
   d The anode is replaced regularly. Why?
3. These terms all describe properties of aluminium. Say what each term means. (Check the glossary?)
   a malleable   b ductile   c low density
   d non-toxic   e resistant to corrosion
4. List six uses of aluminium. For each, say which properties of the metal make it suitable.

13.5 Making use of metals

Properties dictate uses

Think of all the solid things you own, or use, or see around you. Some are made of wood, or plastic, or stone, or cloth. And some are made of metal, or contain metal.

Metals share some properties. Each also has other properties that make it special. How we use the metals depends on their properties. For example you would not use poisonous metals for food containers, or dense soft metals to build aeroplanes.

Here are some common uses of metals:

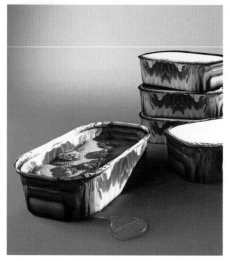

Aluminium foil is used for food cartons like these because it is **non-toxic** (does not damage our health), and resists corrosion.

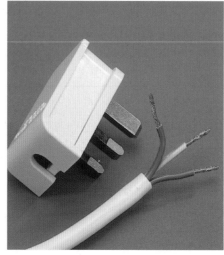

Copper is used for wiring because it is an excellent conductor of electricity, and easily drawn into wires.

Lead is used to hold the glass in stained glass windows because it is soft, easy to cut and bend, and resists attack by air and rain.

Zinc is coated onto steel car bodies, before they are painted, because zinc protects the steel from rusting, by sacrificial protection.

Silver is used for mirrors because it reflects light very well, resists corrosion, and can be deposited on glass as a fine thin coat.

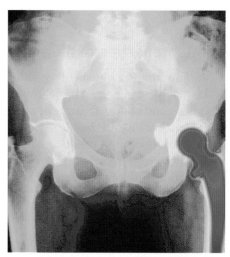

Titanium is used for artificial hip joints like this one, because it is light, hard, strong, non-toxic, and highly resistant to corrosion.

Uses of some metals: a summary

This table summarises the uses of those metals shown opposite.

Metal	Used for ...	Properties that make it suitable
aluminium	overhead electricity cables (with a steel core for strength)	a good conductor of electricity (not as good as copper, but cheaper and much lighter); ductile, resists corrosion
	coating CDs and DVDs	shiny surface reflects the laser beam that 'reads' a CD or DVD
	cooking foil and food cartons	non-toxic, resistant to corrosion, can be rolled into thin sheets
	drinks cans	light, non-toxic, resistant to corrosion
copper	electrical wiring	one of the best conductors of electricity, ductile
	roofing	malleable, develops an attractive protective coating
	saucepan bases	conducts heat well, unreactive, tough
lead	holding the glass in stained-glass windows, and sealing joins in roofs	easy to bend at room temperature, unreactive
	car batteries	gives a current when connected to lead(II) oxide in an electrolyte (dilute sulphuric acid)
zinc	protecting steel from rusting	offers sacrificial protection
	coating or **galvanising** iron and steel	resists corrosion, offers sacrificial protection if coating cracks
	for torch batteries	gives a current when connected to a carbon electrode, in an electrolyte
silver	electrical connections inside mobile phones, keyboards, and photovoltaic (PV) cells	the best metal of all at conducting electricity, ductile
	mirrors, and mirrored sunglasses	reflects light very well, even in a very thin coat
	jewellery	looks good, and resists corrosion
titanium	tooth implants, and replacement hip and knee joints	light, strong, very resistant to corrosion, non-toxic, malleable, and ductile
	exhaust pipes for planes	light, very resistant to corrosion, ductile, high melting point
	pipes and tanks in chemical factories	strong, very resistant to corrosion

(handwritten note: memories underlined)

Making metals more useful: alloys

In this unit the metals were used on their own, not mixed with anything. But very often a metal is much more useful when it is mixed with a small amount of another substance. The mixture is called an **alloy**. You can find out more about alloys in the next unit.

Questions

1 Give two examples of a use for a metal, that depends on the metal being:
   a a good conductor b non-toxic
   c strong d resistant to corrosion
   e light f a good reflector of light

2 Which of the metals above do *you* make use of? What do you use them for?

3 See if you can name other metals you have come across, that are not listed above. What are they used for? What properties make them suitable for this?

13.6 Steels and other alloys

Steels: alloys of iron

Iron is the most widely used metal in the world. But it is almost never used on its own.

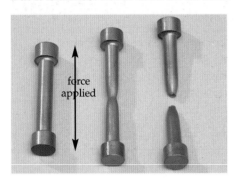

Pure iron is no good for building things, because it is too soft, and stretches quite easily, as you can see from this photo. Even worse, it rusts very easily.

But when a little carbon (0.5%) is mixed with it, the result is **mild steel**. This is hard and strong. It is used for buildings, ships, car bodies, and machinery.

When nickel and chromium are mixed with iron, the result is **stainless steel**. This is hard and rustproof. It is used for car parts, kitchen sinks, and cutlery.

So the properties of iron have been changed by the other substances. **The properties of a metal can be changed by mixing other substances with it. The mixtures are called alloys.**

The other substances can be metals, or non-metals such as carbon or silicon. To make the alloy, you melt the main metal, then dissolve the other substances in it, to give a solution. When this cools and hardens, it is more useful than the pure metal.
Turning a metal into an alloy increases its range of uses.

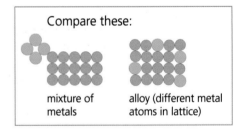

Compare these:

mixture of metals | alloy (different metal atoms in lattice)

Why an alloy has different properties

comes as a diagram Question

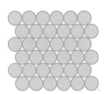

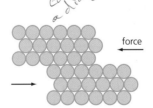

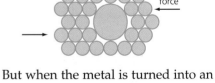

This shows the atoms in a pure metal. They are arranged in a regular lattice. (In fact they are metal ions in a sea of electrons, as you saw on page 52.)

When pressure is applied, for example by hammering the metal, the layers can slide over each other easily. That is why a metal is malleable and ductile.

But when the metal is turned into an alloy, new atoms enter the lattice. The layers can no longer slide easily. So the alloy is stronger than the original metal.

It is not only strength that can change. The alloy may be more resistant to corrosion than the original metal. Its ability to conduct heat and electricity will be different too.

You can add more than one substance, to make the alloy. By trying out different substances, in different amounts, you can design an alloy with exactly the properties you need.

Making steels

This is how steels are made:

1. **First, unwanted impurities are removed from the iron.**
   The molten iron from the blast furnace is poured into an **oxygen furnace**. Calcium oxide is added, and a jet of oxygen turned on. The calcium oxide neutralises any acidic impurities, forming a slag that is skimmed off. The oxygen reacts with the others, burning them away.

   For some steels, *all* the impurities are removed. But many steels are just iron plus a little carbon. Carbon makes steel stronger – but too much makes it brittle, and hard to shape. So the carbon content is checked continually. When it is correct, the oxygen is turned off.

2. **Then other elements may be added.**
   These are measured out carefully, to give steels with the required properties. The table below has one example.

Molten iron being poured into an oxygen furnace.

Some common alloys of iron and other metals

There are thousands of different alloys. Here are just four!

Alloy	Made from	Special properties	Uses
stainless steel *(low carbon)*	70% iron 20% chromium 10% nickel	does not rust	car parts, kitchen sinks, cutlery, tanks and pipes in chemical factories, surgical instruments
aluminium alloy number 7075 TF *(Duralumin)*	90.25% aluminium 6% zinc 2.5% magnesium 1.25% copper	light but very strong	aircraft
brass	70% copper 30% zinc	harder than copper, does not corrode	musical instruments, ornaments, *jewellary*, door knobs and other fittings
bronze	95% copper 5% tin	harder than brass, does not corrode, chimes when struck	statues, ornaments, church bells

Look at the alloy of aluminium above. Aircraft need materials that are light but very strong, and resistant to corrosion. Pure aluminium is light, but not strong enough. So *hundreds* of aluminium alloys have been developed, for aircraft parts.

memories

Questions

1. In the lab, you grind some magnesium powder with iron filings. Is this mixture an *alloy*? Explain.
2. Draw diagrams to show clearly the difference between:
   a. an alloy and a mixture   b. an alloy and a compound
3. An alloy is a type of solution. Explain why.
4. An alloy of aluminium with 1% manganese is much harder and stronger than aluminium alone. Why is this?
5. Steels are alloys. What is the main metal in them?
6. Describe how steels are made.
7. What is brass made of?

13.7 Corrosion

What is corrosion?

When a metal is attacked by air, water, or other substances in its surroundings, the metal is said to **corrode**.

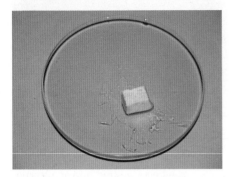

Damp air quickly attacks sodium, turning it into a pool of sodium hydroxide solution. (That is why sodium is stored under oil!)

Iron also corrodes in damp air. So do most steels. This is usually called **rusting**. It happens really quickly beside the sea.

But gold never corrodes. This gold mask of King Tutankhamun was buried in his tomb for over 3000 years. It still looks as good as new.

In general, the more reactive a metal is, the more readily it corrodes.

What does rusting involve?

Rusting needs both air and water, as these tests show.

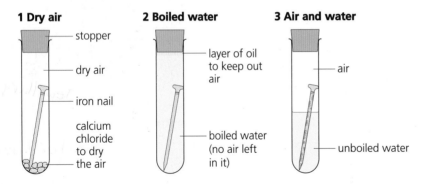

Nails 1 and 2 do not rust. Nail 3 does. The iron is oxidised, like this:

$$4Fe\,(s) + 3O_2\,(g) + 2H_2O\,(l) \longrightarrow 2Fe_2O_3.H_2O\,(s)$$

iron + oxygen + water ⟶ hydrated iron(III) oxide (or rust)

How to prevent rusting

Iron and steel are used for things like ships, oil rigs, bridges, and railways. So rusting can be a huge problem. Replacing rusted items costs billions of pounds a year. There are two ways to prevent rusting:

1 **Coat the metal with something, to keep out air and moisture.**
   You could use:
   Paint. Steel bridges and railings are usually painted. Paints that contain lead or zinc are best at preventing rust. Can you suggest why?

Iron railings painted to keep out air and moisture (which would lead to corrosion).

Grease. Tools and machine parts are coated with grease or oil.

Plastic. Steel is coated with plastic to make things like garden chairs.

Another metal. For example:
- iron for roofs, and steel for car bodies, are coated with **zinc**, by dipping them into molten zinc. This is called **galvanising**.
- food tins are made of steel coated on both sides with **tin**. Tin is used because it is unreactive, and non-toxic. It is deposited on the steel by electrolysis, in a process called **tin plating**.
- car bumpers and bathroom taps are coated with **chromium**, to give a shiny protective surface. The chromium is deposited by electrolysis.

2 **Use sacrificial protection.**
Magnesium is more reactive than iron. So when a bar of magnesium is attached to the side of a steel ship, it corrodes instead of the steel. This is called **sacrificial protection**. The magnesium block must be replaced before it all dissolves away. This costs far less than a new ship!

Zinc can also be used in this way. For more about sacrificial protection, and galvanising, see page 180.

Steel cans for food are plated with tin.

Does aluminium corrode?

Aluminium is more reactive than iron, so you might expect it to corrode faster than iron. In fact it does start to corrode, but soon stops:

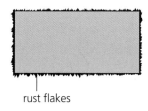
rust flakes

coat of aluminium oxide

When iron corrodes, rust forms in tiny flakes. Damp air can get past them, to attack the metal beneath. In time, it rusts right through.

But when aluminium reacts with air, an even coat of aluminium oxide forms. This acts as a seal, preventing further corrosion.

The layer of aluminium oxide can be made even thicker by electrolysis. This process is called **anodising**. Anodised aluminium is used for cookers, fridges, saucepans, window frames, and so on. The oxide layer can be dyed to bright colours.

Anodised aluminium is used for window frames.

Questions
1 What is *corrosion*?
2 Which two substances cause iron to rust?
3 Name an alloy of iron that doesn't rust. (Kitchen sinks!)
4 Iron that is tin-plated does not rust. Why?
5 a What does the *sacrificial protection* of iron mean?
  b Both magnesium and zinc can be used for it. Why?
  c But copper will not work. Explain why.
6 Why won't aluminium corrode right through?

13.8 Mining, using, and recycling metals

The metal story

This is a summary of how we obtain and use metals – and some of the problems that arise.

1 The metal ore is mined.
We need mining. But it always does some harm to the environment. Forests may be cut down to make way for the mines. Dust, mud and waste liquids from mining can pollute the land, air, and rivers.

2 The ore is separated from the other rock material.
This gives a huge amount of waste: stones, sand, mud, and other material. It can ruin an area, and pollute rivers.

3a The metal is extracted from the ore.
Extraction uses a great deal of energy (electricity and other fuels). It produces waste solids, liquids and gases. Some of these can be harmful – and even deadly.

3b The metal gets further treatment.
For example iron is turned into steel, and copper is purified. This may also use a great deal of energy. And again, there are waste products, and some may be harmful.

4 The metal is turned into useful things.
For example planes, ships, cars, trains, electric cables, fridges, cookers, food containers …

5 These things are used, then discarded.
They may be finished with, or broken, or unsafe – or perhaps merely out of fashion.

8 Some go to rubbish dumps (landfill sites).
Billions of tonnes of scrap metal are dumped every year – tins, old cars and fridges, old computers and so on. The metal is lost forever.

6 Some are collected for recycling.
Food tins and cans are collected from homes. Businesses sell their metal items for recycling. (For example airlines sell old planes.)

7 The scrap metal is recycled.
The scrap is melted, and enters just the final stage of the extraction process. It is made as good as new.

The Bingham Canyon copper mine in Utah, USA. It is 4 km wide and nearly 1 km deep. This is **open cast** or surface mining (not under the ground).

Step **8** is a huge waste. Scrap metal can be recycled in steps **3b** to **7**, again and again. In theory, metals can go on forever!

Recycling metals

The world's population is growing fast. And that means more saucepans are needed, and cookers, computers, cars, trains, planes, electricity cables, ships. So the demand for metals keeps rising. Over 1 billion tonnes of steel are now produced each year.

With growing demand, there is also growing pressure to recycle metals, using steps **6** and **7** in the flowchart.

Why recycling makes sense

1 **Metals are a finite resource.**
   There is only so much of each metal in the Earth's crust. We are digging up ores fast. One day we will have used up all the ores within reach, for some metals. But recycled metals are as good as new.

2 **Recycling helps the environment.**
   Look at steps **1–3a**, opposite. They can all harm the environment. And the more metal we throw away, the more new mines will be needed. But recycling avoids those three steps. Besides, the less we dump, the fewer new landfill sites we will need.

   But note that recycling can also be harmful. For example, metals used in computers can poison people melting them down. So recycling needs to be done safely.

3 **Recycling saves energy.**
   Steps **1 – 3a** need energy. But recycling avoids these steps. Most of the energy we use comes from burning fossil fuels, which are linked to global warming. So the less energy we use, the better.

   However, you do have to collect the scrap metal to recycle it. That uses up energy, in the form of petrol or diesel for trucks.

4 **Recycling saves money.**
   Recycling costs money. Scrap must be collected and brought to the plant, so there will be transport, wages and other costs. But this is usually less than the cost of mining and extracting the same amount of 'new' metal.

Recycling two common metals

Steel About 40% of the 'new' steel produced each year is in fact recycled scrap steel. This is added to the molten iron in the oxygen furnace (page 195).

Aluminium It is one of the easiest metals to recycle. The scrap aluminium is cleaned. Labels and lacquer are scraped off cans. Then the scrap is melted in a furnace, and treated just like the molten aluminium from electrolysis.

'Tin' cans await collection for recycling. The steel in them may be used to make more cans.

Breaking up a ship for scrap metal. Hard work, and it can be very dangerous too. Toxic chemicals are often present in the holds of old cargo ships and oil tankers.

Questions

1 In what ways can our use of metals harm the environment?
2 Scrap metal can be *recycled*. What does that mean?
3 Who might suffer if *more* scrap metal were recycled?
4 Look at the four benefits of recycling. Put them in what you think is their order of importance, with the most important first.
5 What can *you* do to ensure more metal gets recycled?

Questions on Chapter 13

1 Only a few elements are found uncombined in the Earth's crust. Gold is one example. The rest occur as compounds, and have to be extracted from their ores. This is usually carried out by heating with carbon, or by electrolysis.
Some information about the extraction of three different metals is shown below.

metal	formula of main ore	method of extraction
iron	Fe_2O_3	heating with carbon
aluminium	Al_2O_3	electrolysis
sodium	NaCl	electrolysis

   a Give the chemical name of each ore.
   b Arrange the three metals in order of reactivity.
   c i How are the two more reactive metals extracted from their ores?
      ii Explain why this is a reduction of the ores.
   d i How is the least reactive metal extracted from its ore?
      ii Explain why this is a reduction of the ore.
      iii Why can't this method be used for the more reactive metals?
   e Which of the methods would you use to extract:
      i potassium? ii lead? iii magnesium?
   f Gold is a metal found native in the Earth's crust. Explain what *native* means.
   g Where should gold go, in your list from b?
   h Name another metal that occurs native.

2 Many metals are more useful when mixed with other elements, than when they are pure.
   a What name is given to the mixtures?
   b i Which metals are found in these mixtures?
        brass   bronze   stainless steel
      ii Describe the useful properties of the mixtures.
      iii Give one use for each of these mixtures.
   c Name another mixture of metals and the useful properties it has.

3 *aluminium   gold   iron   tin   magnesium*
   mild steel   calcium   stainless steel
   a In the above list of metals and alloys, only four are resistant to corrosion.
      i Which ones are they?
      ii Explain why each is resistant to corrosion.
   b Which one of the other metals or alloys in the list will corrode the fastest? Give reasons for your choice.

4 a Draw a diagram of the blast furnace. Show clearly on your diagram:
      i where air is 'blasted' into the furnace
      ii where the molten iron is removed
      iii where a second liquid is removed
   b i Name the three raw materials used.
      ii What is the purpose of each material?
   c i Name the second liquid that is removed from the furnace.
      ii When it solidifies, does it have any uses? If so, name one.
   d i Name a waste gas from the furnace.
      ii Does this gas have a use? If so, what?
   e i Write an equation for the chemical reaction that produces the iron.
      ii Explain why this is a reduction of the iron compound.
      iii What acts as the reducing agent for it?

5 Zinc and lead are obtained from ores that contain only the metal and sulphur, in the molar ratio 1:1.
   a Name the compounds in these ores.
   b Write down their formulae.
   In the extraction of the metal, the compounds are roasted in air to obtain the oxide of the metal. The sulphur forms sulphur dioxide.
   c i Write equations for the roasting of the ores.
      ii Which type of reaction is this?
      iii Care must be taken in disposing of the sulphur dioxide produced. Explain why. (See page 206?)
   Then the oxide can be heated with coke (carbon) to obtain the metal and carbon monoxide.
   d i Write equations for the reactions with carbon.
      ii Which substances are reduced, in the reactions?
   e Which gives the greater mass of metal, 1000 g of lead ore or 1000 g of zinc ore? Explain your answer. (A_r: Zn = 65, Pb = 207.)

6 Aluminium is the most abundant metal in the Earth's crust. Even so, more of it is being recycled.
   a i What is the main ore of aluminium called?
      ii Which aluminium compound is it?
   b How is aluminium extracted from this ore?
   c Aluminium is used in thin sheets of 'foil', for covering things cooking in the oven. Give two properties that make it suitable for this use.
   d Why is it worth recycling aluminium? Give at least two reasons.

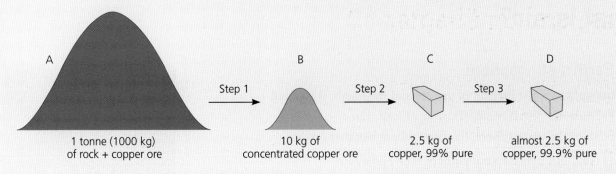

1 tonne (1000 kg) of rock + copper ore — 10 kg of concentrated copper ore — 2.5 kg of copper, 99% pure — almost 2.5 kg of copper, 99.9% pure

7 The diagram above shows stages in obtaining copper from a low-grade ore. This ore contains copper(II) sulphide, CuS. It may also contain small amounts of silver, gold, platinum, iron, zinc, cadmium, and arsenic.
  a What is an *ore*?
  b What is a *low-grade* ore?
  c i How much waste rock is removed per tonne in **step 1**?
   ii What percentage of the ore is finally extracted as pure copper metal?
  d Why might it be economic to extract copper from a low-grade ore like this?
  e i Which type of chemical reaction occurs in **step 2**?
   ii With what could the copper ore be reacted, to obtain the metal?
  f i What process is carried out at **step 3** to purify the metal? (Hint: page 140.)
   ii What will the main cost in this process be?
   iii As well as pure copper, this process may produce other valuable substances. Explain why, and where they will be found.
  g List some of the environmental problems that may arise in going from A to D.

8 In an experiment to investigate the corrosion of iron, a nail is placed in water.
  a i What name is given to the corrosion of iron?
   ii Which substances are involved in the corrosion of iron?
   iii Write an equation for the corrosion reaction.
  b The corrosion of iron is a redox reaction.
   i What is oxidised in the reaction?
   ii What is reduced?
   iii Write a half-equation for the electron transfer that takes place at the surface of the iron, in this redox reaction.
  c i Which one of these metals could be put with the nail to prevent it from corroding?
   silver   lead   magnesium
   ii Explain how this metal prevents the iron from corroding.

9 Aluminium is the most abundant metal in the Earth's crust. It makes up about 8% of the crust. Iron is next, making up about 6%.
Iron and aluminium are extracted from their ores in large quantities. The table below summarizes the two extraction processes.

	Iron	Aluminium
Chief ore	haematite	bauxite
Formula of main compound in ore	Fe_2O_3	Al_2O_3
Energy source used	burning of coke in air (exothermic reaction)	electricity
Other substances used in the extraction	limestone	carbon (graphite) cryolite
Temperature at hottest part of reactor /°C	1900	1000
How the metal separates from the reaction mixture	melts and collects at the bottom	melts and collects at the bottom
Other products	carbon dioxide sulphur dioxide slag	carbon dioxide

  a In each extraction, is the metal oxide *oxidised* or *reduced*, when it is converted to the metal?
  b Explain *why* each of these substances is used:
   i limestone in the extraction of iron
   ii carbon in the extraction of aluminium
   iii cryolite in the extraction of aluminium.
  c Describe any two similarities in the extraction processes for aluminium and iron.
  d Give a use for the slag that is produced as a by-product in the extraction of iron.
  e Aluminium costs over three times as much per tonne as iron. Suggest two reasons why aluminium is more expensive than iron, even though it is more abundant in the Earth's crust.
  f Most of the iron produced is converted into steels. **i** Why? **ii** How is this carried out?
  g Both steel and aluminium are recycled. Why is it important to recycle these metals?

14.1 What is air?

The Earth's atmosphere

The **atmosphere** is the blanket of gas around the Earth. It is about 700 km thick, and has four layers.

- We live in the **troposphere**, the layer where weather occurs.
- We take flight into the **stratosphere** now and again.
- Only a few people – the astronauts – have passed through the **mesosphere**.
- The **ionosphere** is by far the deepest layer. It consists mainly of charged particles.

Because of gravity, the atmosphere is densest at sea level. As you rise above the Earth, it quickly thins out. About 50% of the gas lies below 5.5 km. 30 km above us, the atmosphere is mainly just lighter gas particles, and charged particles.

Down here on the Earth's surface, we call the atmosphere around us **air**.

What's in air?

This pie chart shows the gases that make up clean air:

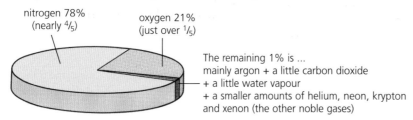

- nitrogen 78% (nearly ⁴/₅)
- oxygen 21% (just over ¹/₅)
- The remaining 1% is ... mainly argon + a little carbon dioxide + a little water vapour + a smaller amounts of helium, neon, krypton and xenon (the other noble gases)

But the composition changes very slightly from day to day and place to place. There is more water vapour in the air around you on a damp day. Over busy cities and industrial areas there are pollutants such as carbon monoxide and sulphur dioxide. But air is continually on the move, so pollutants soon get spread around too.

The ozone layer

The ozone layer is about 25 km above sea level, in the stratosphere. Ozone has the formula O_3. It forms when the ultraviolet radiation from the sun causes oxygen molecules to break into atoms. These then react with other oxygen molecules:

$$O_2 \xrightarrow{\text{ultraviolet light}} 2O \quad \text{then} \quad 2O_2 + 2O \longrightarrow 2O_3$$

At the same time, some ozone molecules absorb ultraviolet radiation and break down again to oxygen. And that is how the ozone layer protects us. Radiation that would harm us if it reached Earth is absorbed instead by the ozone layer, and used to break bonds.

The energy given out when new bonds form is given out as heat. So the ozone layer is warmer than the air below it, as the drawing above shows.

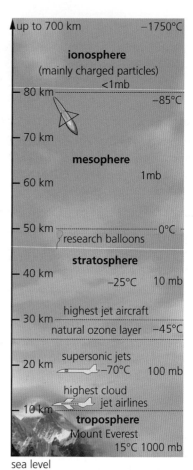

The Earth's atmosphere.

Compared with the Earth's diameter, the atmosphere is just a very thin layer.

Allotropes

Oxygen (O_2) and ozone (O_3) are allotropes of oxygen – two different forms of the same element.

Measuring the percentage of oxygen in air

The first person to realize that air was a mixture of gases was the French chemist Lavoisier. He experimented by heating mercury in air in a closed vessel. The mercury turned into a red powder, and about one-fifth of the air got used up. (What was the red powder?)

No matter how much mercury Lavoisier started with, or how much he heated it, only one-fifth of the air got used up. He tested the remaining gas and found that a candle wouldn't burn in it. He put a mouse in it and the mouse died. Lavoisier concluded that the gas that got used up was also the one that supported life. He named it **oxygen**.

You too can do an experiment along the same lines as Lavoisier's.

The apparatus A tube of hard glass is connected to two gas syringes A and B. The tube is packed with small pieces of copper wire. One syringe contains 100 cm^3 of air. The other is empty:

Lavoisier (1743–1794) in his lab. He met his death on the guillotine during the French Revolution.

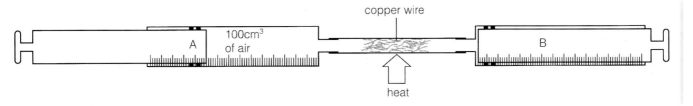

The method These are the steps:
1. The tube is heated by a Bunsen burner. When A's plunger is pushed in, the air is forced through the tube and into B. The oxygen in it reacts with the hot copper, turning it black. When A is empty, B's plunger is pushed in, forcing the air back to A. This cycle is repeated several times.
2. Heating is stopped after about 3 minutes, and the apparatus allowed to cool. Then all the gas is pushed into one syringe and its volume measured. (It is now less than 100 cm^3.)
3. Steps 1 and 2 are repeated until the volume of the gas remains steady. This means that all the oxygen has been used up. The final volume is noted.

The results Starting volume: 100 cm^3. Final volume: 79 cm^3.
So the volume of oxygen in 100 cm^3 air is 21 cm^3.

The percentage of oxygen in air is therefore $\frac{21}{100} \times 100 = \mathbf{21\%}$.

Air dissolved in water
- Oxygen is more soluble than nitrogen, in water.
- If you boiled the air out of water, and measured the percentage of oxygen in it, you'd get **33%**.

Questions
1. What percentage of air is made up of:
   a nitrogen?   b oxygen?   c nitrogen + oxygen?
2. About how much more nitrogen is there than oxygen in air, by volume?
3. What is the combined percentage of all the other gases in air?
4. Which do you think is the most reactive gas in air?
5. Lavoisier obtained a red powder, during his experiments with mercury and oxygen. Give its name and formula.
6. a Write down the name and formula of the black substance that forms in the experiment above.
   b Suggest a way to turn it back into copper. (Page 104!)

14.2 Making use of air

Separating gases from the air

Air is a mixture of gases. Most of them are useful to us. But to use them, we must first separate them from each other.

So how can we separate gases? There is a very clever way. First the air is cooled until it turns into a liquid. Then the liquid mixture is separated using a method you met on page 20: **fractional distillation**.

The fractional distillation of liquid air

This works because the gases in air have different boiling points, as shown in the table on the right. So, when liquid air is heated up, the gases boil at different temperatures, and can be collected one by one.

The boiling points of the gases in air (°C)	
carbon dioxide	− 32
xenon	−108
krypton	−153
oxygen	−183
argon	−186
nitrogen	−196
neon	−246
helium	−269

increasing ↑

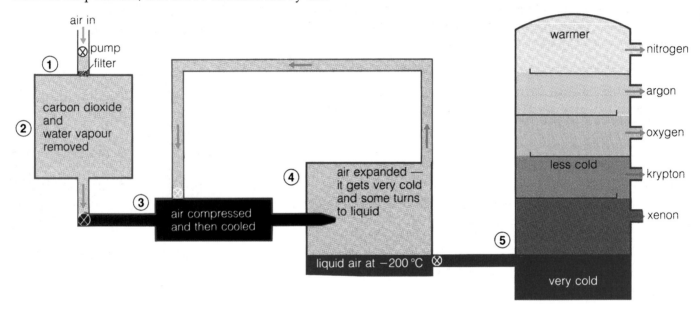

The steps The diagram above shows the steps in the process.

1 Air is pumped into the plant, and filtered to remove dust particles.
2 Next, water vapour, carbon dioxide, and pollutants are removed (since these would freeze later and block the pipes). Like this:
   - First the air is cooled until the water vapour condenses to water.
   - Then it is passed over beds of adsorbent beads to trap the carbon dioxide, and any pollutants in it.
3 Now the air is forced into a small space, or **compressed**. That makes it hot. It is cooled down again by recycling cold air, as the diagram shows.
4 The cold, compressed air is passed through a jet, into a larger space. It expands rapidly, and this makes it very cold.

Steps 3 and 4 are repeated several times. The air gets colder each time. By −200 °C, it is liquid, except for neon and helium. These gases are removed. They can be separated from each other by adsorption on charcoal.

5 The liquid air is pumped into the fractionating column. There it is slowly warmed up. The gases boil off one by one, and are collected in tanks or cylinders. Nitrogen boils off first. Why?

Liquid nitrogen.

Making use of oxygen

1. Astronauts carry oxygen with them, and so do deep-sea divers. Planes also carry their own oxygen supply.
2. In hospitals, patients with breathing problems are given oxygen through an **oxygen mask,** that covers the nose and mouth, or in an **oxygen tent.** This is a plastic tent that fits over a bed and is tucked in at the sides. Oxygen-rich air is pumped into it.
3. In steel works, a jet of oxygen is blown through molten iron to remove the carbon (coke) used to extract the iron from its ore.
   The oxygen oxidises the carbon to carbon dioxide and carbon monoxide. But the oxygen flow is controlled so that a little carbon is left, since this helps to make steel hard.
4. A mixture of oxygen and the gas **ethene** (or acetylene, C_2H_2) is used as the fuel in **oxy-acetylene torches** for cutting and welding metal. When this gas mixture burns, the flame can reach 6000 °C.
   To **cut** metals, you just move the flame along the line you want to cut. The metal melts. (Steel melts at around 3150 °C.)
   Welding is where you join two pieces of metal. You press the edges together, and melt them with the flame. The puddle of melted metal cools and hardens, giving a solid join.

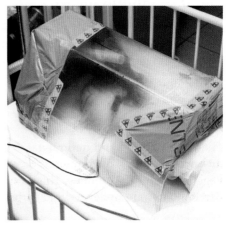
An infant in an oxygen tent.

Making use of nitrogen

1. Liquid nitrogen is very cold (–196°C). So it is used to quick-freeze food in food factories, and to freeze liquid in cracked pipes before repairing them.
2. It is also used to **shrink-fit** metal machine parts. One part is dipped into liquid nitrogen. It shrinks as it cools. It can then be fitted easily into another part. When it warms up again, the result is a very tight fit.
3. Nitrogen is also flushed through food packaging to remove oxygen and keep the food fresh.

Welding with an oxy-acetylene flame.

Making use of the noble gases

The noble gases are unreactive or inert, and glow when a current is passed through them at low pressure. So they have many uses.

1. Argon provides the inert atmosphere in ordinary tungsten light bulbs. (In air, the tungsten filament would quickly burn away.)
2. Neon is used in advertising signs because it glows red when a current is passed through it.
3. Helium is used to fill balloons, since it is very light, and safe.

For futher examples of their uses, see page 67.

Liquid nitrogen is used to freeze liquid in damaged pipes, before repairs.

Questions

1. In the separation of air into its gases:
   a. why is the air compressed and then expanded?
   b. why is argon obtained *before* oxygen?
   c. what do you think is the biggest expense? Why?
2. Write down three uses of oxygen gas.
3. Why does a mixture of oxygen and ethyne burn better than a mixture of air and ethyne?
4. Nitrogen is used to keep food frozen during transportation. Which properties make it suitable for this?
5. Give three uses for noble gases. (Check page 67 too.)

14.3 Pollution alert!

The air: a dump for waste gases

Everyone likes clean fresh air. But for the last 250 years, ever since the world began to industrialize, we have been pumping harmful gases into the air. From engines, and factories, and power stations, and homes.

Most of the pollutants in the air come from burning **fossil fuels**.

The fossil fuels

These are **coal**, **oil** and **natural gas**. They are called fossil fuels because they were formed, over millions of years, from the remains of dead vegetation and sea animals. They consist of **hydrocarbons** – compounds that contain only hydrogen and carbon.

The world depends on the fossil fuels – and especially on oil. They provide energy for cooking, and heating, and generating electricity. Petrol and diesel from oil are used as car fuel. But the trouble is, they produce many harmful compounds when they burn.

Don't breathe in!

The main air pollutants

This table shows the main pollutants found in air, and the harm they do:

Pollutant	How is it formed?	What harm does it do?
Carbon monoxide, CO colourless gas, insoluble, no smell	Forms when the carbon compounds in fossil fuels burn in too little air. For example inside car engines and furnaces.	Poisonous even in low concentrations. It reacts with the **haemoglobin** in blood, and prevents it from carrying oxygen around the body – so you die from oxygen starvation.
Sulphur dioxide, SO_2 colourless acidic gas, sharp smell, soluble	Sulphur occurs naturally in the fossil fuels (mainly in coal and oil) and forms sulphur dioxide when it burns.	Irritates the eyes and throat, and causes **respiratory** (breathing) problems. Dissolves in rain to form **acid rain** (page 123), which damages buildings, trees, and plants, and kills fish and other river life.
Nitrogen oxides, NO and NO_2 acidic gases	Form when the nitrogen and oxygen in air react together inside hot car engines, and in hot furnaces.	Cause respiratory problems. Dissolve in rain to form acid rain.
Lead compounds	A compound called tetra-ethyl lead was added to petrol to help it burn smoothly in engines. On burning, it produced particles of other lead compounds, mainly lead halides such as PbBrCl.	Lead damages children's brains. It damages the kidneys and nervous system in adults.

Reducing air pollution

Here are three of the things being done to cut down air pollution.
- The exhausts of new cars are fitted with **catalytic converters**, in which harmful gases are converted to harmless ones. See page 155.
- In modern power stations and factories, waste gases containing sulphur dioxide and nitrogen oxides are treated with slaked lime (calcium hydroxide). It removes the acidic oxides by neutralising them.
- Most countries have now banned lead from petrol, and people use lead-free petrol instead. This has made a big difference to the amount of lead pollution. (But there is still some, for example from lead extraction plants and battery factories.)

Power stations are responsible for much of the pollution in the atmosphere.

Is carbon dioxide a pollutant?

Air naturally contains some carbon dioxide. You could not live without it. Plants use it to make glucose, and other compounds. Then you and other living things eat plants, and plant-eating animals, for food.

Carbon dioxide helps us in another big way too. It is a **greenhouse gas**. That means it traps heat around the Earth, and stops it escaping into space. Without carbon dioxide and other greenhouse gases we would freeze to death at night, when the sun was not there to warm the land up.

But now we are pumping so much carbon dioxide into the air, from burning fossil fuels, that the Earth is getting warmer. This is called **global warming**.

How will it affect us? Climates all over the world will change. (This is already happening.) The ice at the poles will melt. Sea levels will rise, and may drown low-lying coasts. Experts predict more storms, droughts, floods, and famine. It will mean disaster for millions of people.

Cutting back on carbon dioxide

To try to slow down global warming, we need to cut back on the amount of carbon dioxide we pump out. Here are some ways to do that:
- Use less fossil fuel. For example cut down on car use. Use public transport instead.
- Switch to clean sources of power. For example use solar power to generate electricity, and hydrogen as a fuel for cars. (See page 213.)
- Try to find ways to store carbon dioxide, and not let it escape into the atmosphere. Some scientists are working on ways to store it under the sea!

Wearing a mask against traffic fumes.

Questions

1. Look at the pollutants in the table on page 206.
   a Which come from petrol burned in car engines?
   b Which come from air? How and why do these form?
2. Natural gas or methane is a fossil fuel. In a plentiful supply of air, it burns to give carbon dioxide and water. Write a balanced equation to show this.
3. If methane burns in a poor supply of air it will give carbon monoxide and water instead. See if you can write a balanced equation to show this.
4. In both **2** and **3**, a product of the reaction causes problems. For each, say which product it is, and what the problem is.
5. Why has tetra-ethyl lead been banned from petrol?

14.4 Water supply

Everyone needs water

We all need water.
- At home we need it for drinking, cooking, washing things (including ourselves) and flushing toilet waste away.
- On farms it is needed as a drink for animals, and to water crops.
- In industry, they use it as a solvent, and to wash things, and to keep hot reaction tanks cool. (Cold water pipes are curled around the tanks.)
- Power stations use it to make steam. The steam then drives the turbines that generate electricity.

In many places, our water supply is pumped from rivers. The water is cleaned up, the germs are killed, and then it is pumped to homes.

So where does the water come from?

Much of the water we use is taken from rivers. But some is pumped up from below ground, where water that has drained down through the soil lies trapped in rocks.

This underground water is called **ground water**. Often a large area of rock holds a lot of ground water, like a sponge. This rock is called an **aquifer**.

Is it clean?

River water is not clean – even if it looks clean! There will be mud in it, and animal waste, and bits of dead vegetation. But worst of all there will be **microbes**: bacteria and other tiny organisms that can make us ill.

Over 1 billion people around the world have no access to clean water. They depend on dirty rivers for their drinking water. And over 2 million people, most of them children, die each year from **diarrhoea** and diseases such as **cholera** and **typhoid**, caused by drinking infected water.

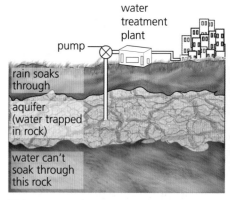

Using an aquifer as a water supply.

Providing a water supply on tap

No matter where in the world you are, the steps in providing a clean safe water supply, on tap, are the same:

1. Find a clean source – a river or aquifer – to pump water from.

2. Remove as many solid particles from the water as you can.
   - You could make fine particles stick together and skim them off.
   - You could filter the water through clean gravel or sand.

3. Add something to kill the microbes in it. (Usually chlorine.)

4. Store the water in a clean covered reservoir, ready for pumping to taps.

How well you can clean the water up depends on how dirty it is, and what kind of treatment you can afford!

A modern treatment plant

This diagram shows a modern water treatment plant. Follow the numbers to see how particles are removed and microbes killed.

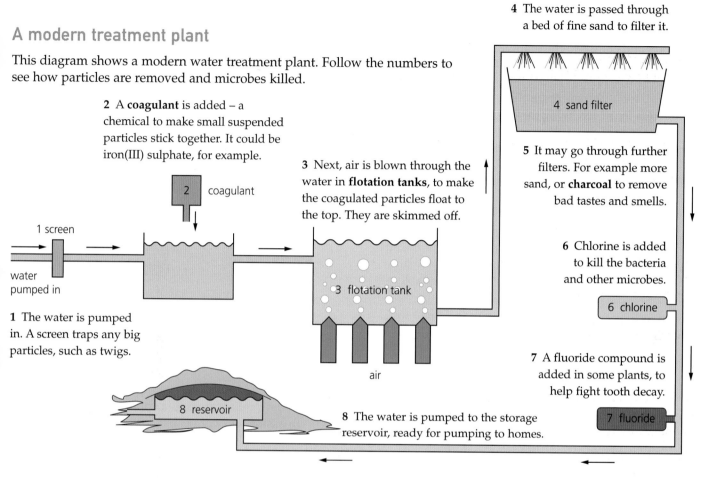

4 The water is passed through a bed of fine sand to filter it.

2 A **coagulant** is added – a chemical to make small suspended particles stick together. It could be iron(III) sulphate, for example.

3 Next, air is blown through the water in **flotation tanks**, to make the coagulated particles float to the top. They are skimmed off.

5 It may go through further filters. For example more sand, or **charcoal** to remove bad tastes and smells.

1 The water is pumped in. A screen traps any big particles, such as twigs.

6 Chlorine is added to kill the bacteria and other microbes.

7 A fluoride compound is added in some plants, to help fight tooth decay.

8 The water is pumped to the storage reservoir, ready for pumping to homes.

This treatment can remove even the tiniest particles. And chlorine can kill all the microbes. But the water may still have harmful substances dissolved in it. For example, nitrates from fertilisers, that can make babies ill.

It is possible to remove dissolved substances, using special membranes. But that is very expensive, and is not usually done. The best solution is to find the cleanest source you can, for your water supply.

If there is only dirty water to drink ...

- Leave it to sit in a container for a while, to let particles settle.
- Scoop out the clearer water from the top of the container, and boil it for several minutes to kill the microbes.
- If you are not able to boil it, leave it sitting in a clear plastic container in the sun for several hours. This will kill most microbes!

Questions

1. What is: **a** ground water? **b** an aquifer? **c** a microbe? (Check the glossary?)
2. What is a *coagulant* used for, in water treatment plants?
3. Why is chlorine such an important part of the treatment?
4. A fluoride compound may be added to water. Why?
5. Some water can be harmful even after treatment. Explain.
6. You need a drink of water – but there is only dirty river water. What will you do to clean it?

Questions on Chapter 14

1. Copy and complete the following paragraph:
   Air is a .... of different gases. 99% of it consists of the two elements .... and .... One of these, ..., is needed for respiration, which is the process by which living things obtain the .... they need. The two elements above can be .... from liquid air by ...., because they have different .... Much of the .... obtained is used to make nitric acid and fertilisers.
   Some of the remaining 1% of air consists of two compounds, ... and .... One of these is important because it is taken in by plants, in the presence of ...., to make .... The rest of the air is made up of elements called the .... These are all members of Group .... of the Periodic Table.

2. The bar charts below show the composition of the atmospheres on Venus, Earth, and Mars.

   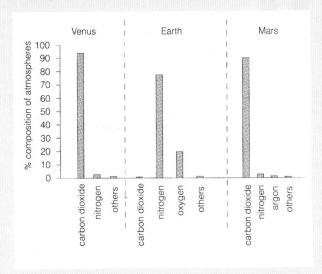

   a. Write the formula of the named gases that make up the atmospheres of the three planets.
   b. What are the main differences between the atmospheres of the Earth, Mars, and Venus?
   c. The Earth's atmosphere was once similar to that of Venus. Then green plants appeared on the Earth, using photosynthesis to create their food.
      i. Explain how, and why, this changed the composition of the Earth's atmosphere. (Page **266** may help.)
      ii. What can you conclude about Venus?
   d. The planet Mercury is the smallest planet in our solar system, and closest to the Sun. Mercury has practically no atmosphere. Suggest a reason.

3. Oxygen and nitrogen, the two main gases in air, are both slightly soluble in water.
   Using this apparatus, a sample of water was boiled until 100 cm^3 of dissolved air had been collected:

   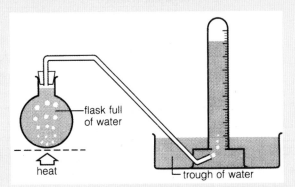

   This air was then passed over heated copper. Its final volume was 67 cm^3.
   a. Does air get *more* soluble in water, or *less*, as the temperature rises? How can you tell?
   b. The copper reacts with the oxygen in the dissolved air. Write an equation for this.
   c. Draw a diagram of the apparatus used for passing air over heated copper.
   d. What volume of oxygen was present in 100 cm^3 of dissolved air?
   e. Calculate the approximate percentages of oxygen and nitrogen in dissolved air.
   f. What are the percentages of these gases in atmospheric air?
   g. Explain why the answers are different, for parts **e** and **f**.
   h. Which gas is more soluble in water, nitrogen or oxygen?
   i. What is the biological importance of the air dissolved in water?

4. Air is a *mixture* of different gases.
   a. Which gas makes up about 78% of the air?
   b. Only one gas in the mixture will allow things to burn in it. Which gas is this?
   c. How are the gases in the mixture separated from each other, in industry?
   d. Which noble gas is present in the greatest quantity, in air?
   e. i. Which gas containing sulphur is a major cause of air pollution?
      ii. What harmful effect does this gas have?
   f. Name two other gases that contribute to air pollution, and say what harm they do.

5 The rusting of iron wool gives a way to find the percentage of oxygen in air. This is the apparatus:

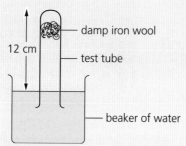

After five days the water level had risen by 2.5 cm, and the iron had rusted to iron(III) oxide.

a Write a balanced equation for the reaction between iron wool and oxygen.
b Why was the iron wool dampened with water before the experiment?
c Why does the water rise up the tube?
d What result does the experiment give, for the percentage of oxygen in air?
e What sort of result would you get if you doubled the amount of iron wool?
f Comment on the accuracy of this method compared with the method on page 203.

6 In the catalytic converters fitted to new cars, carbon monoxide and oxides of nitrogen in the exhaust gas are converted to other substances. (See page 155.)
a i Why is carbon monoxide removed?
  ii Give one harmful effect of nitrogen dioxide.
b What is meant by a *catalytic* reaction?
c In one reaction in a catalytic converter, nitrogen monoxide (NO) reacts with carbon monoxide to form nitrogen and carbon dioxide.
  Write a balanced equation for this reaction.
d i Name one exhaust gas that is not removed in the catalytic converter, and is causing us a big problem.
  ii How could this problem be tackled?

7 Nitrogen and oxygen are separated from air by fractional distillation. Oxygen boils at −183 °C and nitrogen at −196 °C.
a Write the chemical formulae of these two gases.
b Is air a mixture or a compound?
c What state must the air be in, before fractional distillation can be carried out?
d Very low temperatures are required for **c**. How are these achieved?
e Explain, using their boiling points, how the the nitrogen and oxygen are separated.
f Name one other gas that is also obtained from the fractional distillation of air.

8 Oxides of three non-metals, carbon, sulphur and nitrogen, are among the common pollutants in air.
a i Oxides of one of the non-metals form when air gets so hot that gases in it react together. Which of the three non-metals is this?
  ii Where does air get hot enough for this?
b Which oxide is formed from the *incomplete* combustion of fossil fuels? Name it and give its formula.
c i Name the oxide containing sulphur that causes acid rain, and give its formula.
  ii What is the main source of this oxide?
d i Which of the oxides above are found in the exhaust gases from car engines?
  ii Manufacturers of exhaust systems try to remove these pollutants. Explain how.
e *Metal* oxides do not contribute significantly to air pollution. Suggest a reason for this.
f i Lead compounds in the air are also pollutants. What damage can they do?
  ii Lead pollution is less of a problem today than it was twenty years ago. Why is this?

9 This diagram shows one stage in the treatment of water to make it ready for piping to homes:

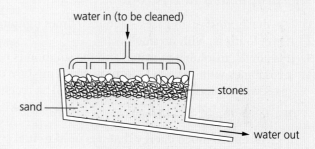

a Name the process being carried out here.
b The water entering this stage has already been treated with a *coagulant*, aluminium sulphate. What does a coagulant do?
c Which kinds of impurities will the above process: i remove? ii fail to remove?
d A layer of carbon may be added below the sand. What is it for?
e The next stage in treatment is *chlorination*.
  i What does this term mean?
  ii Why is this process carried out?
f In some places the water is too acidic to be piped to homes. What could be added to reduce the acidity level?
g At the end of treatment, another element may be added to water, for dental reasons, in the form of a salt. Which element is this?

15.1 Hydrogen

The lightest element

Hydrogen is the lightest of all the elements. It is so light that there is none in the air: it has escaped from the Earth's atmosphere.

But out in space, it is the most common element in the universe. And your life depends on it. We would die without the sun's light and heat – and these are the result of a nuclear reaction inside the sun, where hydrogen atoms fuse to form helium atoms.

Sunshine, thanks to hydrogen!

Making hydrogen in the lab

Hydrogen is made in the lab by using a metal to displace it from dilute acid. (See the panel on the right below.) Zinc and dilute sulphuric acid are the usual choice. The apparatus is shown on the right.

The acid is dripped into the flask. Bubbles of hydrogen form around the zinc, while it dissolves. Hydrogen flows to the gas jar. It is collected over water since it is almost insoluble. The equation for the reaction is:

$$Zn\,(s) + H_2SO_4\,(aq) \longrightarrow ZnSO_4\,(aq) + H_2\,(g)$$
zinc | dilute sulphuric acid | zinc sulphate | hydrogen

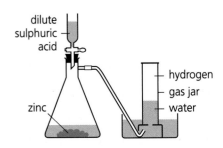

One way to make hydrogen in the lab.

The properties of hydrogen

1. It is the lightest of all gases – about 20 times lighter than air.
2. It is colourless, with no smell.
3. It is almost insoluble in water, and the solution is neutral.
4. It combines with oxygen to form water. A mixture of the two gases will explode when lit. So take care! The reaction is:

   $$2H_2\,(g) + O_2\,(g) \longrightarrow 2H_2O\,(l)$$

   It gives out so much energy that it is used to fuel space rockets.
5. Hydrogen is more reactive than copper, as you can see from the panel on the right. So it will grab oxygen from copper(II) oxide, reducing it to copper. The hydrogen is itself oxidised to water. The reaction is:

   $$CuO\,(s) + H_2\,(g) \longrightarrow Cu\,(s) + H_2O\,(l)$$

The test for hydrogen

If you think a gas might be hydrogen:
1. Collect a sample of it in a test tube. Trap it with your thumb.
2. Remove your thumb and hold a lighted splint to the mouth of the tube.
3. If the gas is hydrogen, it will burn with a squeaky pop.

Uses of hydrogen

- It is used to 'harden' vegetable oils to make margarine (page 105).
- It is used in the manufacture of many important chemicals, including ammonia, nitric acid, hydrochloric acid, and methanol.
- It is growing more important as a fuel, in hydrogen fuel cells.

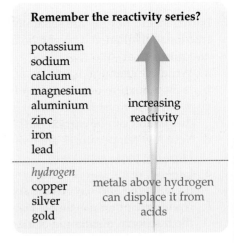

The hydrogen fuel cell

Its job is to provide a current. So it has a negative pole that pushes electrons into a circuit, a positive pole that accepts them, and an electrolyte. The negative pole is surrounded by hydrogen, and the positive pole by oxygen (in air). The electrolyte contains OH⁻ ions:

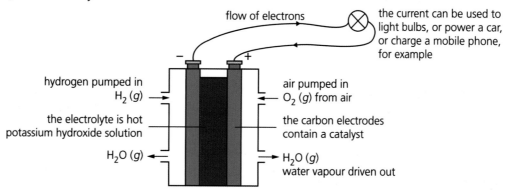

At the negative pole	At the positive pole
Hydrogen loses electrons to the OH⁻ ions. It is oxidised: $$2H_2(g) + 4OH^-(aq) \longrightarrow 4H_2O(l) + 4e^-$$ The electrons flow through the wire to the positive pole, as a current. On the way the current lights your house, or powers your car, or charges your mobile phone.	The electrons are accepted by oxygen molecules. Oxygen is reduced to OH⁻ ions: $$O_2(g) + 2H_2O(l) + 4e^- \longrightarrow 4OH^-(aq)$$ But the concentration of OH⁻ ions in the electrolyte does not change. Can you see why?

Adding the two half-equations above gives the full equation:

$$2H_2(g) + O_2(g) \longrightarrow 2H_2O(l)$$

So the overall reaction is that hydrogen and oxygen combine to form water.

Advantages of the hydrogen fuel cell

- Only water is formed. No pollutants! No nitrogen oxides form because air does not get heated up.
- The reaction gives out plenty of energy. 1kg of hydrogen gives about 2.5 times as much energy as 1 kg of natural gas (methane).
- We won't run out of hydrogen. It can be made by the electrolysis of water with a little acid added. (See the electrolysis of dilute hydrochloric acid, on page 137.) Solar power could give cheap electricity for this.

But there is a drawback. Hydrogen is highly flammable so it must be stored with great care. A spark will cause a mixture of hydrogen and air to explode.

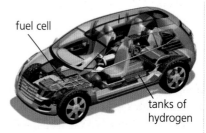

Model of a car with a hydrogen fuel cell. Look at the tanks of hydrogen. There is no oxygen tank. Why not?

Questions
1. How is hydrogen made in the lab?
2. Why and how does hydrogen react with copper(II) oxide?
3. Explain how the hydrogen fuel cell works.
4. Write down the two half-equations given above. Then show how you would add them to find the overall reaction.
5. Where do you think oxygen comes from, for the car above?

15.2 Nitrogen and ammonia

Nitrogen

Nitrogen is a colourless, odourless, unreactive gas, that makes up nearly four-fifths of the air. You breathe it in – and breathe it right out again, unchanged.

But you also take in nitrogen in the **proteins** in your food. Your body uses these to build muscle, bone, skin, hair, blood, and other tissues. In fact you are about 2.6% nitrogen, by weight!

The properties of nitrogen

1. It is a colourless gas, with no smell.
2. It is only slightly soluble in water.
3. It is very unreactive compared with oxygen.
4. But it will react with hydrogen to form ammonia, with the help of high pressure, a moderate temperature, and a catalyst. The reaction is reversible:

    $$N_2(g) + 3H_2(g) \rightleftharpoons 2NH_3(g)$$

    This reaction is the first step in making nitric acid, and nitrogen fertilisers. You can find out more about it in the next unit.
5. It also combines with oxygen at high temperatures to form nitrogen monoxide (NO) and nitrogen dioxide (NO_2). The reactions occur naturally in the air during lightning – and also inside hot car engines, and power station furnaces. The nitrogen oxides are acidic, and cause air pollution, and acid rain. (See page 123.)

The main chemical use of nitrogen

The main chemical use of nitrogen is in the manufacture of ammonia, which is then used to make nitric acid, and fertilisers. The nitrogen in air is used for this.

Other uses of nitrogen

Nitrogen is separated from air by fractional distillation, and used in many other ways. These uses rely on the fact that it is inert, non-toxic, very cold when liquid (it boils at –196°C), and abundant. For example:

Nitrogen gas is used ...
- to flush air out of food packages, in order to keep the food fresh. (Food goes 'off' in air.)
- to force oil out of oil wells; it is pumped in and drives the oil upwards.
- to flush air out of tanks of chemicals that might catch fire in air.

Liquid nitrogen is used ...
- to freeze liquid in damaged pipes, making it easier to repair them.
- to quick-freeze foods such as pizzas.
- to chill containers used for transporting food.
- to shrink-fit machine parts. One part is put into liquid nitrogen and it shrinks, so it slides into the other part more easily.
- to freeze tissue samples, for medical research.

We live in a sea of nitrogen.

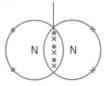

The bonding in nitrogen. Since three pairs of electrons are shared, the bond is a triple bond. It can be as shown as $N \equiv N$.

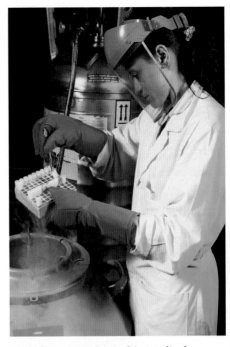

Liquid nitrogen is used in medical research, to keep tissue samples frozen.

Ammonia

Ammonia is a gas with the formula NH_3. It is a very important compound, since it is used to make fertilisers. It is made in industry by reacting nitrogen with hydrogen. The details are in the next unit.

Making ammonia in the lab

You can make ammonia in the lab by heating *any* ammonium compound with a strong base. The base drives out or *displaces* ammonia from the compound. You could use ammonium chloride and calcium hydroxide, for example. The equation for the reaction is:

$$2NH_4Cl\,(s) + Ca(OH)_2\,(s) \longrightarrow CaCl_2\,(s) + 2H_2O\,(l) + 2NH_3\,(g)$$
ammonium chloride + calcium hydroxide → calcium chloride + water + ammonia

This reaction can also be used as a test. If a compound gives off ammonia when heated with a strong base, it must be an ammonium compound. You can confirm that the gas is ammonia using the reaction in **3** below.

The properties of ammonia

1. It is a colourless gas with a strong, choking smell.
2. It is less dense than air.
3. It reacts with hydrogen chloride gas to form a white smoke. The smoke is made of tiny particles of solid ammonium chloride:

   $$NH_3\,(g) + HCl\,(g) \longrightarrow NH_4Cl\,(s)$$

   This reaction can be used to test whether a gas is ammonia.
4. It is very soluble in water. (It shows the fountain effect.)
5. The solution in water is alkaline – it turns red litmus blue. So it must contain hydroxide ions. This is because ammonia acts as a base, accepting protons from water molecules:

   $$NH_3\,(aq) + H_2O\,(l) \longrightarrow NH_4^+\,(aq) + OH^-\,(aq)$$

   But only a small % of its molecules do this, so ammonia is a weak base. See page 117 for more.
6. Since ammonia solution is a base, it reacts with acids to form salts. For example with nitric acid it forms ammonium nitrate:

   $$NH_3\,(aq) + HNO_3\,(aq) \longrightarrow NH_4NO_3\,(aq)$$

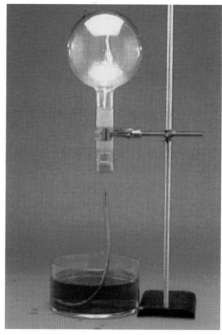

The **fountain experiment**. The flask contains ammonia. It dissolves in the first drops of water that reach the top of the tube, so a fountain of water rushes up to fill the vacuum. (Any very soluble gas will show this effect.)

Questions

1. What is the main chemical use of nitrogen?
2. Nitrogen has many other uses too. Choose three, and for each, explain why it suits that use.
3. A solution of ammonia is alkaline. Explain why.
4. Ammonia is a *weak* base. Why?
5. How would you make solid ammonium sulphate, as pure as you can, starting with ammonia solution? Write an equation for the reaction.
6. How would you test a compound to prove it was an ammonium compound? (Check out page 291 too.)

15.3 Making ammonia in industry

It's a key chemical

Ammonia is one of the world's most important chemicals, because it is needed to make fertilisers. We depend on fertilisers to grow enough food.

Ammonia is made from nitrogen and hydrogen. Nitrogen is unreactive, and its reaction with hydrogen is reversible. So making ammonia is not easy. The process below is called the **Haber process**, after the German chemist, Fritz Haber, who developed it in 1908. It is still used today.

The Haber process

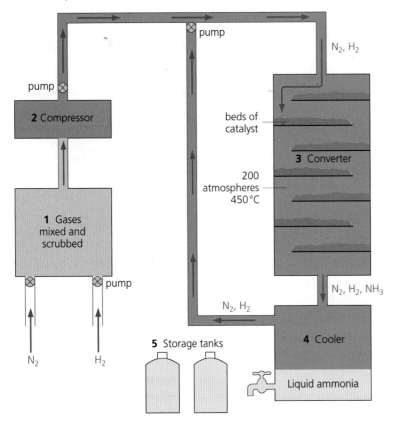

An ammonia plant. Ammonia plants are often close to oil refineries, to make use of the hydrogen from cracking.

1. The raw materials are hydrogen and nitrogen. The panel on the right shows how they are obtained. The two gases are mixed, and **scrubbed** (cleaned) to remove impurities.
2. The gas mixture is **compressed**. More and more gas is pumped in, until the pressure reaches 200 atmospheres.
3. The compressed gas flows to the **converter** – a round tank with beds of hot iron at 450°C. The iron catalyses this reversible reaction:

   $$N_2(g) + 3H_2(g) \rightleftharpoons 2NH_3(g)$$

   But only 15% of the mixture leaving the converter is ammonia.
4. The mixture is cooled until the ammonia condenses to a liquid. The nitrogen and hydrogen are **recycled** to the converter for another chance to react. Steps **3** and **4** are continually repeated.
5. The ammonia is run into tanks, and stored as a liquid under pressure.

The raw materials

Nitrogen
Air is nearly 80% nitrogen and 20% oxygen, with just small amounts of other gases.

The oxygen is removed by burning hydrogen in the air.

$$2H_2(g) + O_2(g) \rightarrow 2H_2O(l)$$

That leaves mainly nitrogen.

Hydrogen
It is usually made from natural gas or **methane**, and water (as steam):

$$CH_4(g) + 2H_2O(g) \xrightarrow{catalyst} CO_2(g) + 4H_2(g)$$

Compounds in oil can also be **cracked** to give hydrogen. For example:

$$C_2H_6(g) \xrightarrow{catalyst} C_2H_4(g) + H_2(g)$$
ethane → ethene

Improving the yield of ammonia

The reaction between nitrogen and hydrogen is reversible, and exothermic – it gives out heat.

Since it is reversible, a mixture of the two gases will *never* react completely. So the yield will never be 100%.

But all of the following will help to improve the yield, as Unit 11.5 explained:

- high pressure
- low temperature
- removing the ammonia as it forms.

This graph on the right shows how the yield changes with temperature and pressure.

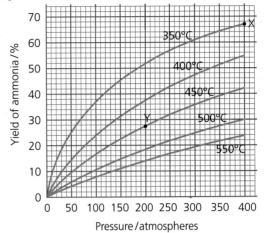

The yield of ammonia at different temperatures and pressures

The final reaction conditions

The temperature and pressure The highest yield on the graph above is at **X**, at 350 °C and 400 atmospheres. But in fact the Haber process uses 450 °C and 200 atmospheres, at **Y** on the graph. Why?

Because at 350 °C, the reaction is too slow. 450 °C gives a better rate.

And second, a pressure of 400 atmospheres needs very powerful pumps, and very strong and sturdy pipes and tanks, and a lot of electricity. 200 atmospheres is safer, and saves money.

So the conditions in the converter do not give a high yield. But the ammonia is removed, so that more will form. And the unreacted gases are recycled, to give them another chance to react. So the final overall yield is high.

The catalyst Iron speeds up the reaction. But it does not change the yield!

The raw materials

The panel on the opposite page shows how the raw materials are produced. Air and water are easy to find. But you need natural gas (methane), or compounds from oil, to make hydrogen.

So ammonia plants are usually built close to natural gas terminals, or oil refineries. In fact many oil and gas companies now make ammonia, as a way to increase their profits.

Making ammonia: a summary

$$N_2 + 3H_2 \underset{\text{endothermic}}{\overset{\text{exothermic}}{\rightleftharpoons}} 2NH_3$$

4 molecules → 2 molecules

To improve the yield:
- high pressure
- low temperature
- remove ammonia

To get a decent reaction rate:
- raise temperature (compromise!)
- use a catalyst

Questions

1. Ammonia is made from nitrogen and hydrogen.
   a. How are the nitrogen and hydrogen obtained?
   b. What is the process for making ammonia called?
   c. Write an equation for the reaction.
2. Look at the catalyst beds in the diagram on page 216.
   a. What is in them?
   b. Why are they arranged this way?
3. a. Explain *why* high pressure and low temperature helps the yield, in making ammonia. (Check Unit 11.5?)
   b. 400 atmospheres and 250 °C would give a high yield. Why are these conditions *not* used in the Haber process?
   c. What is the maximum possible % yield of ammonia at 200 atmospheres and 450 °C? (Use the graph.)
   d. What happens to the unreacted gases?

15.4 Fertilisers

What plants need

A plant needs carbon dioxide, light, and water. It also needs several different elements. The main ones are **nitrogen**, **potassium**, and **phosphorus**.

It needs nitrogen for chlorophyll, and all its other proteins.

Potassium helps it to produce proteins, and to resist disease.

Phosphorus helps roots to grow, and helps crops to ripen.

Plants obtain these elements from compounds in the soil, which they take in through their roots, as solutions. The most important one is nitrogen, which every plant cell needs. Plants take it in as **nitrate** ions and **ammonium** ions.

Fertilisers

Every crop a farmer grows takes compounds from the soil. Some get replaced naturally. But in the end the soil gets worn out, and new crops won't grow well. So the farmer has to add **fertilisers**.

A fertiliser is any substance added to the soil to make it more fertile.

Animal manure is a natural fertiliser. **Artificial fertilisers** are salts made in factories, and sprinkled or sprayed on fields. Here are some examples.

ammonium nitrate, NH_4NO_3 ammonium sulphate, $(NH_4)_2SO_4$
potassium sulphate, K_2SO_4 ammonium phosphate, $(NH_4)_3PO_4$

Getting ready to apply fertiliser to fields.

Ammonia: the key to nitrogenous fertilisers

The air is rich in nitrogen. But plants need it in the soil! The Haber process gives a way to turn it into ammonia, which can then be passed on to plants.

However, nitrogen reaches plants faster as nitrate ions than as ammonium ions. And here again ammonia is the key. It is the starting point for making nitric acid, from which nitrates are made.

From ammonia to nitric acid and nitrates

Ammonia is oxidised to nitric acid with the help of a catalyst of platinum and rhodium. Several reactions take place. The *overall* reaction is:

$$NH_3\,(g) + 2O_2\,(g) \longrightarrow HNO_3\,(aq) + H_2O\,(l)$$
ammonia + oxygen → nitric acid + water

Fitting the platinum/rhodium gauze (the catalyst) to a reactor where ammonia will be made into nitric acid.

The nitric acid can then be neutralised to give nitrates, for use as fertilisers. For example the fertiliser ammonium nitrate is made by neutralising nitric acid with ammonia solution:

$$HNO_3\,(aq) + NH_3\,(aq) \longrightarrow NH_4NO_3\,(aq)$$
nitric acid + ammonia → ammonium nitrate

This is then added to soil to help crops grow strong and healthy.

It's not all good news

Fertilisers help to feed the world. We could not grow enough crops without them. But there are drawbacks – as usual!

Fertilisers can kill fish Fertilisers can seep into rivers from farmland. There they help tiny water plants called algae to grow. These grow so well that they cover the water like a carpet. When they die, bacteria feed on them. In the process the bacteria use up all the oxygen dissolved in the water. So fish die.

Fertilisers can cause health problems in infants Nitrate ions from fertilisers can get into drinking water. They are converted to nitrites in the body. These combine with the haemoglobin in blood, which means the blood can't carry so much oxygen around the body. This can cause oxygen starvation in young infants. Their skin takes on a blue tinge. They don't usually die, but their development may be affected.

What % of it is nitrogen?

Ammonium nitrate is rich in nitrogen. What % of it is nitrogen? Find out like this:

Formula: NH_4NO_3
A_r: N = 14, H = 1, O = 16
M_r: $(14 \times 2) + (4 \times 1) + (16 \times 3)$
    = 80

% of this that is nitrogen
    = $\dfrac{28}{80} \times 100\%$
    = **35%**

Questions

1. What are *artificial fertilisers*? Why are they needed? Which key elements do they provide for plants?
2. *Nitrogenous fertilisers* are fertilisers that contain nitrogen. Name three nitrogenous fertilisers.
3. Why is the Haber process so important to farming?
4. Fertilisers can harm river life. Explain how.
5. The salt ammonium nitrate is an important fertiliser. Write a balanced equation for the reaction used to make it. Which type of reaction is this?
6. a Find the % of nitrogen in ammonium sulphate. (S = 32.)
   b How does it compare with the % in ammonium nitrate, given in the panel above?

Questions on Chapter 15

1. The reaction that takes place when hydrogen is passed over iron(II) oxide is:

   FeO (s) + H$_2$ (g) ⟶ Fe (s) + H$_2$O (l)

   a Complete the sentences below.
   In the reaction the iron(II) oxide is .... to iron and the hydrogen is .... to water. So hydrogen is acting as a ....
   b Draw a diagram of the apparatus that could be used for this reaction, in the laboratory.
   c Suggest a reason why carbon monoxide is used instead of hydrogen, to extract iron from iron oxide in industry. (Hint: which oxide is it?)
   d Give one use of hydrogen, in industry.

2. This paragraph is about the element nitrogen. Rewrite it, choosing the correct item from each pair in brackets.
   Air is a (mixture/compound) which contains ($\frac{1}{5}$/$\frac{4}{5}$) nitrogen. The symbol for nitrogen is (N/N$_2$) and the gas is made of (molecules/atoms) represented by the formula (N/N$_2$). It is therefore a (monatomic/diatomic) gas. Nitrogen is a (colourless/brown) gas which is (very/slightly) soluble in water. It is a very (reactive/unreactive) element. Nitrogen is needed by plants to make (proteins/sugars). We can provide nitrogen for plants in the form of (fertilisers/limestone). For these, the nitrogen is first reacted with (hydrogen/oxygen) to give the (gas/solid) ammonia. This is then turned into (sulphuric/nitric) acid. The acid is then (oxidised/neutralised) to give fertilisers such as ammonium nitrate.

3. The apparatus on the right can be used to prepare a solution of ammonia in water *safely*. Ammonia gas is given off when the test tube is heated.
   a i Name two compounds that could be heated together in the test tube.
     ii Write an equation for the reaction.
   b Why would it be dangerous to dip the glass tube into the water, *without* using the filter funnel? (Hint: fountain experiment, page 215.)
   c If a few drops of litmus solution is added to the water in the trough, what colour change will you see during the experiment?
   d i If the water is replaced by dilute nitric acid, what salt will be formed in the trough?
     ii Write the equation for the reaction in d i.

4. Ammonia can be made using the apparatus below. X and Y are two gas syringes, connected to combustion tube A.

   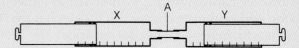

   At the start, X contained 75 cm^3 of hydrogen, and Y contained 25 cm^3 of nitrogen.
   a Copy the diagram, labelling the gases.
   b How would you make the two gases mix?
   c Where should the apparatus be heated? Show this on your diagram.
   d i The reaction needs a catalyst. Why?
     ii Name a suitable catalyst.
     iii Where should it be placed? Show it on your diagram.
   e Write a balanced equation for the reaction.
   f How could you confirm that some ammonia had been formed?

5. Hydrogen can be made in the laboratory by dripping dilute sulphuric acid onto zinc in a flask.
   a Write the chemical equation for the reaction.
   b What will you see in the flask?
   c The gas can be collected by bubbling it through water, into a gas jar. What does that tell you about it?
   Hydrogen will remove oxygen from the compound nickel oxide (NiO).
   d i Would you place hydrogen above nickel or below it, in the reactivity series? Explain.
     ii Write the equation for this reaction.
     iii Which type of reaction is it?

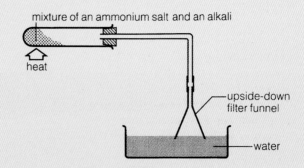

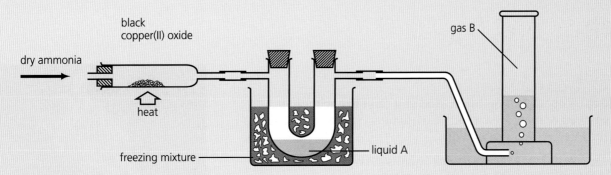

6 Using the above apparatus, dry ammonia is passed over heated copper(II) oxide. The gases given off are passed through a cooled U-tube. A liquid (A) forms in the U-tube and a colourless gas (B) collects in the gas jar.
  a The copper(II) oxide is reduced to copper. What will you see as the gas passes over the heated copper(II) oxide?
  b Why is the U-tube surrounded by a freezing mixture?
  c The liquid A is found to turn blue cobalt chloride paper pink, and to have a boiling point of 100°C. Identify it.
  d Is gas B soluble or insoluble in water?
  e Identify gas B. (Hint: look at the other chemicals in this reaction.)
  f Write a word equation for the reaction.
  g The copper(II) oxide is *reduced* by the ammonia. Explain what this means.
  h How will the mass of the heated tube and its contents change, during the reaction?

7 The manufacture of ammonia is an important industrial processes.
  a Name the raw materials used.
  b Which two gases react together?
  c Why are the two gases scrubbed?
  d Why is the mixture passed over iron?
  e What happens to the *unreacted* nitrogen and hydrogen?
  f Why is the manufactured ammonia stored at high pressure?

8 An NPK fertiliser contains the three main elements that plants need, for healthy growth.
  a Name the three elements.
  b Describe *how* each element helps plants.
  c Which of the three elements are provided by the following fertilisers?
    i ammonium phosphate
    ii potassium nitrate
    iii ammonium sulphate
  d Write a formula for each fertiliser in **c**.

9 Nitrogen and hydrogen are converted to ammonia in the Haber process:

$$N_2 (g) + 3H_2 (g) \rightleftharpoons 2NH_3 (g)$$

  a The energy level diagram for the reaction is shown below. What does this diagram tell you about the reaction?

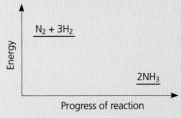

  b The reaction is *reversible*, and *reaches equilibrium*.
    i Explain clearly what the two terms in italics mean.
    ii What effect does a catalyst have on an equilibrium reaction?
    iii Which catalyst is used in the Haber process?
    iv What effect does this catalyst have on the % yield of ammonia?
  c Much of the ammonia from the Haber process is then used to make nitric acid (HNO_3).
    i Write the overall equation for the reaction used to make nitric acid.
    ii Which type of reaction is it?
    iii What is the oxidising agent in this reaction?
    iv Ammonia is a *raw material* for the manufacture of nitric acid. What does the term in italics mean?
  d i Some nitric acid is used to make ammonium nitrate. Write a word equation for this reaction.
    ii Which type of reaction is this?
    iii Give one use for ammonium nitrate.

10 Ammonium chloride (NH_4Cl) is a fertiliser, often used for growing coconut, oil palm, and kiwi fruit.
  a Why is it able to help plant growth?
  b i How would you make it from ammonia?
    ii Write the equation for the reaction.
  c What is the % of nitrogen in that fertiliser? (A_r : H = 1, N = 14, Cl = 35.5.)

16.1 Oxygen

A gas for life

Oxygen makes up about 20% of the air. And you can't live without it. If you are deprived of it for even a couple of minutes, you will have brain damage. A little longer, and you will die.

Oxygen was given its name by the French scientist Lavoisier, in 1778. You can read about the experiments he did, on page 203.

Producing oxygen in industry

In air, oxygen is part of a mixture of gases. But often it is more useful on its own – for example for welding metals.

So in industry, oxygen is extracted from air by **fractional distillation**. The air is cooled until it turns liquid, then warmed up again. Oxygen boils off at – 183 °C. You can find out more about this, and what the oxygen is used for, in Unit 14.2.

Don't forget your oxygen supply!

Producing oxygen in the lab

In the lab, you can obtain oxygen from hydrogen peroxide, a colourless liquid. The apparatus is shown on the right.

Hydrogen peroxide has the formula H_2O_2. It decomposes like this:

$$2H_2O_2\,(aq) \longrightarrow 2H_2O\,(l) + O_2\,(g)$$

It decomposes very very slowly on its own. But manganese(IV) oxide is a catalyst for this reaction, so it is added to the flask.

Oxygen is only slightly soluble in water. How can you tell, from this apparatus?

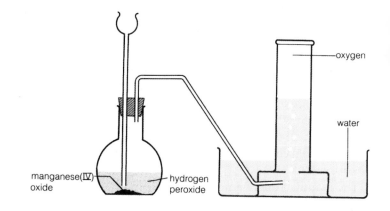

Producing oxygen in the lab.

The properties of oxygen

1 It is a clear, colourless gas, with no smell.
2 It is only slightly soluble in water. About 0.01 g of it will dissolve in 1 litre of water at 20 °C. But this is enough to keep fish alive. As the water passes through their gills, the oxygen diffuses into their blood.
3 It is a very reactive element. It reacts with many substances to produce **oxides**. The reactions usually give out a lot of energy.
For example when oxygen is mixed with hydrogen, and the mixture is lit, there is an explosive reaction, giving water:

$$2H_2\,(g) + O_2\,(g) \longrightarrow 2H_2O\,(l)$$

4 When oxygen reacts with other substances, we say it **oxidises** them. (Remember, one definition of oxidation is *gain of oxygen*.) So in the reaction above, hydrogen is oxidised. Which substance is reduced?

For us humans, the two most important reactions of oxygen are **respiration** and the **burning of fuels**. These are very similar, as you will see next.

Like us, fish can't live without oxygen.

Respiration

Respiration is the reaction that keeps you alive. During respiration, oxygen reacts with glucose in your body cells, like this:

$$C_6H_{12}O_6\,(aq) + 6O_2\,(g) \longrightarrow 6CO_2\,(g) + 6H_2O\,(l) + \text{energy}$$
$$\text{glucose} + \text{oxygen} \longrightarrow \text{carbon dioxide} + \text{water} + \text{energy}$$

The glucose comes from digested food, and the oxygen from air:

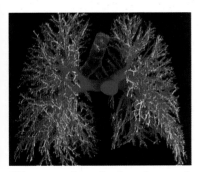

When you breathe in, air goes into your lungs. It goes along tiny tubes to their surface. From there, the oxygen diffuses into your blood.

Your blood carries the oxygen to the millions of cells in your body, along with glucose from digested food. Respiration takes place in each cell.

The energy from respiration keeps your heart and muscles working, and keeps you warm. Without it, no body reactions could go on. You would die.

Respiration goes on in every cell of every living thing. Fish use the oxygen dissolved in water. Like us, plants take oxygen from the air.

The burning of fuels

Fuels are substances we burn in air to obtain energy – usually as heat. When we burn the fuel natural gas, or **methane**, in air, the reaction is:

$$CH_4\,(g) + 2O_2\,(g) \longrightarrow CO_2\,(g) + 2H_2O\,(l) + \text{energy}$$
$$\text{methane} + \text{oxygen} \longrightarrow \text{carbon dioxide} + \text{water} + \text{energy}$$

Compare this with the equation for respiration. What do you notice?

The test for oxygen

Things burn much faster in pure oxygen than in air. So, to test for oxygen:
1. Light a wooden splint, then blow the flame out. The splint keeps glowing because it is reacting with the oxygen in air.
2. Plunge the splint into the unknown gas.
3. If the gas is oxygen, the splint will burst into flame.

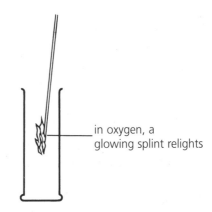

The test for oxygen.

Questions
1. Draw a labelled diagram, showing how you would obtain oxygen from hydrogen peroxide solution.
2. How would you show that the gas you collected in **1** was in fact oxygen?
3. What is *respiration*? Where does it take place?
4. How do fish obtain oxygen for respiration?
5. What is a *fuel*? See if you can give four examples.
6. In what ways is respiration like the burning of methane?
7. *Food is fuel for the body*. Do you agree? Give reasons.
8. *Why* do things burn faster in pure oxygen than in air?

16.2 Oxides

What are oxides?

Oxides form when oxygen reacts with other elements, or compounds. There are four types of oxide: basic, acidic, amphoteric, and neutral. Find out more below.

Basic oxides

Look how these metals react with oxygen:

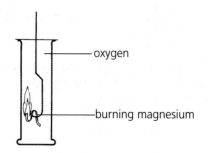

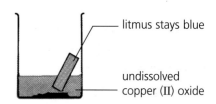

Magnesium ribbon is lit over a Bunsen flame, and plunged into a jar of oxygen. It burns with a brilliant white flame, leaving a white ash, **magnesium oxide**:

$2Mg\ (s) + O_2\ (g) \longrightarrow 2MgO\ (s)$

Hot iron wool is plunged into a gas jar of oxygen. It glows bright orange, and throws out a shower of sparks. A black solid is left in the gas jar. It is **iron(III) oxide**:

$4Fe\ (s) + 3O_2\ (g) \longrightarrow 2Fe_2O_3\ (s)$

Copper is too unreactive to catch fire in oxygen. But when it is heated in a stream of the gas, its surface turns black. The black substance is **copper(II) oxide**:

$2Cu\ (s) + O_2\ (g) \longrightarrow 2CuO\ (s)$

As you can see, the most reactive metal reacts the most vigorously.

The copper(II) oxide produced in the last reaction above is insoluble in water. But it does dissolve in dilute acid:

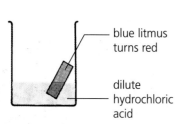
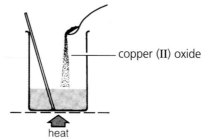
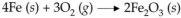

This is dilute hydrochloric acid. It turns blue litmus paper red, like all acids do.

Copper(II) oxide dissolves in it, when it is warmed. But after a time, no more will dissolve.

The resulting liquid has no effect on blue litmus. So the oxide has **neutralised** the acid.

Copper(II) oxide is called a **basic oxide** since it can neutralise an acid. The products are a salt and water:

    base   +   acid $\longrightarrow$ salt   +   water
    $CuO\ (s)\ +\ 2HCl\ (aq) \longrightarrow CuCl_2\ (aq)\ +\ H_2O\ (l)$

Iron(III) oxide and magnesium oxide behave in the same way – they too can neutralise acid, so they are basic oxides.

In general, metals react with oxygen to form basic oxides.
Basic oxides belong to the larger group of compounds called **bases**.

Acidic oxides

Now look how these non-metals react with oxygen:

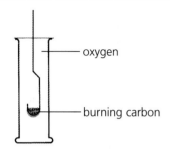 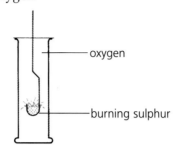 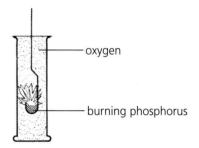

Powdered carbon is heated over a Bunsen until red-hot, and then plunged into a jar of oxygen. It glows bright red, and the gas **carbon dioxide** is formed:

$$C\ (s) + O_2\ (g) \longrightarrow CO_2\ (g)$$

Sulphur catches fire over a Bunsen burner, and burns with a blue flame. In pure oxygen it burns even brighter. The gas **sulphur dioxide** is formed:

$$S\ (s) + O_2\ (g) \longrightarrow SO_2\ (g)$$

Phosphorus bursts into flame in air or oxygen, without heating. (So it is stored under water!) A white solid, **phosphorus pentoxide**, is formed:

$$P_4\ (s) + 5O_2\ (g) \longrightarrow P_4O_{10}\ (s)$$

Carbon dioxide is slightly soluble in water. The solution will turn litmus red: it is **acidic**. The weak acid carbonic acid has formed:

$$CO_2\ (g) + H_2O\ (l) \longrightarrow H_2CO_3\ (aq)$$

Sulphur dioxide and phosphorus pentoxide also dissolve in water to form acids. So they are all called **acidic oxides**.

In general, non-metals react with oxygen to form acidic oxides.

Amphoteric oxides

Aluminium is a metal, so you would expect aluminium oxide to be a base. In fact it is both acidic *and* basic. It reacts as a base with hydrochloric acid:

$$Al_2O_3\ (s) + 6HCl\ (aq) \longrightarrow 2AlCl_3\ (aq) + 3H_2O\ (l)$$

But it reacts as an acidic oxide with sodium hydroxide, giving a compound called sodium aluminate:

$$Al_2O_3\ (s) + 6NaOH\ (aq) \longrightarrow 2Na_3AlO_3\ (aq) + 3H_2O\ (l)$$

So it is called an **amphoteric oxide**. Zinc oxide is also amphoteric.

Neutral oxides

Some oxides are neither acidic nor basic: they are **neutral**. Carbon monoxide, CO, and dinitrogen oxide, N_2O are neutral. (Other nitrogen oxides are acidic.)

Zinc oxide: an amphoteric oxide. It will dissolve in both acid and alkali.

Questions
1. How could you show that magnesium oxide is a base?
2. Copy and complete: Metals usually form .... oxides while non-metals form .... oxides.
3. See if you can arrange carbon, phosphorus and sulphur in order of reactivity, using their reaction with oxygen.
4. What colour change would you see, on adding litmus solution to a solution of phosphorus pentoxide?
5. What is an *amphoteric* oxide? Give two examples.
6. Dinitrogen oxide is a *neutral* oxide. It is quite soluble in water. How could you prove it is neutral?

16.3 Sulphur and sulphur dioxide

Where is sulphur found?

Sulphur is a non-metal. It is quite a common element in the Earth's crust.

- It is found, as the *element*, in large underground beds in several countries, including Mexico, Poland and the USA. It is also found around the rims of volcanoes.
- It occurs as a compound in many metal ores. For example in the ore **galena**, which is lead(II) sulphide, PbS.
- Sulphur compounds also occur naturally in the fossil fuels: coal, oil and natural gas.

A molecule of sulphur. It has 8 atoms – so the molecular formula of sulphur is S_8. But it is just called S in equations.

Extracting the sulphur

From oil and gas Most sulphur is now obtained from the sulphur compounds found in oil and natural gas. (These compounds are removed from oil and gas to help prevent air pollution.)

For example natural gas is mainly methane. But it can have as much as 30% **hydrogen sulphide**. This is separated from the methane. Then it is reacted with oxygen, with the help of a catalyst, to give sulphur:

$$2H_2S\ (g)\ +\ O_2\ (g)\ \longrightarrow\ 2S\ (s)\ +\ 2H_2O\ (l)$$
hydrogen sulphide + oxygen ⟶ sulphur + water

From sulphur beds Only about 5% of the sulphur we use comes from the underground sulphur beds. Superheated water is pumped down to melt the sulphur and carry it to the surface. (It melts at 115 °C.)

The properties of sulphur

1. It is a brittle yellow solid.
3. It has two different forms or **allotropes**, as shown on the right.
4. Because it is molecular, it has quite a low melting point.
5. Like other non-metals, it does not conduct electricity.
6. Like most non-metals, it is insoluble in water.
7. It reacts with metals to form sulphides. For example with iron it forms iron(II) sulphide:

    $Fe\ (s) + S\ (s) \longrightarrow FeS\ (s)$

8. It burns in oxygen to form sulphur dioxide:

    $S\ (s) + O_2\ (g) \longrightarrow SO_2\ (g)$

Uses of sulphur

- Most sulphur is used to make sulphuric acid.
- Some is added to rubber, for example for car tyres, to toughen it. This is called **vulcanizing** the rubber.
- Some is used to make drugs, pesticides, matches, and paper.
- Some is added to cement to make **sulphur concrete**. This is not attacked by acid. So it is used for walls and floors in factories where acid is used.

This is a crystal of **rhombic** sulphur, the allotrope that is stable at room temperature.

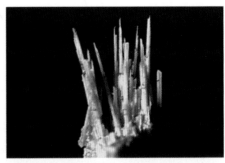

If you heat rhombic sulphur slowly to above 96°C, the molecules rearrange themselves. The result is needle-shaped crystals of **monoclinic sulphur**.

Sulphur dioxide

Sulphur dioxide is a gas, with the formula SO_2. It forms when sulphur burns in air.

1 It is a colourless gas, heavier than air, with a strong, choking smell.
2 Like most non-metal oxides, it is an acidic oxide. It dissolves in water, forming **sulphurous acid**, H_2SO_3:

$$H_2O\,(l) + SO_2\,(g) \longrightarrow H_2SO_3\,(aq)$$

This breaks down easily again to sulphur dioxide and water.

3 It acts as a bleach when it is damp or in solution. That's because it removes the colour from coloured compounds by **reducing** them.

Sulphur dioxide as a pollutant

Coal, oil, and petrol all contain some sulphur compounds (even after oil has been treated to remove some). In particular, coal can contain a high % of sulphur.

When these fuels are burned in power stations, and car engines, and factory furnaces, the sulphur compounds are oxidised to sulphur dioxide.

This gas escapes into the air, where it can cause a great deal of harm. It can attack your lungs, giving breathing problems. It also dissolves in rain to give **acid rain**. This attacks buildings and metal structures, and can kill fish and plants. (See page 123.)

Uses of sulphur dioxide

- Its main use is in the manufacture of sulphuric acid.
- It is used to bleach wool, silk, and wood pulp for making paper.
- It is used as a sterilizing agent in making soft drinks and jam, and in drying fruit. It stops the growth of bacteria and moulds.

Death by sulphur dioxide. These trees were killed by acid rain.

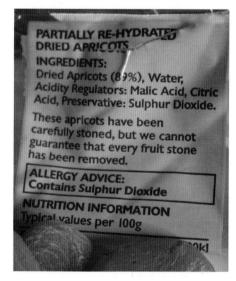

Sulphur dioxide is used to preserve dried fruit (like these apricots) and other foods.

Questions

1 Name three sources of sulphur in the Earth's crust.
2 Sulphur has quite a low melting point. Why is this?
3 Sulphur has two *allotropes*. What does that mean?
4 Sulphur reacts with iron to form iron(II) sulphide. Is this a redox reaction? Explain your answer.
5 a Sulphur dioxide is an *acidic* oxide. Explain.
  b In streets with lots of traffic, the air often has a high level of sulphur dioxide. Why? What harm can it do?
6 Sulphur dioxide is a heavy gas. Do you think this contributes to air pollution? Explain your answer.

16.4 Sulphuric acid

Making sulphuric acid by the Contact process

More sulphuric acid is made than any other chemical!
Most of it is made by the **Contact process**. The raw materials are:

- **sulphur**, air, and water ... or
- **sulphur dioxide**, air, and water. The sulphur dioxide is obtained when sulphide ores, such as lead and zinc ores, are roasted in air to extract the metal from them.

Starting with sulphur, the steps in the Contact process are:

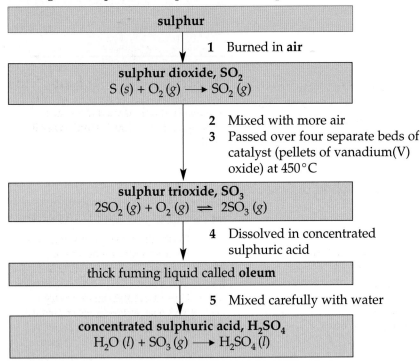

Part of a sulphuric acid plant in the UK. Since it is such a dangerous chemical, safety is a big concern.

You will see this sign on tanks of sulphuric acid. What is the message?

Things to note about the Contact process

- The reaction in step 3 is **reversible**. The sulphur trioxide continually breaks down again. So the mixture is passed over four separate beds of catalyst, to give the reactants further chances to react.
- Sulphur trioxide is removed between the last two beds of catalyst (using step 4) in order to increase the yield.
- The reaction in step 3 is **exothermic**. So yield rises as temperature falls. But the catalyst will not work below 400°C, and it works better as the temperature rises. So 450°C is a compromise.
- To keep the temperature down to 450°C in step 3, heat must be removed from the catalyst beds. So pipes of cold water are coiled around them to carry heat away. The heat makes the water boil. The steam may be used to generate electricity for the plant, or sold to nearby buildings for heating.
- In step 4, the sulphur trioxide is dissolved in concentrated acid rather than water, because water would cause a thick, dangerous mist of acid to form.

Pressure in the Contact process

In step 3:

$$2SO_2 + O_2 \rightleftharpoons 2SO_3$$

3 molecules   2 molecules

- So increasing the pressure will increase the yield of sulphur trioxide. (Page 169 explains why.)
- But in practice, the pressure is increased only a little (to less than two atmospheres).

Uses of sulphuric acid

Sulphuric acid is one of the world's most important chemicals. Around 160 million tonnes of it are made every year. It has thousands of uses in industry. Its main uses are in making:
- fertilisers such as ammonium sulphate
- paints, pigments, and dyestuffs
- fibres and plastics
- soaps and detergents.

It is also the acid used in car batteries.

Concentrated sulphuric acid

Concentrated sulphuric acid is a thick oily liquid. It is a **dehydrating agent**: it removes water from other substances. It removes the water of crystallisation from blue copper(II) sulphate crystals, giving the **anhydrous** salt:

$$CuSO_4 \cdot 5H_2O \;(s) \xrightarrow[-5H_2O \;(l)]{conc \; H_2SO_4 \;(l)} CuSO_4 \;(s)$$
blue → white

It can remove water from flesh too. So it must be handled with great care.

Concentrated sulphuric acid was added to two teaspoons of sugar – and this is the result. It dehydrated the sugar, leaving carbon behind. Think what it could do to flesh!

Dilute sulphuric acid

It turns blue litmus red, and shows the usual reactions of dilute acids:
1. acid + metal ⟶ metal salt + hydrogen
2. acid + metal oxide or hydroxide ⟶ metal salt + water
3. acid + metal carbonate ⟶ metal salt + water + carbon dioxide

Its salts are called **sulphates**. Reactions **2** and **3** are neutralisations. You can tell this because water is produced.

Note that the dilute acid should be made by adding the concentrated acid to water. And never the other way round – because so much heat is produced that the acid could splash out and burn you.

Sulphuric acid for drying gases

- Concentrated sulphuric acid can be used to dry damp gases. It removes the water vapour from them.
- For example, it is used to dry hydrogen chloride, when this gas is made in the lab. (Why might this gas in particular be damp?)
- But you would *not* use it to dry ammonia. Why not?

Questions
Unit 11.4 will help you answer these questions.
1. For making sulphuric acid, name: **a** the process **b** the raw materials **c** the catalyst
2. The reaction between sulphur dioxide and oxygen is *reversible*. What does that mean?
3. **a** Is the *breakdown* of sulphur trioxide to sulphur dioxide endothermic, or exothermic?
   b Will *more* sulphur trioxide break down, or *less*, as the temperature rises? Why?
4. **a** Suggest a reason why a catalyst is needed.
   b At 500°C, the catalyst makes sulphur trioxide form even faster. Why isn't this temperature used?
5. Explain how this helps to increase the yield of sulphur trioxide, in the Contact process:
   a Several beds of catalyst are used.
   b The sulphur trioxide is removed by dissolving it.
6. Identify two *oxidation* reactions in the manufacture of sulphuric acid.
7. Write down three uses of sulphuric acid.
8. Three reactions of dilute acids are listed above.
   a Write word equations for the reactions of zinc metal, zinc oxide (ZnO) and zinc carbonate (ZnCO$_3$) with dilute sulphuric acid.
   b Now write a balanced equation for each reaction in **a**.

16.5 Chlorine

What is chlorine?

Chlorine is a green-yellow gas. It is a non-metal. It is very reactive, so is not found on Earth as the free element. It occurs mainly as the compound **rock salt**, or sodium chloride.

How chlorine is made in industry

Chlorine is made in industry by the electrolysis of:
- molten sodium chloride, or
- **brine**, which is a concentrated solution of sodium chloride (page 138).

Some properties of chlorine

1. It is a greenish-yellow poisonous gas, with a choking smell.
2. It is heavier than air.
3. It is soluble in water. The solution is called **chlorine water**.
   It is acidic, because chlorine reacts with water to form *two* acids:

   $$Cl_2\,(g) + H_2O\,(l) \longrightarrow \underset{\text{hydrochloric acid}}{HCl\,(aq)} + \underset{\text{hypochlorous acid}}{HOCl\,(aq)}$$

   The hypochlorous acid slowly decomposes, giving off oxygen:

   $$2HOCl\,(aq) \longrightarrow 2HCl\,(aq) + O_2\,(g)$$

4. Chlorine water acts as a bleach. This is because the hypochlorous acid can lose its oxygen to other substances – it **oxidises** them. Many coloured substances lose their colour when oxidised.

5. Hydrogen burns in chlorine to form hydrogen chloride. The reaction can be explosive:

   $$H_2\,(g) + Cl_2\,(g) \longrightarrow 2HCl\,(g)$$

6. Chlorine combines with most metals, forming metal chlorides. For example, it combines with heated aluminium like this:

   $$2Al\,(s) + 3Cl_2\,(g) \longrightarrow 2AlCl_3\,(s)$$

A displacement reaction of chlorine

When chlorine gas is bubbled through a colourless solution of potassium bromide, the solution goes orange, because of this reaction:

$$\underset{\text{colourless}}{Cl_2\,(g)} + 2KBr\,(aq) \longrightarrow 2KCl\,(aq) + \underset{\text{orange}}{Br_2\,(aq)}$$

Bromine has been displaced. The half-equations for the reaction are:

chlorine: $Cl_2\,(g) + 2e^- \longrightarrow 2Cl^-\,(aq)$ (reduction)
bromide ion: $2Br^-\,(aq) \longrightarrow Br_2\,(aq) + 2e^-$ (oxidation)

So chlorine has acted as an oxidising agent. It has oxidised the bromide ion. Combining these half-equations gives the ionic equation for the reaction:

$$Cl_2\,(g) + 2Br^-\,(aq) \longrightarrow 2Cl^-\,(aq) + Br_2\,(aq)$$

This shows clearly that the oxidised substance has been displaced.

Chlorine was used as a weapon in World War I. This soldier is ready for a chlorine gas attack.

Remember!

Oxidation is ...
gain of oxygen
loss of hydrogen
loss of electrons

Reduction is ...
loss of oxygen
gain of hydrogen
gain of electrons

Why is chlorine so reactive?

Why is chlorine so reactive? Because its atoms are just one electron short of a full shell – so they have a strong drive to gain an electron.

In chlorine molecules, two chlorine atoms gain full shells by sharing electrons. But they 'prefer' to take an electron over completely. So in reaction **6** on the last page, chlorine reacts with aluminium in order to take electrons from it. The aluminium loses electrons: it is oxidised.

Chlorine acts as an oxidising agent, in order to gain electrons.

All the halogens are reactive, and act as oxidising agents, for the same reason.

Why does chlorine displace bromine?

Chlorine is *more* reactive than bromine because its atoms are smaller than bromine atoms – so the nucleus can attract an electron more strongly:

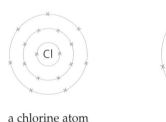

a chlorine atom
(three shells)

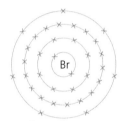

a bromine atom
(four shells)

Like all bleaching agents, chlorine kills bacteria and other germs. So it is added to the water in swimming pools.

So, in a competition for electrons, chlorine wins. It oxidises bromide ions, displacing bromine. All the halogens behave in the same way:

A more reactive halogen will always displace a less reactive halogen from solutions of its compounds. This is a redox reaction.

Uses of chlorine

- to make the plastic polyvinyl chloride (PVC)
- to make hydrochloric acid
- to make solvents such as tetrachloroethane (for dry cleaning)
- in making medical drugs, bleaches, disinfectants, and insecticides
- to sterilize drinking water, and water in swimming pools.

Group 7

chlorine
bromine
iodine

size of atoms ↓   reactivity and oxidising power ↑

Questions

1. Explain why a solution of chlorine in water is:
   a acidic   b able to bleach things
2. Chlorine reacts quite violently with heated aluminium. Show that this is a redox reaction, and say what is oxidised, and what is reduced.
   (See if you can use the term *oxidation state* in your answer. Look back at Unit 7.4.)
3. a When chlorine gas is bubbled through a colourless solution of potassium iodide, the solution turns red. Give a reason for the colour change.
   b Write an equation for the reaction.
   c What is being: i oxidised? ii reduced?
   d It is a d_____ reaction. What is the missing word?
4. Which uses of chlorine have benefited *you*?

16.6 Carbon and its compounds

Carbon

Some carbon occurs as the free element, in the Earth's crust, in the form of **diamond** and **graphite**. Both of them contain only carbon atoms, but they are quite different. They are called **allotropes** of carbon. Diamond is a very hard, clear solid. Graphite is a dark and greasy solid.

Carbon dioxide

The gas carbon dioxide (CO_2) occurs naturally in air. It is also a product in all these reactions:

1 Carbon compounds burning in plenty of air. For example, when natural gas (methane) burns in plenty of air, the reaction is:

$$CH_4\,(g) + 2O_2\,(g) \longrightarrow CO_2\,(g) + 2H_2O\,(l)$$

2 The reaction between glucose and oxygen, in your body cells:

$$C_6H_{12}O_6\,(aq) + 6O_2\,(g) \longrightarrow 6CO_2\,(g) + 6H_2O\,(l)$$

This is called **respiration**. You breathe out the carbon dioxide.

3 The reaction between dilute acids and carbonates. For example between hydrochloric acid and marble chips (calcium carbonate):

$$CaCO_3\,(s) + 2HCl\,(aq) \longrightarrow CaCl_2\,(aq) + CO_2\,(g) + H_2O\,(l)$$

Charcoal is a form of graphite. It is made by heating coal or wood in just a little air.

Properties of carbon dioxide

1 It is a colourless gas, with no smell.
2 It is much heavier than air.
3 Things will not burn in it. We say it **does not support combustion**.
4 It is slightly soluble in water, forming carbonic acid, H_2CO_3. This is a weak acid (page 116).

Uses of carbon dioxide

1 It is used in fire extinguishers. (Why?)
2 It is put in soft drinks, to make them fizzy. It is bubbled into them under pressure, to make more dissolve. When the bottles are opened it escapes, causing the 'fizz'.
3 Solid carbon dioxide (dry ice) is used to keep food frozen.

The 'fizz' in this soft drink is caused by carbon dioxide escaping.

A test for carbon dioxide

This uses lime water, a solution of calcium hydroxide in water.
When carbon dioxide is bubbled into lime water, the lime water at first turns milky, because a fine white precipitate of calcium carbonate forms:

$$Ca(OH)_2\,(aq) + CO_2\,(g) \longrightarrow CaCO_3\,(s) + H_2O\,(l)$$

But when more carbon dioxide is bubbled through, the precipitate disappears again. This is because it reacts with the carbon dioxide to form calcium hydrogen carbonate, which is soluble:

$$CaCO_3\,(s) + CO_2\,(g) + H_2O\,(l) \longrightarrow Ca(HCO_3)_2\,(aq)$$

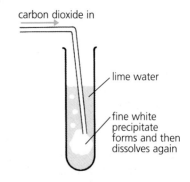

The test for carbon dioxide: first the lime water goes milky, then it goes clear again.

Carbon monoxide

Carbon monoxide (CO) forms when carbon compounds burn in too little oxygen. For example, for methane:

$$2CH_4\,(g) + 3O_2\,(g) \longrightarrow 2CO\,(g) + 4H_2O\,(l)$$

It is a deadly poisonous gas, with no smell. It binds to the haemoglobin in red blood cells, and prevents it from carrying oxygen around the body. So victims die from oxygen starvation.

Gas-fuelled water heaters and boilers should be checked regularly. Every year, hundreds of people are killed by carbon monoxide from faulty burners.

Carbonates

Carbonates are compounds that contain the carbonate ion, CO_3^{2-}. One example is calcium carbonate, $CaCO_3$, which occurs naturally as limestone, chalk and marble.

These are the main properties of carbonates:

1. They are insoluble in water – except for sodium, potassium, and ammonium carbonates, which are soluble.
2. They react with acids to form a salt, water, and carbon dioxide.
3. Most of them break down on heating, to an oxide and carbon dioxide:

$$\underset{\substack{\text{calcium carbonate}\\(\text{limestone})}}{CaCO_3\,(s)} \rightleftharpoons \underset{\substack{\text{calcium oxide}\\(\text{quicklime})}}{CaO\,(s)} + \underset{\text{carbon dioxide}}{CO_2\,(g)}$$

But the carbonates of sodium and potassium do not break down.

Organic compounds

Organic compounds are the carbon compounds found in, or derived from, living things. There are thousands of them – far more than all the other compounds put together. They include:

- the proteins, carbohydrates, and fats in your body
- the hundreds of different compounds in crude oil and coal
- the plastics and medical drugs made from the compounds in crude oil.

The study of these carbon compounds is called **organic chemistry**. The last two chapters in this book are all about organic chemistry.

Built up from organic compounds, and at least 20% carbon!

Questions

1. What are *allotropes*?
2. Name two allotropes of carbon, and give *two* properties of each, to show how different they are.
3. Gas boilers should be checked regularly, to make sure air flows through the burner properly. Explain why.
4. What is an *organic* compound?
5. Name an organic compound that contains just carbon and hydrogen, is a gas, and is used as a fuel.
6. Sodium carbonate is an *inorganic* compound.
   a. What do you think that means?
   b. Name another inorganic compound.
7. On heating, most carbonates break down to an oxide and carbon dioxide. Name this type of reaction.
8. Write an equation to show what happens when:
   a. lead(II) carbonate is heated
   b. limestone reacts with dilute hydrochloric acid

16.7 Limestone

Limestone: from sea creatures

Most of the creatures that live in the sea have shells or skeletons made of calcium carbonate. When they die, their remains fall to the sea floor. Slowly, over millions of years, pressure has turned the layers of shells and bones into limestone rock.

Over millions of years, powerful forces have then raised some sea beds upwards, draining and folding them to form mountains. That explains why limestone is found inland, miles from the sea!

Around 5 billion tonnes of limestone are quarried from the Earth's crust every year. This is what it is used for:

A limestone quarry in the Dominican Republic.

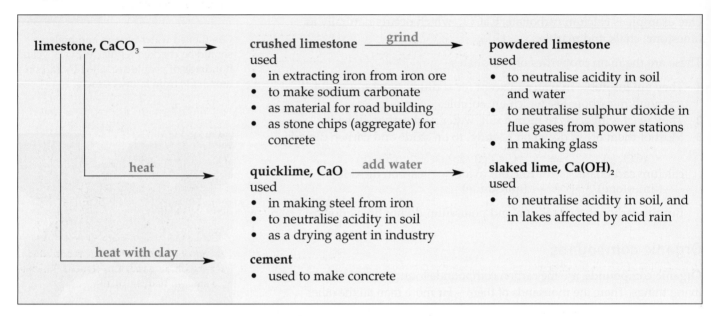

Substances made from limestone

Quicklime When limestone is heated, it breaks down to quicklime:

$$CaCO_3\,(s) \rightleftharpoons CaO\,(s) + CO_2\,(g)$$
calcium carbonate calcium oxide carbon dioxide
(limestone)   (quicklime)

This reaction is an example of **thermal decomposition**.

The drawing shows a quicklime kiln. The kiln is heated. Limestone is fed in one end. Quicklime comes out the other end.

The decomposition is reversible. So the calcium oxide and carbon dioxide *could* combine again. But a strong current of air is passed through the kiln, to carry the carbon dioxide away before it has a chance to react.

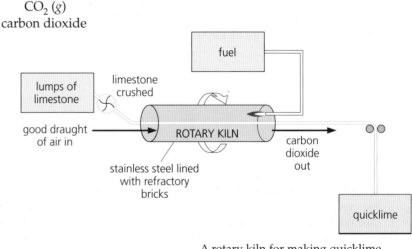

A rotary kiln for making quicklime.

Slaked lime This forms when water is added to quicklime. The reaction is exothermic, so the mixture hisses and steams. Conditions are controlled so that the slaked lime forms as a fine powder:

$$CaO\,(s) + H_2O\,(l) \longrightarrow Ca(OH)_2\,(s)$$
calcium oxide → calcium hydroxide
(quicklime) (slaked lime)

Slaked lime is used to neutralise acidity in soil, and in lakes.

Cement This is made by mixing limestone with clay, heating the mixture strongly in a kiln like the one below, adding gypsum (calcium sulphate), and grinding up the final solid to give a powder.

Plenty of cement in use!

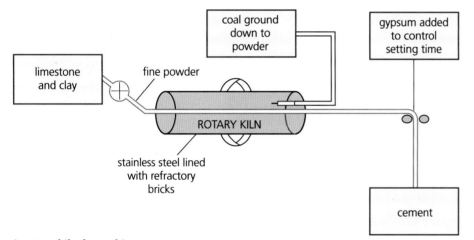

A rotary kiln for making cement.

The art of plastering. Plaster is a mixture of slaked lime, sand, and cement. It dries to a smooth hard surface.

Concrete For this, cement is mixed with **aggregate** (a mixture of small stone chips and sand) and water. As the concrete 'sets', the compounds in the cement bond to form a hard mass of hydrated crystals – that is, crystals with water molecules bonded inside them.

The setting process is exothermic. If the concrete sets too fast, the heat could crack it. It is also a big nuisance if you mix concrete, and it sets before you are ready to use it. And that is why gypsum is added to cement: it slows the setting process down.

Questions

1. How was limestone formed?
2. a How is quicklime made? Write the equation for the reaction.
   b Why is it important to remove the carbon dioxide?
   c How is the carbon dioxide removed, in a lime kiln?
3. How is slaked lime made? Write the equation.
4. Give two uses each for quicklime and slaked lime.
5. a What is *cement*, and how is it made?
   b Why is gypsum added to it?
6. Which ingredients are mixed, to make concrete?

Questions on Chapter 16

1. Look at each description below. Say whether it fits oxygen, or sulphur, or chlorine.
   a. quite soluble in water
   b. solid at room temperature
   c. reacts with metals to form oxides
   d. exists in two forms (like diamond and graphite)
   e. when damp, removes the colour from dyes
   f. burns in air with a blue flame
   g. reacts with hydrogen to form water
   h. a poisonous gas
   i. added to rubber to make it tough and strong
   j. relights a glowing splint
   k. colourless
   l. reacts with other elements to form chlorides
   m. burns to form an oxide which causes acid rain.

2. The following diagrams represent molecules of different substances that contain sulphur.
   a. Write the chemical formula for each substance, then name the substance.

   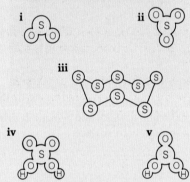

   b. How would you convert:
   substance **iii** into substance **i**?
   substance **i** into substance **ii**?
   substance **i** into substance **v**?
   substance **ii** into substance **iv**?

3. Oxygen reacts with many different elements. Combustion experiments are carried out on some elements, using gas jars of oxygen.
   a. Copy and complete this table of results.

Element	What you would see in the gas jar
iron	
carbon	
phosphorus	
magnesium	
copper	
sulphur	

   b. Which element does not need heating?
   c. Which element is the least reactive?
   d. In the case of phosphorus, the product is soluble in water. Name one of the other products that is soluble in water.
   e. Which of the products give acidic solutions?
   f. Which of the elements form basic oxides?

4. Sulphuric acid is made by the Contact process. The first stage is to make sulphur trioxide, like this:
   $$2SO_2\,(g) + O_2\,(g) \rightleftharpoons 2SO_3\,(g)$$
   The energy change in the reaction is -97 kJ/mol
   a. Name the catalyst used in this reaction.
   b. Is the reaction exothermic, or endothermic?
   c. What are the reaction conditions for making sulphur trioxide?
   d. Will the yield of sulphur trioxide increase, decrease, or stay the same, if the temperature is raised? Explain your answer.
   e. Describe how sulphur trioxide is changed into concentrated sulphuric acid.

5. Below is a flow chart for the Contact process:

   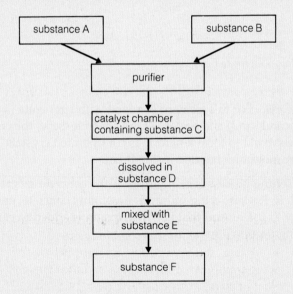

   a. Name the substances **A**, **B**, **C**, **D**, **E**, and **F**.
   b. Why is a catalyst used?
   c. Write a chemical equation for the reaction that takes place on the catalyst.
   d. The production of substance **F** is very important. Why? Give three reasons.
   e. Copy out the flow chart, and write in the full names of the different substances.

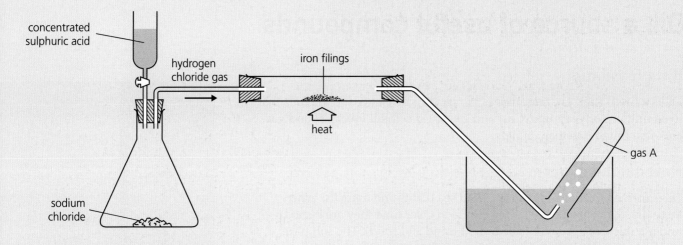

6 a Copy the diagram below. Then fill in:
   i the common names of the substances
   ii their chemical formulae

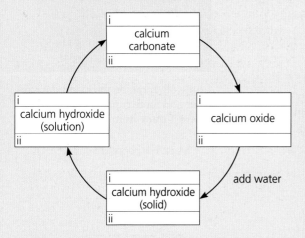

   b Beside each arrow say how the change is carried out. One example is shown.
   c Give three reasons why limestone is an important raw material.

7 Limestone is calcium carbonate, $CaCO_3$. It is quarried on a huge scale.
   a Which elements does it contain?
   b Much of the quarried limestone is turned into quicklime (CaO) for the steel industry.
      i What is the chemical name for quicklime?
      ii Describe how it is made from limestone.
   c Powdered limestone is used to improve the water quality in acidified lakes.
      i How might the water in the lakes have become acidified?
      ii How does limestone help it?
      iii Why is the limestone used in powdered form, rather than as lumps? See if you can think of more than one reason.
   d List other important uses of limestone.

8 The apparatus above was used to investigate the reaction between hydrogen chloride and iron.
   a Write the formulae for the two chemicals from which hydrogen chloride is made.
   b The reaction between iron and hydrogen chloride is exothermic. What would you see in the combustion tube?
   c Gas A burns with a squeaky pop. What is it?
   The iron compound that forms is iron(II) chloride.
   d For that compound, give the oxidation state of:
      i the iron   ii the chlorine
   e In a formula, the oxidation states add up to zero. Use this idea to write the correct formula for iron(II) chloride.
   f i Write the equation for the reaction above.
      ii Explain why the reaction is a redox reaction.
      iii Write the half-equation for the iron.

9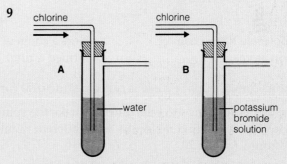

Chlorine gas is passed through two solutions, as shown above.
   a What will you *see* in each test tube?
   b What chemical(s) are formed in each tube?
   c Write a chemical equation for the reaction in each test tube.
   d i Which of the reactions is a displacement?
      ii Write the ionic equation for this reaction.
      iii Identify the oxidising and reducing agents.
   e Write a word equation for another displacement reaction, using chlorine.

17.1 Oil: a source of useful compounds

It's a fossil fuel

Oil is a **fossil fuel**. That means it is the remains of plants and animals that lived millions of years ago. Coal and gas are also fossil fuels. Oil and gas are often found together.

How did the oil form?

Oil is formed from the remains of dead sea plants and animals. When these die, their remains fall to the sea floor. Over time they are buried under a thick, heavy layer of **sediment** (sand and mud). The soft parts can't rot away, because there is no oxygen down there. Instead, they slowly turn into **oil** and **gas**, which gather in wells.

Later, movements in the Earth's crust may cause the sea floor to rise. The water drains away, turning the sea floor into land. The oil and gas are still there. But now they are under the ground, rather than under water. That is why you find oil wells on land *and* at sea.

Crude oil. It can be thick and dark or paler and more runny, depending on where it came from.

What is in crude oil?

Crude oil is a smelly mixture of hundreds of different compounds. They are **organic compounds**, which means they started off in living things. Most are also **hydrocarbons** – they contain *only* carbon and hydrogen.

Compare these hydrocarbon molecules:

This molecule has a chain of 5 carbon atoms.

Here a chain of 6 carbon atoms has formed a ring.

Here 6 carbon atoms form a **branched** chain.

A formula drawn out like these is called a **structural formula**.

Carbon atoms are joined to each other, to make the spine of each molecule. In crude oil you will find molecules with different numbers of carbon atoms, from 1 to over 70.

It's black gold

The compounds in crude oil have many different uses. They are so useful that over 13 billion litres of oil are used up every day!

In particular, most of the compounds can be used as fuels – which we burn to give energy – or turned into fuels. Over a third of the world's energy needs are met by oil.

Since it is so useful, countries that do not have oil are prepared to pay a lot of money for it. And countries that have oil to sell become wealthy.

Almost a half of the world's known oil deposits are in the Middle East. Saudi Arabia has around one-fifth of known deposits.

How we use oil

This shows the three main uses of oil.

Around half the oil pumped from oil wells is used for transport. It provides the fuel for cars, trucks, planes, and ships. You won't get far without it!

Most of the rest is burned for heat, in factories, homes and power stations. In a power station (above), the heat is used to turn water to steam, to drive turbines.

A small % is used as the starting chemicals to make many other things: plastics, shampoo, paint, thread, fabric, detergents, makeup, medical drugs, and more.

Many of the things you use every day were probably made from oil. Toothbrush, comb, and shampoo just for a start!

A non-renewable resource

Crude oil is still forming, very slowly, under the oceans. But we are using it up much faster than it can form, which means it will run out one day.

So oil is called a **non-renewable** resource.

It is hard to tell when it will run out. But at the present rate of use, some experts say that our oil reserves will last for about 40 more years.

What will we do then?

A platform for pumping oil from a well under the North Sea.

Questions
1. Why is crude oil called a fossil fuel?
2. What is a *hydrocarbon*?
3. What is in crude oil?
4. Explain why oil is such a valuable resource.
5. Oil is called a *non-renewable* resource. Why?
6. What do you think we will do for fuel when oil runs out?

17.2 Separating oil into fractions

Refining oil

Crude oil contains *hundreds* of different hydrocarbons. A big mixture of different compounds is not very useful. So the first step is to separate the oil into groups of compounds that have molecules of similar size. This is called **refining** the oil. It is carried out by **fractional distillation**.

Oil refining in the lab

The apparatus on the right can be used to refine oil in the lab.

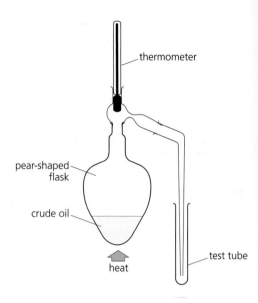

1. As you heat the oil, the compounds start to evaporate. The ones with smaller lighter molecules go first, since it takes less energy to free these from the liquid.
2. As the hot vapours rise, so does the thermometer reading. The vapours condense in the cool test tube.
3. When the thermometer reading reaches 100 °C, replace the first test tube with an empty one. The liquid in the first test tube is your first **fraction** from the distillation.
4. Collect three further fractions in the same way, replacing the test tube at 150 °C, 200 °C, and 300 °C.

Comparing the fractions

Now compare the fractions – how runny they are, how easily they burn, and so on. You can burn samples on a watch glass, like this:

fraction 1	fraction 2	fraction 3	fraction 4

It catches fire easily. The flame burns high, which shows that the liquid is **volatile** – it evaporates easily.	This catches fire quite easily. The flame burns less high – so this fraction is less volatile than fraction 1.	This seems less volatile than fraction 2. It does not catch fire so readily or burn so easily – it is not so **flammable**.	This one does not ignite easily. You need to use a wick to keep it burning. It is the least flammable of the four.

This table summarizes all the trends you will notice:

Fraction	Boiling point range	How easily does it flow?	How volatile is it?	How easily does it burn?	Size of molecules
1	up to 100°C	very runny	volatile	very easily	small
2	100°C to 150°C	runny	less volatile	easily	
3	150°C to 200°C	not very runny	even less volatile	not easily	
4	200°C to 300°C	viscous (thick and sticky)	least volatile	only with a wick	large

What those results tell you

Those results show that, the larger the molecules in a hydrocarbon:

- the higher its boiling point will be
- the less volatile it will be
- the less easily it will flow (or the more viscous it will be)
- the less easily it will burn.

Fractional distillation in the oil refinery

In the refinery, the fractional distillation is carried out in a tower that is kept very hot at the base, and cooler towards the top. Look at the drawing below.

Crude oil is pumped in at the base. The compounds start to boil off. Those with the smallest molecules boil off first, and rise to the top of the tower. Others rise only part of the way, depending on their boiling points, and then condense.

The table shows the fractions that are collected.

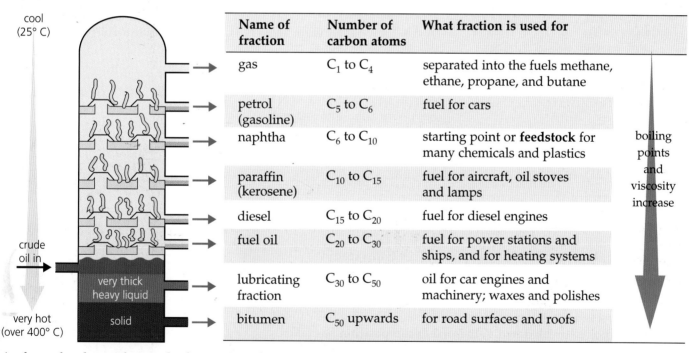

Name of fraction	Number of carbon atoms	What fraction is used for
gas	C_1 to C_4	separated into the fuels methane, ethane, propane, and butane
petrol (gasoline)	C_5 to C_6	fuel for cars
naphtha	C_6 to C_{10}	starting point or **feedstock** for many chemicals and plastics
paraffin (kerosene)	C_{10} to C_{15}	fuel for aircraft, oil stoves and lamps
diesel	C_{15} to C_{20}	fuel for diesel engines
fuel oil	C_{20} to C_{30}	fuel for power stations and ships, and for heating systems
lubricating fraction	C_{30} to C_{50}	oil for car engines and machinery; waxes and polishes
bitumen	C_{50} upwards	for road surfaces and roofs

As the molecules get larger, the fractions get less runny, or more viscous: from gas at the top of the tower to solid at the bottom. They also get less flammable. So the last two fractions in the table are not used as fuels.

Questions
1. Which two opposite processes take place during fractional distillation?
2. A group of compounds collected together during fractional distillation is called a ____?
3. What does it mean? **a** volatile **b** viscous
4. List four ways in which the properties of different fractions differ.
5. Name the oil fraction that: **a** is used for petrol **b** has the smallest molecules **c** is the most viscous **d** has molecules with 20 to 30 carbon atoms

17.3 Cracking hydrocarbons

After fractional distillation ...

Crude oil is separated into fractions by fractional distillation.
But that's not the end of the story. The fractions all need further treatment before they can be used.

1. They contain impurities – mainly sulphur compounds. If these are left in fuel, they will burn to form harmful sulphur dioxide gas.
2. Some fractions are separated further into single compounds, or smaller groups of compounds. For example the gas fraction is separated into methane, ethane, propane, and butane.
3. Part of a fraction may be **cracked**. Cracking breaks molecules down into smaller ones.

Cracking a hydrocarbon in the lab

The experiment below was carried out using a hydrocarbon from oil. The product is a gas, collected over water in the inverted test tube:

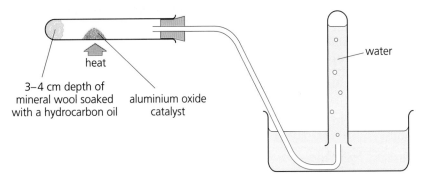

A heap of sulphur that has been removed from oil, outside the oil refinery.

	The reactant	The product
Appearance	thick colourless liquid	colourless gas
Smell	no smell	pungent smell
Flammability	difficult to burn	burns readily
Reactions	few chemical reactions	many chemical reactions

So the product is quite different from the reactant. Heating has caused the hydrocarbon to break down. A **thermal decomposition** has taken place. Note that:

- the reactant had a high boiling point and was not flammable – which means it had large molecules, with long chains of carbon atoms.
- the product has a low boiling point and is very volatile – so it must have small molecules, with short carbon chains.
- the product must also be a hydrocarbon, since nothing new was added during the reaction.

So the molecules of the starting hydrocarbon have been cracked.
And since the product is more reactive, it could be a useful chemical.

Cracking in the oil refinery

In an oil refinery, cracking is carried out in much the same way. The long-chain hydrocarbon is heated to vaporize it, and the vapour is usually passed over a hot catalyst. Thermal decomposition takes place. When a catalyst is used, it is called **catalytic cracking**.

Why cracking is so important

1. Cracking helps you to make the best use of your oil. Suppose you have plenty of the naphtha fraction, but not enough petrol. You can crack some naphtha to get molecules the right size for petrol.
2. Cracking *always* produces short-chain compounds with a carbon–carbon double bond. This bond makes the compounds **reactive**. So they can be used to make plastics and other substances.

Example 1: Cracking the naphtha fraction This is the kind of reaction that takes place when you crack naphtha:

A cracking plant at an oil refinery in the UK.

The decane has been broken down into smaller molecules. Look at the propene and ethene molecules. They have carbon–carbon double bonds. They are called **alkenes**, and are very reactive.

Example 2: Cracking ethane Ethane already has very short molecules – but even it can be cracked, to give ethene:

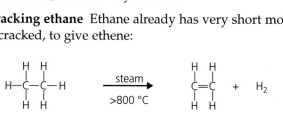

The hydrogen from this can be used to make ammonia – see page 216.

Questions
1. What happens during cracking?
2. Cracking is a *thermal decomposition*. Explain why.
3. Describe the usual conditions needed for cracking a hydrocarbon in an oil refinery.
4. What is *always* produced in a cracking reaction?
5. Explain why cracking is so important.
6. a A straight-chain hydrocarbon has the formula C_5H_{12}. Draw a diagram of one of its molecules.
   b Now show what might happen when the compound is cracked.

17.4 The alkanes

What are alkanes?

There are *thousands* of organic compounds, obtained from oil and other sources. They can be put into families of compounds with similar properties.

The **alkanes** are the simplest family.

The alkanes are hydrocarbons: they contain only carbon and hydrogen. This table shows the first four members of the family:

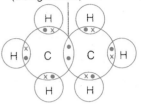

The bonding in ethane.

Name	methane	ethane	propane	butane
Formula	CH_4	C_2H_6	C_3H_8	C_4H_{10}
Structure of molecule	H–C–H with H above and below	H–C–C–H with H's	H–C–C–C–H with H's	H–C–C–C–C–H with H's
Number of carbon atoms in chain	1	2	3	4
Boiling point	–164 °C	–87 °C	–42 °C	–0.5 °C

→ boiling point increases with chain length

They form a homologous series

The compounds above fit the general formula C_nH_{2n+2}, where n is a number. So the family forms a **homologous series**. For methane n is 1, giving the formula CH_4. For ethane n is 2, giving C_2H_6, and so on.

In a homologous series:

- all the compounds fit the same general formula
- the chain length increases by 1 each time
- as the chain gets longer, the compounds show a gradual change in properties, as shown in the panel on the right.

In a homologous series …

As the chain gets longer:
- melting and boiling points increase
- viscosity increases
- flammability decreases.

Things to remember about the alkanes

1. They are found in oil and natural gas. The alkanes in oil have from 1 to 70 carbon atoms. Natural gas is mainly methane, with small amounts of ethane, propane and butane.
2. The first four alkanes are gases at room temperature. The next twelve are liquids. The rest are solids. Boiling points increase with chain length because attraction between the molecules increases – so it takes more energy to separate them.
3. In alkanes, each carbon atom forms four single covalent bonds. This makes alkanes quite unreactive, except for the reactions in **4** and **6**.
4. Alkanes burn well in a good supply of oxygen, forming carbon dioxide and water vapour, and giving out plenty of heat. So they are used as **fuels**. Methane burns the most easily. Like this:

$$CH_4\,(g) + 2O_2\,(g) \longrightarrow CO_2\,(g) + 2H_2O\,(l) + \text{heat energy}$$

The bottled gas for homes is butane.

5 But with less oxygen, alkanes undergo **incomplete combustion,** giving poisonous carbon monoxide, or even just carbon (soot):

$$2CH_4 (g) + 3O_2 (g) \rightarrow 2CO (g) + 4H_2O (l) + \text{less heat energy, or}$$

$$CH_4 (g) + O_2 (g) \rightarrow C (s) + 2H_2O (l) + \text{even less heat energy}$$

6 Alkanes react with chlorine in sunlight. For example:

methane + Cl₂ → chloromethane + HCl

A chlorine atom takes the place of a hydrogen atom. So this is called a **substitution** reaction. If there is enough chlorine, all four hydrogen atoms will be replaced, one by one. Look at the panel on the right.

The reaction can be explosive in sunlight. But it will not take place at all in the dark, because light energy is needed to break the bonds in the chlorine molecules, to start the reaction off.

Chlorine and methane

The hydrogen atoms can be replaced one by one:

chloromethane CH_3Cl
dichloromethane CH_2Cl_2
trichloromethane $CHCl_3$
tetrachloromethane CCl_4

All four are used as solvents. But they can cause health problems, so are being used less and less.

Isomers

butane

2-methylpropane

Look at these molecules. Both have the formula C_4H_{10}. But they have different structures. The first one has a **straight** or **unbranched** chain. The second is **branched**. They are called **isomers**.
Isomers have the same formula, but different structures.

The more carbon atoms in a compound, the more isomers it has. There are 75 isomers with the formula $C_{10}H_{22}$! And since isomers have different structures, they also have slightly different properties. For example:

- branched isomers have lower boiling points, because the branches stop the molecules getting close. So they cannot attract each other so strongly.

- branched isomers are also less flammable.

Boiling points of some isomers (°C)

C_4H_{10}
straight chain   0
branched chain −10

C_5H_{12}
straight chain 36
1 branch   28
2 branches 9.5

Questions

1 The fifth alkane in the alkane family is pentane.
   a What is its formula?   b Draw a pentane molecule.
   c In what state is pentane at room temperature?
2 Why are alkanes such as methane and butane used as fuels? Give *three* reasons.
3 a The reaction of chlorine with methane is called a *substitution* reaction. Why?
   b What special condition is needed, for this reaction?
4 Ethane reacts with chlorine, in a substitution reaction.
   a Draw the structural formula for each compound that can form, as the reaction proceeds.
   b Write the formula for each compound in **a**.
5 The compound C_5H_{12} has *three* isomers.
   a Draw the structures of these three isomers.
   b The panel above shows their boiling points. Match these to your drawings, and explain your choice.

17.5 The alkenes

The **alkenes** are a family of **hydrocarbons**. So they contain only carbon and hydrogen. But unlike the alkanes, they have a **double bond** between carbon atoms. Look at the first three alkenes:

Name	ethene	propene	but-1-ene
Formula	C_2H_4	C_3H_6	C_4H_8
Structure of molecule	H₂C=CH₂	H₂C=CH−CH₃	H₂C=CH−CH₂−CH₃
Number of carbon atoms in chain	2	3	4
Boiling point	−102 °C	−47 °C	−6.5 °C

← boiling point increases with chain length →

- The alkenes fit the general formula C_nH_{2n}. Do you agree? So they form a homologous series, and show a gradation in properties.
- The carbon–carbon double bond largely dictates how the alkenes react. So it is called the **functional group** of the alkenes.

Things to remember about the alkenes

1 They are made from alkanes by **cracking** (page 243).
2 Because they have a double bond, they are called **unsaturated**. (Alkanes have no double bonds, so are **saturated**.)
3 They are much more reactive than alkanes. This is because the double bond can break to form single bonds, and add on other atoms. For example, ethene reacts with hydrogen like this:

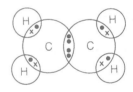

The bonding in ethene.

$$CH_2=CH_2 \text{ (g)} + H_2 \text{ (g)} \xrightarrow[\text{a catalyst}]{\text{heat, pressure}} CH_3-CH_3 \text{ (g)}$$
ethene → ethane

The hydrogen just adds on, so this is called an **addition** reaction.

4 Alkenes also do an addition reaction with steam, to form compounds called **alcohols**, that have an OH group. For example:

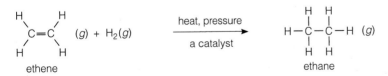

ethene → ethanol

5 The alkenes are highly flammable, and burn readily in air. For example when ethene burns in enough oxygen, the reaction is:

$$C_2H_4 \text{ (g)} + 3O_2 \text{ (g)} \longrightarrow 2CO_2 \text{ (g)} + 2H_2O \text{ (l)}$$

6 But note that an alkene has a higher % of carbon than the corresponding alkane does. So alkenes are more likely to undergo incomplete combustion, and to burn with a sooty flame.

7 Finally, alkene molecules can add on *to each other* to form compounds with very long carbon chains. These compounds are called **polymers**. The reaction is called **polymerisation**. For example ethene polymerises to form **poly(ethene)** or **polythene**. The reaction can be shown like this:

$$n \begin{pmatrix} H \\ \diagdown \\ C=C \\ \diagup \\ H H \end{pmatrix} \xrightarrow[\text{a catalyst}]{\sim 50\,°C,\ 3\ \text{or}\ 4\ \text{atmospheres pressure,}} \begin{pmatrix} H & H \\ | & | \\ -C-C- \\ | & | \\ H & H \end{pmatrix}_n$$

where *n* can be thousands. (Poly- means *many*.) See Unit 18.2 for more.

A fire at an oil well. Crude oil contains many unsaturated hydrocarbons, and these don't burn cleanly.

A test for unsaturation

How do you tell whether a hydrocarbon has double bonds? You test it with bromine! For example, ethene reacts like this:

$$\underset{\substack{\text{ethene}\\\text{(colourless)}}}{H_2C=CH_2}\,(g) + \underset{(\text{orange})}{Br_2\,(aq)} \longrightarrow \underset{\substack{\text{1,2-dibromoethane}\\\text{(colourless)}}}{BrCH_2-CH_2Br}\,(l)$$

The bromine is dissolved in water, giving orange **bromine water**. When this is shaken with *any* unsaturated hydrocarbon, an addition reaction occurs, and the orange colour disappears. But alkanes show no reaction. Why not?

Are there isomers in the alkene family?

The answer is yes. The chains can branch in different ways, *and* the double bond can be in different positions, as here:

but-1-ene   but-2-ene   2-methylpropene

All three compounds above have the formula C_4H_8, but each has a different structure. So they are isomers.

Questions

1 What is the general formula of the alkenes?
2 Write a formula for an alkene with 10 carbon atoms.
3 a Name the two simplest alkenes.
   b Now draw their structures.
4 What makes the alkenes behave so differently from the alkanes?
5 Propene reacts with hydrogen to form propane.
   a What is this type of reaction called?
   b Write an equation for the reaction.
6 Propene can *polymerise*. What does that mean?
7 a Propene is *unsaturated*. What does that mean?
   b Write an equation for its reaction with bromine.

17.6 The alcohols

What are alcohols?

The **alcohols** are a third family of organic compounds. They all contain the **OH** group. This table shows the first five members:

Formula	CH$_3$OH	C$_2$H$_5$OH	C$_3$H$_7$OH	C$_4$H$_9$OH	C$_5$H$_{11}$OH
Name	methanol	ethanol	propan-1-ol	butan-1-ol	pentan-1-ol
Structure of molecule	(structure)	(structure)	(structure)	(structure)	(structure)
Number of carbon atoms	1	2	3	4	5
Boiling point (°C)	65	78	97	117	137

- The alcohols form a homologous series. The compounds above all fit the general formula $C_nH_{2n+1}OH$. Check it out!
- The **OH group** is the functional group for the alcohols. Thanks to it, they all react in a similar way.
- Three of the names above have -1- in. This tells you that the OH group is attached to a carbon atom at one end of the chain.

Ethanol, an important alcohol

- Ethanol is the alcohol in alcoholic drinks.
- It is a good solvent. It dissolves many things that don't dissolve in water.
- It evaporates easily. That makes it a suitable solvent to use in things like glues, printing inks, perfumes, and aftershave.
- It is the starting point for many chemicals. For example, for liquids called **esters** that are used as flavours and scents. (See page 253.)
- It is used as a car fuel – added to, or instead of, petrol.

Ethanol is used as a solvent for perfume. Why?

Things to remember about ethanol

1. It is a clear, colourless liquid that boils at 78°C.
2. It is **miscible** with water – it mixes completely with it.
3. It burns well in oxygen, giving out plenty of heat:

   $C_2H_5OH\ (l) + 3O_2\ (g) \longrightarrow 2CO_2\ (g) + 3H_2O\ (l)$ + heat

4. Ethanol can be **dehydrated** to ethene, by passing its vapour over heated aluminium oxide, which acts a catalyst:

   (structural equation: ethanol $\xrightarrow[\text{Al}_2\text{O}_3,\ \text{heat}]{-\text{H}_2\text{O}}$ ethene)

   Compare this with reaction 4 on page 246. What do you notice?

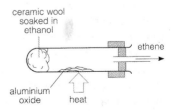

The dehydration of ethanol to give ethene. The gas is collected over water.

5 If ethanol is left standing in air, it will be **oxidised** with the help of bacteria, forming **ethanoic acid.** This has a sour taste, which is why wine left open in air goes sour:

```
    H   H                           H   O
    |   |                           |   ‖
H — C — C — OH    [O]→    H — C — C
    |   |                           |   \
    H   H                           H    O—H
    ethanol                         ethanoic acid
```

But ethanol is oxidised to the acid much faster by warming it with an oxidising agent such as acidified potassium dichromate solution. The dichromate ion is reduced in the reaction. (See page 111.) So the solution changes colour from orange to green.

Ethanol in alcoholic drinks

Alcoholic drinks are just a solution of ethanol. The ethanol is produced by the fermentation of grapes or other plant material. Alcohol has these effects on the body:

- Just one drink will affect a person's co-ordination and judgement. That is why drunk drivers are so dangerous.
- It makes people aggressive, so they get into fights easily.
- Heavy drinking causes depression and other mental disorders.
- It can lead to ulcers, high blood pressure, brain and liver damage, and cancer of the mouth, throat, and gullet. Heavy drinkers who also smoke are at even greater risk from these cancers.

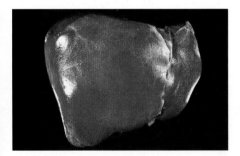

This is a healthy human liver …

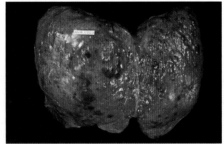

… and this is one damaged by alcohol.

Questions
1 Why do alcohols all react in a similar way?
2 Write the formula and draw the structure for:
   a methanol   b propan-1-ol.
3 a In *propan-1-ol*, what does the 1 show?
   b There is another isomer with the same formula as propan-1-ol. Draw its structure, and suggest a name.
4 See if you can draw an isomer for ethanol.
5 Look at the boiling points in the table on the opposite page. What do you notice? Suggest a reason.
6 Are alcohols *hydrocarbons*? Explain your answer.
7 a Write an equation for the combustion (burning) of:
      i methanol   ii methane
   b Compare the equations. What do you notice?
8 Suggest a way to turn ethanol into ethene – and then back to ethanol again. Give equations.
9 Give three uses of ethanol.
10 a Which alcohol is found in alcoholic drinks?
   b Give three harmful effects of this alcohol.

17.7 The manufacture of ethanol

Ethanol is growing important as a car fuel. So how is it made? In two ways. One starts with plant material, and the other with ethene, obtained from oil.

1 Ethanol by fermentation

Look what happens when you mix glucose (a sugar) and yeast in the absence of air:

$$C_6H_{12}O_6\,(aq) \xrightarrow[\text{in yeast}]{\text{enzymes}} 2C_2H_5OH\,(aq) + 2CO_2\,(g) + \text{energy}$$

glucose → ethanol + carbon dioxide

Enzymes in the yeast break the glucose down to ethanol and carbon dioxide. The process is called **fermentation**. It is exothermic.

Any substance that contains sugar, starch, or cellulose can be fermented. (Starch and cellulose break down to give glucose.) So ethanol can be made from sugar cane, corn, potatoes, wood, waste paper, and even grass. For example, you could make ethanol from corn, like this:

- Grind the corn, treat it with enzymes to break its starch down to glucose, and then add yeast.
- Leave the material to ferment for 20–60 hours.
- The fermented liquid contains a mixture of substances. Separate the ethanol from it by fractional distillation (page 20).

Methylated spirits is ethanol plus a little methanol and some colouring. Methanol is poisonous, and even low doses can cause blindness.

2 Ethanol from ethene

Ethanol can also be made by the hydration of ethene:

$$CH_2{=}CH_2\,(g) + H_2O\,(g) \xrightleftharpoons[\text{a catalyst (phosphoric acid)}]{570\,°C,\ 60\text{–}70\ \text{atm}} CH_3CH_2OH\,(l)$$

ethene → ethanol

- The ethene is obtained by cracking long-chain alkanes from oil.
- The hydration is exothermic and reversible. High pressure and a low temperature would give the best yield. But in practice, it is carried out at 570 °C, to give a decent rate of reaction.

Batch versus continuous

The two processes above are also very different *types* of process:

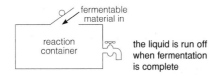

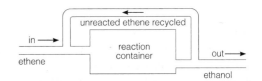

For fermentation, prepare a **batch** of material and leave it to ferment. Then when fermentation is complete, remove the liquid. It contains ethanol mixed with other things.

But the hydration of ethene is **continuous**: feed ethene in one end of the reaction container, and pure ethanol comes out the other end. Unreacted ethene is recycled.

Comparing the two methods

Ethanol by fermentation	Ethanol from ethene
Advantages	*Advantages*
• It uses renewable resources like corn and sugar cane.	• The reaction is fast.
• It's a good way to use waste organic material.	• The plant runs continuously. No time is wasted shutting down and restarting.
	• It doesn't need such large reaction vessels.
Disadvantages	• It gives pure ethanol.
• You need a lot of material to make even 1 litre of ethanol. So the fermentation tanks must be big, and you need lots of them.	*Disadvantages*
• Fractional distillation uses energy, so costs money.	• Ethene is made from oil, a non-renewable resource.
• Fermentation is slow, and the tanks must be shut down between batches.	• Energy is needed to make steam, and get the righ temperature and pressure for the reaction.
• When ethanol reaches a certain concentration, the yeast stops working. This limits the amount of ethanol you can get per batch.	• Under the reaction conditions, a lot of the ethene remains unreacted. You have to keep recycling it.

Which method is best to use?

- In countries where oil is cheap, they'll probably go for ethene.
- But fermentation is growing more important in countries like Brazil and the USA, that depend on imported oil. The more ethanol they get from fermentation, the less oil they need to import.
- Ethanol is a 'cleaner' fuel than petrol – it causes less pollution. The oxygen in it helps to promote complete combustion, so less carbon monoxide is produced. And it contains no sulphur compounds, so sulphur dioxide is not a problem.

Many countries now encourage the use of ethanol for cars, by charging less tax on it. And car companies make **flexible-fuel cars**, that will run on petrol, ethanol, or a mixture of the two.

Making ethanol in the drinks industry

The ethanol in alcoholic drinks is *always* made by fermentation. Grapes are used for wine, and barley for beer. After fermentation, the liquid is filtered off, and got ready for drinking. The mixture of things in the liquid gives the drink its taste and smell.

Gasohol is a mixture of petrol and ethanol. It is sold in several countries such as Thailand, Denmark and the USA.

Questions

1. a What is *fermentation*? b Write an equation for it.
2. Fermentation is a *batch process*. Explain.
3. The hydration of ethene is a *continuous process*. The reaction is *reversible*.
   a Write an equation for the hydration reaction.
   b Now explain the words in italics.
4. Give two advantages and two disadvantages of making ethanol:
   a by fermentation
   b from ethene.
5. What is *gasohol*?
6. Write down two advantages of ethanol as a fuel.

17.8 Carboxylic acids, and esters

The carboxylic acids

Now we look at a family of organic acids: the carboxylic acids.
Here are the first four members of the family:

Name of acid	methanoic	ethanoic	propanoic	butanoic
Formula	HCOOH	CH_3COOH	C_2H_5COOH	C_3H_7COOH
Structure of molecule	H–C(=O)OH	H–C(H)(H)–C(=O)OH	H–C(H)(H)–C(H)(H)–C(=O)OH	H–C(H)(H)–C(H)(H)–C(H)(H)–C(=O)OH
Number of carbon atoms	1	2	3	4
Boiling point	101 °C	118 °C	141 °C	164 °C

- Again, the family forms a homologous series. Its general formula is $C_nH_{2n+1}COOH$. Do you agree?
- This time the func.tional group is **COOH**, or the **carboxyl group**. Because of it, the carboxylic acids all behave in a similar way.
- But, as usual, the properties change a little with chain length.

Things to remember about the carboxylic acids

1. They can be made by the oxidation of alcohols. (See page 249.)
2. Their solutions turn litmus red (just as inorganic acids do).
3. Like inorganic acids, their solutions contain H^+ ions. That is what gives them their acidity. Ethanoic acid dissociates in water like this:

$$CH_3COOH\ (aq) \rightleftharpoons CH_3COO^-\ (aq) + H^+\ (aq)$$

ethanoic acid → ethanoate ion + hydrogen ion

or $CH_3COOH\ (aq) \rightleftharpoons CH_3COO^-\ (aq) + H^+\ (aq)$

This is a reversible reaction. At any given time, only some of the acid molecules are dissociated into ions. So carboxylic acids are **weak** acids. (See page 116).

3. Like inorganic acids, they react with metals, and bases, to form salts. Ethanoic acid reacts with sodium hydroxide like this:

ethanoic acid + sodium hydroxide → sodium ethanoate (a salt) + water

or $CH_3COOH\ (aq) + NaOH\ (aq) \longrightarrow CH_3COONa\ (aq) + H_2O\ (l)$

Vinegar: a solution of ethanoic acid in water.

Sodium ethanoate. It is used as a preservative, in the food industry. What will the sodium salt of propanoic acid be called?

5 Carboxylic acids react with alcohols to give compounds called **esters**. The reaction is reversible. Sulphuric acid acts a catalyst. Here the alcohol is shown with its OH group first, to help you see what is happening:

ethanoic acid + propanol ⇌ propyl ethanoate (an ester) + water (conc H_2SO_4)

this group is called an **ester** link

Propyl ethanoate smells of pears. (Many esters smell of fruits or flowers.)

Note that two molecules have joined to make a larger molecule, with the loss of a small molecule, water. This type of reaction is called a **condensation reaction**. You can write the reaction like this:

$$CH_3COOH\ (l) + C_3H_7OH\ (l) \underset{}{\overset{conc.H_2SO_4}{\rightleftharpoons}} CH_3COOC_3H_7\ (l) + H_2O\ (l)$$

Compare it with the reaction of ethanoic acid and sodium hydroxide in **4**. The reactions look similar. But an ester is a covalent compound, while sodium ethanoate is an ionic salt.

6 In naming esters, the alcohol part comes *first* in the name, but *second* in the formula (just like the sodium comes first in sodium ethanoate).

The smell and flavour of the apple come from natural esters. Artificial esters are used in the shampoo.

More about esters

You have probably tasted, and smelled, lots of esters!
- The tastes and smells of fruit and vegetables are due to natural esters.
- The oils and fats obtained from plants and animals are a mixture of many esters. Palm oil is an example.
- Thousands of 'artificial' esters are made in industry. They are added to soaps, shampoos, and perfumes for their smells. And to ice cream, foods and soft drinks for their flavour.

Here are some examples of artificial esters used in sweets and ice cream:

Alcohol	Acid	Ester	Smells of ...
pentanol	ethanoic acid	pentyl ethanoate	banana
pentanol	pentanoic acid	pentyl pentanoate	apple
methanol	benzoic acid	methyl benzoate	marzipan

All these oils are mixtures of natural esters. Are they mostly saturated or unsaturated? How can you tell?

Questions
1 What is the functional group of the carboxylic acids?
2 Copy and complete. (Page 119 may help!)
   carboxylic acid + metal → _____ + _____
   carboxylic acid + alkali → _____ + _____
   carboxylic acid + alcohol ⇌ _____ + _____
3 Carboxylic acids are *weak* acids. Explain why.
4 Draw a 'structural' equation for the reaction between ethanol and ethanoic acid, and name the products.
5 What is a *condensation reaction*?
6 Esters are important compounds in industry. Why?

Questions on Chapter 17

1 This diagram represents the process used to separate crude oil into different fractions.

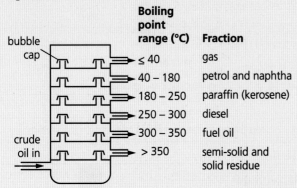

   a i   Name this process.
      ii It makes use of the fact that different compounds have different _____ _____. What are the missing words?
   b i   Is naphtha just one compound, or a group of compounds? Explain.
      ii Using the terms *evaporation* and *condensation*, explain how naphtha is produced.
      iii How do the bubble caps help?
   c   Give one use for each fraction obtained.
   d i   A hydrocarbon has a boiling point of 200 °C. In which fraction will it be?
      ii Are its carbon chains shorter, or longer, than those found in naphtha?
      iii Is it more viscous, or less viscous, than the compounds found in naphtha?

2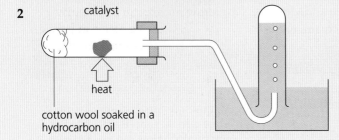

A hydrocarbon can be cracked in the lab using the apparatus above.
   a   What is *cracking*?
   b   Which two things are needed to crack hydrocarbons?
   c   The first tube of gas that is collected should be discarded. Why? (What else is in the heated tube?)
   d   Ethane, C_2H_6, can be cracked to give ethene, C_2H_4, and hydrogen. Write an equation for this reaction.

3 The saturated hydrocarbons form a homologous series with the general formula C_nH_{2n+2}.
   a   What is a *homologous series*?
   b   Explain what the term *saturated* means.
   c   Name the series described above.
   d i   Give the formula and name for a member of this series with two carbon atoms.
      ii Draw its structural formula.
   e i   Name a homologous series of *unsaturated* hydrocarbons, and write a general formula for it.
      ii Give the formula and name for a member of this series with two carbon atoms.
      iii Draw the structural formula for the compound.

4 Answer these questions about the alkanes.
   a   Which two elements do alkanes contain?
   b   Which alkane is the main compound in natural gas?
   c   After butane, the next two alkanes in the series are *pentane* and *hexane*. How many carbon atoms are there in a molecule of:
      i   pentane?   ii hexane?
   d   Give the formulae for pentane and hexane.
   e i   Draw the structural formulae for pentane and hexane (as in the table on page 244).
      ii Now draw an isomer for each.
   f   Is pentane solid, liquid, or a gas, at room temperature?
   g   Suggest a value for the boiling point of pentane, and explain your answer.
   h   Will pentane react with bromine water? Explain.
   i   Alkanes burn in a good supply of oxygen. Name the gases formed when they burn.
   j   Write a balanced equation for the complete combustion of pentane in oxygen.
   k i   Name a harmful substance formed during *incomplete* combustion of pentane in air.
      ii Write a balanced equation for this reaction.

5 Octane (C_8H_{18}) is a liquid hydrocarbon. It is generally unreactive, but does react with chlorine under certain conditions.
   a   Which type of hydrocarbon is octane?
   b   Why is octane generally unreactive?
   c i   What special condition is required for octane to react with chlorine?
      ii Which type of reaction occurs?
      iii Write one equation for the reaction.

6 Hex-1-ene is an unsaturated hydrocarbon. It melts at −140 °C and boils at 63 °C. Its empirical formula is CH_2. Its relative molecular mass is 84.
   a i   To which family of hydrocarbons does hex-1-ene belong?
      ii What is its molecular formula?
      iii What does the 1 in its name mean? (A bullet point about alcohols at the top of page 248 will give you a clue.)
      iv Draw the structural formula for it.
   b i   In what state is hex-1-ene at room temperature?
      ii Make a guess at the boiling point of hept-1-ene, the next member of the series.
   c i   Hex-1-ene reacts with bromine water. Write an equation to show this reaction.
      ii What is this type of reaction called?
      iii What would you *see* during the reaction?

7 Three hydrocarbons share the formula C_5H_{12}. Their boiling points are 36 °C, 28 °C, and 10 °C.
   a To which family of hydrocarbons do these compounds belong?
   b What name is given to different compounds that have the same molecular formula?
   c i   Draw the structural formulae for the three hydrocarbons.
      ii Now assign a boiling point to each, and explain your choice.
   d Would you expect the three hydrocarbons to have the same *chemical* properties? Explain.

8 When ethanol vapour is passed over heated aluminium oxide, a dehydration reaction occurs, and the gas ethene is produced.
   a Draw a diagram of suitable apparatus for carrying out this reaction in the lab.
   b What is meant by a *dehydration reaction*?
   c Write an equation for this reaction, using the structural formulae.
   d i   What will you see if the gas that forms is bubbled through bromine water?
      ii You will *not* see this if ethanol vapour is passed through bromine water. Why not?
   e In industry, ethanol is made by a reaction which is the reverse of this dehydration reaction. Steam and ethene are reacted together in the presence of phosphoric acid, at high temperature and pressure.
      i   Write an equation for this reaction.
      ii Which type of reaction is it?
      iii Why is the phosphoric acid used?
   f Which other way of manufacturing ethanol is becoming more important? Why?

9 Ethanol is a member of a homologous series.
   a Give two general characteristics of a homologous series.
   b i   To which homologous series does ethanol belong?
      ii What is the general formula for the series?
      iii What does *functional group* mean?
      iv What is the functional group in ethanol's homologous series?
   c Write down the formula of ethanol.
   d Ethanol reacts with sodium.
      i   Bubbles of gas form around the sodium. What gas is this?
      ii Write a balanced equation for the reaction.
   e i   Draw the structural formula for the fifth member of the series, pentan-1-ol.
      ii Draw the structural formula for an isomer of pentan-1-ol.
      iii Describe how pent-1-ene could be made from pentan-1-ol.
      iv Name the organic product formed when pentan-1-ol is oxidised using acidified potassium dichromate(VI).

10 Ethanoic acid is a member of the homologous series with general formula $C_nH_{2n+1}COOH$.
   a Name this series.
   b What is the functional group of the series?
   c Ethanoic acid is a *weak* acid. Explain what this means, using an equation to help you.
   d Ethanoic acid reacts with carbonates.
      i   What would you *see* during this reaction?
      ii Write a balanced equation for the reaction with sodium carbonate.
   e i   Name the member of the series for which $n = 3$, and draw its structural formula.
      ii Give the equation for the reaction between this compound and sodium hydroxide.

11 Ethanoic acid reacts with ethanol in the presence of concentrated sulphuric acid.
   a Name the organic product formed.
   b Which type of compound is it?
   c How could you tell quickly that it had formed?
   d What is the function of the sulphuric acid?
   e The reaction is *reversible*. What does this mean?
   f Write an equation for the reaction.
   g Name the two chemicals you would use to make methyl propanoate.
   h i   Reactions between organic acids and alcohols are important in industry. Why?
      ii Give three examples of things you buy that contain products from these reactions.

18.1 Introducing polymers

What is a polymer?

A polymer is any substance containing very large molecules, formed when lots of small molecules join together.

For example, look what happens when ethene molecules join:

This test tube contains the colourless gas ethene, C_2H_4. When ethene is heated to 50 °C, at a few atmospheres pressure, and over a special catalyst …

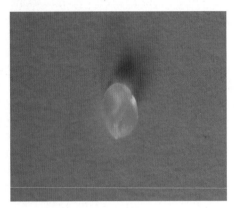

… it turns into a liquid that cools to a waxy white solid. This is found to contain very long molecules, made by the ethene molecules joining.

And it is really useful. It can be used to make toys, dustbins, tables and chairs, water pipes, buckets, crates, washing-up bowls and so on.

The reaction that took place is:

```
    H   H   H   H   H   H   H   H   H   H   H   H
     \ /     \ /     \ /     \ /     \ /     \ /
      C = C   C = C   C = C   C = C   C = C   C = C
     / \     / \     / \     / \     / \     / \
    H   H   H   H   H   H   H   H   H   H   H   H
```
ethene molecules (monomers)

↓ polymerisation

part of a polythene molecule (a polymer)
```
    H H H H H H H H H H H H
    | | | | | | | | | | | |
   —C—C—C—C—C—C—C—C—C—C—C—C—
    | | | | | | | | | | | |
    H H H H H H H H H H H H
```

The drawing shows six ethene molecules adding together. In fact thousands add together, giving molecules with very long chains. These very large molecules are called **macromolecules**.

A polymer is a substance made of macromolecules.

The polymer made from ethene is called **poly(ethene)** or **polythene**. Poly- means *many*. The reaction is called a **polymerisation**.

In a polymerisation reaction, thousands of small molecules join to give macromolecules. The product is called a polymer.

Synthetic polymers

Polythene is a **synthetic polymer**. *Synthetic* means it is made in a factory. Other synthetic polymers include nylon, Terylene, lycra, chewing gum, and plastics such as polystyrene and perspex. Hair gels and shower gels contain water-soluble polymers.

Hair gel: a water-soluable polymer. When you put it on, the water in it evaporates so the gel gets stiff.

Natural polymers

Polythene was first made in 1935. But for billions of years, nature has been busy making natural polymers. Look at these examples:

Starch is a polymer made by plants. The starch molecules are built from molecules of **glucose**, a sugar. We eat plenty of starch in rice, bread, and potatoes.

Plants also use glucose to make another polymer called **cellulose**. Cotton T-shirts and denim jeans are almost pure cellulose, made by the cotton plant.

Your skin, hair, nails, bones and muscles are mostly polymers, made of macromolecules called **proteins**. Your body builds these up from **amino acids**.

The wood in trees is about 50% cellulose. Paper is made from wood pulp, so this book is mainly cellulose. The polymer in your hair and nails, and in wool and silk, and animal horns and claws, is called **keratin**. The polymer in your skin and bones is called **collagen**.

So – you contain polymers, you eat polymers, you wear polymers, and you use polymers. Polymers play a big part in your life!

The reactions that produce polymers

All polymers, natural and synthetic, consist of macromolecules, formed by small molecules joined together.

But these macromolecules are not all made in the same way. There are two types of reaction: **addition polymerisation** and **condensation polymerisation**. You can find out more about these in the next two units.

Questions

1 What is:
   a   a macromolecule?   b   a polymer?
   c   a natural polymer?   d   a synthetic polymer?
   e   polymerisation?

2 Name the natural polymer found in:
   a   your hair   b   this book

3 Name at least three objects you own, that are made of polymers.

18.2 Addition polymerisation

Another look at the polymerisation of ethene

Here again is the reaction that produces polythene:

ethene molecules (monomers)

↓ polymerisation

part of a polythene molecule (a polymer)

The reaction can be shown in a short form like this:

$$n \begin{pmatrix} H & H \\ C=C \\ H & H \end{pmatrix} \xrightarrow{\text{heat, pressure, a catalyst}} \begin{pmatrix} H & H \\ -C-C- \\ H & H \end{pmatrix}_n$$

where n stands for a large number. It could be many thousands. The catalyst for the reaction is usually a mixture of titanium and aluminium compounds.

It's an addition reaction

The reaction above takes place because the double bonds in ethene break, allowing the molecules to add on to each other. So this is called **addition polymerisation**.
In addition polymerisation, double bonds in molecules break and the molecules add on to each other.

The monomer

The small starting molecules in an polymerisation are called **monomers**. So in the reaction above, ethene is the monomer.
For addition polymerisation, the monomers always have double bonds.

The chain lengths in polythene

In the polythene above, all the macromolecules are long chains of carbon atoms, with hydrogen atoms attached. So they are all similar. But they are *not* all identical. The chains are not all the same length. That is why we can't write an exact formula for polythene.

By changing the reaction conditions, the *average* chain length can be changed. But the chains will never be all the same length.

The relative atomic mass (M_r) of an ethene molecule is 28.
The *average* M_r of the macromolecules in a sample of polyethene can be 500 000 or more. In other words, when making polythene, at least 17 000 ethene molecules join, on average!

Pellets of polythene. Each pellet is enough to make ...

... one of these plastic bottles. The pellets are melted and the polythene is blown into a mould.

Making other polymers by addition

The polymers below are all made by addition polymerisation. Compare them:

The monomer	Part of the polymer molecule	The equation for the reaction
chloroethene (vinyl chloride) — $CH_2=CHCl$	poly(chloroethene) or polyvinyl chloride (PVC)	n stands for a large number!
tetrafluoroethene — $CF_2=CF_2$	poly(tetrafluoroethene) or Teflon	
phenylethene (styrene) — $C_6H_5CH=CH_2$	poly(phenylethene) or poly(styrene)	

Identifying the monomer

If you know the structure of the addition polymer, you can work out what the monomer was. Like this:

- Identify the repeating unit. (It has two carbon atoms side by side, in the main chain.) You could draw brackets around it.
- Then draw the unit, but put a double bond between the two carbon atoms. That is the monomer.

For example this is **poly(propene)**. The brackets show the repeating unit.

So this is the monomer. It is the alkene **propene**.

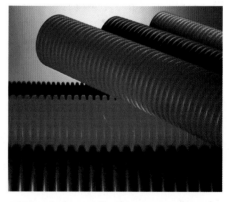

PVC is light and flexible so is widely used for hoses and water pipes, and as an insulating cover for electrical wiring.

Questions

1. Why was *addition polymerisation* given that name?
2. a What is a *monomer*?
   b Could methane (CH_4) be used as a monomer for addition polymerisation? Explain your answer.
3. It is not possible to give an exact formula for the macromolecules in polythene. Why not?
4. Draw a diagram to show the polymerisation of:
   a ethene   b chloroethene   c phenylethene
5. A polymer has the general formula shown on the right. Draw the monomer that was used to make it.

18.3 Condensation polymerisation

Condensation polymerisation

In addition polymerisation, there is only one monomer. Double bonds break, allowing the monomer molecules to join together.

But in condensation polymerisation:

- two *different* monomers join.
- double bonds do not break. Instead, the monomers join by eliminating (getting rid of) small molecules.

Let's look at two examples.

1 Making nylon

Below are the two monomers used in making nylon. We will call them A and B, for convenience:

A 1,6-diaminohexane

B hexan-1,6-dioyl chloride

Only the functional groups take part in the polymerisation reaction. So here we show the rest of the molecules as coloured blocks, to help you see what is happening:

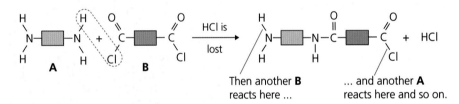

Then another **B** reacts here ... ... and another **A** reacts here and so on.

No double bonds break. Instead, single bonds break, and new single bonds form. The monomers are able to join to each other by eliminating a small molecule: hydrogen chloride.

The reaction continues at each end of A and B. Thousands of molecules join, giving a macromolecule. This shows just part of a macromolecule in nylon:

this linking group of atoms is called an **amide link**

Note the **amide** link where the monomers have joined. Because of this link, nylon is called a **polyamide**. (You will come across the same link later, in **proteins**.)

Nylon can be drawn into tough strong fibres that do not rot away. So it is used for thread, ropes, fishing nets, car seat belts, and carpets.

Making nylon in the school lab. **A** is dissolved in water. **B** is dissolved in an organic solvent that does not mix with water. Nylon begins to form as soon as the solutions meet in the beaker.

Nylon thread: tough, strong, great for flying kites.

2 Making Terylene

Like nylon, Terylene is made by condensation polymerisation, using two different monomers. This time we call them C and D:

C benzene-1,4-dicarboxylic acid

D ethane-1,2-diol

Once again, only the functional groups (the acid and alcohol groups) at each end of the molecules take part in the polymerisation reaction. So we can show the rest of the molecules as coloured blocks:

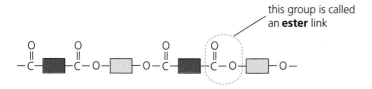

Then another **D** reacts here ... ... and another **C** reacts here and so on.

Fibres of nylon and Terylene are made by pumping the melted polymer through a spinneret, which is like a shower head. As it comes out through the holes, the polymer hardens into long threads.

This time, the monomers join by eliminating a water molecule. The reaction continues at each end of C and D. Thousands of molecules join up, giving a macromolecule. This shows just part of a macromolecule in Terylene:

this group is called an **ester** link

Note the **ester** link where the monomers have joined. Compare it with the esters you met on page 253. Because of this link, Terylene is called a **polyester**.

Terylene is used for shirts and other clothing, and for bedlinen. It is usually woven with cotton. The resulting fabric is more hard-wearing than cotton, and does not crease so easily. Terylene is also sold as polyester thread.

Shirts made from Terylene woven with cotton.

Questions

1. How many products are there, in a condensation polymerisation?
2. In what ways is condensation polymerisation different from addition polymerisation?
3. Draw a diagram to show the reaction that produces nylon. You can show the carbon chains as blocks.
4. a Nylon is called a *polyamide*. Why?
   b Make a drawing of just the polyamide linkage.
5. a Look back at the reaction that gives esters, on page 253. Now compare it with the reaction to make Terylene, above. What do you notice?
   b Draw the ester functional group.

261

18.4 Making use of synthetic polymers

Plastics are synthetic polymers

Synthetic polymers are usually called **plastics**. (*Plastic* means *can be moulded into shape without breaking*, and this is true of all synthetic polymers while they are being made.) But when they are used in fabrics, and for thread, we still call them synthetic polymers.

Most plastics are made from chemicals found in the naphtha fraction of oil (pages 241 and 243). They are usually quite cheap to make.

The properties of plastics

Most plastics have these properties:

1. They do not usually conduct electricity or heat.
2. They are unreactive. Most are not affected by air, water, acids or other chemicals. This means they are usually safe for storing things in, including food.
3. They are usually light to carry – much lighter than wood, or stone, or glass, or most metals.
4. They don't break when you drop them. You have to hammer most rigid plastics quite hard, to get them to break.
5. They are strong. This is because their long molecules are attracted to each other. Most plastics are hard to tear or pull apart.
6. They do not catch fire easily. But when you heat them, some soften and melt, and some char (go black as if burned).

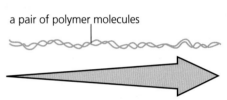

as molecules get longer, the force of attraction between them increases

Changing the properties

By choosing monomers and reaction conditions carefully, or mixing other chemicals with the monomers, you can make plastics with exactly the properties you want.

For example, look at how you can change the properties of polythene:

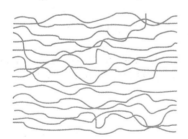

At about 50°C, 3 or 4 atmospheres pressure, and using a catalyst, you get long chains like these. They are packed close together so the polythene is quite **dense**.

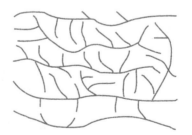

At about 200°C, 2000 atmospheres pressure, and with a little oxygen present, the chains will branch. Now they can't pack closely, so the polythene is far less dense.

So by choosing the right conditions, you can change the density of the polythene, and make it 'heavy' or 'light' to suit your needs.

The **high-density** polythene is hard and strong, which is why it is used for things like bowls and dustbins. The **low-density** polythene is ideal for things like plastic bags, and 'cling film' for wrapping food. Because the chains are not closely packed, you can see through it.

Uses for synthetic polymers

Given all those great properties, it is not surprising that plastics have thousands of uses. Here are some examples.

Polymer	Examples of uses
polythene	plastic bags and gloves, clingfilm (*low density*) mugs, bowls, chairs, dustbins (*high density*)
polychloroethene (PVC)	water pipes, wellingtons, hoses, covering for electricity cables
polypropene	crates, ropes
polystyrene	used as expanded polystyrene in fast-food cartons, packaging, and insulation for roofs and walls (to keep homes warm)
Teflon	coated on frying pans to make them non-stick, fabric protector, windscreen wipers, flooring
nylon	ropes, fishing nets and lines, tents, curtains
Terylene	clothing (especially mixed with cotton), thread

She is filling the spaces in the box with polystyrene chips. They will help to protect the contents, on the way to the customer.

Teflon is a very slippery polymer. It is used to coat frying pans and woks, to stop food sticking to them during cooking.

Questions

1 Look at the properties of plastics, on page 262. Which *three* properties do you think are the most important for:
   a plastic bags?   b kitchen bowls and utensils?
   c water pipes?   d fishing nets?
   e hair dryers?   f polystyrene fast-food containers?
2 What is *low-density* polythene, and how is it made?
3 Teflon is used to coat frying pans, to make them non-stick. So what properties do you think Teflon has? List them.
4 a What is expanded polystyrene?
   b Give three uses of this material.

5 a Now make a table with these headings:

Item	Properties of the plastic in it	Disadvantages of this plastic	Name of this plastic

   b i Fill in the first column of your table, giving three or four plastic items you own or use.
   ii In the second column, give the properties you observe, for that plastic. (You are a scientist!) For example is the plastic rigid? Or flexible?
   iii In the third column give any disadvantages you notice, for this plastic.
   iv Then see if you can name it. If you can, well done!

18.5 Plastics: here to stay?

Plastics: the problem

There were only a few plastics around before the 1950s. Since then, dozens of new ones have been developed, and more are on the way.

Now it is hard to imagine life without them. They are used in homes, schools, shops, offices, hospitals, and factories. And under the ground, for pipes that carry water, oil, gas, sewage, and electricity cables.

One big reason for their success is their unreactivity. But this is also a problem. They do not break down or rot away. Most of the plastics thrown out in the last 50 years are still around – and may still be here 50 years from now. A mountain of waste plastic is growing.

Polythene: the biggest problem

Polythene is the biggest problem. It is the most-used plastic in the world, thanks to its use in plastic bags and food packaging. Around 5 trillion polythene bags are made every year. (That's 5 million million.) Most are used only once or twice, then thrown away.

In many places, rubbish is collected and brought to **landfill sites**. Plastic bags just fill up these sites. In other places, rubbish is not collected. So the plastic bags lie around and cause many problems. For example:

- they choke birds, fish and other animals that try to eat them. Or they fill up the animals' stomachs so that they can't eat proper food, and starve to death. (Animals can't digest plastics.)
- they clog up drains, and sewers, and cause flooding.
- they collect in rivers, and get in the way of fish. Some river beds now contain a thick layer of plastic.
- they blow into trees and onto beaches. So the place looks a mess. Tourists are put off – and many places depend on tourists.

Because of these problems, plastic bags have been banned in many places. For example in Bangladesh, Rwanda and several states in India.

Her polythene shopping bags could still be lying undecomposed, in landfill sites, 50 years from now.

Nice beach?

Cattle – and vultures – at risk from plastic.

Recycling plastics

Some waste plastics do get reused. For example:

- some are melted down and made into new plastic bags, and things like soles for shoes, and fleeces.
- some are melted and their long chains cracked, to make small molecules that can be polymerised into new plastics.
- some are burned, and the heat is used to produce electricity.

But only a small % of waste plastic is reused in these ways. One problem is the many different types of plastic. These must be separated before reusing them, but that is not easy to do. Burning also poses problems, since some plastics give off poisonous gases.

Degradeable plastics

Degradeable polythene is already here. Some is **biodegradeable**: it contains additives such as starch that bacteria can feed on. Some is **photodegradeable**: it contains additives that break down in sunlight. The result is that the polythene breaks down into tiny flakes.

The amount of additive can be varied for different purposes – for example to make rubbish sacks that will break down within weeks.

Bio-polymers: the future?

In future, the plastics used in your mobile phone, toothbrush, trainers and back pack could be **bio-polymers** – grown inside plants, or made by bacteria living in tanks, instead of from oil compounds.

For example, one strain of bacteria is able to feed on sugar from crops such as corn, to produce polyesters. And some plants have already been genetically modified to produce plastics in their cells. When the plants are harvested, the plastic is extracted using a solvent. Then the solvent is evaporated.

Work on bio-polymers is still at an early stage. But when oil runs out, we may be glad of bio-polymers. And they have two advantages for the environment: they are a renewable resource, and biodegradeable.

A degradeable plastic bag: it will break down along with the leaves.

Questions

1. Describe some negative effects of plastics on the environment.
2. Polythene is responsible for most of the environmental problems caused by plastics. Explain why.
3. Explain what these are, in your own words:
   a photodegradable polythene b bio-polymers
4. See if you can come up with some ideas, to help prevent pollution by plastic bags.

18.6 The macromolecules in food (I)

What's in your food?

No matter what kind of food you eat, or where you eat it, its main ingredients are the same: **carbohydrates**, **proteins** and **fats**. All three are made up of macromolecules. And plants can make them all.

Plants: the polymer factories

1 Plants take in carbon dioxide from the air and water from the soil, as raw materials.
2 Using energy from sunlight, and **chlorophyll** as a catalyst, they turn them into glucose, in a process called **photosynthesis**:

$$6CO_2\,(g) + 6H_2O\,(l) \longrightarrow C_6H_{12}O_6\,(s) + 6O_2\,(g)$$
carbon dioxide   water   glucose (a sugar)   oxygen

3 Then they turn the glucose molecules into macromolecules of **starch** and **cellulose**, by polymerisation. These natural polymers are called **carbohydrates**. Plants use cellulose to build stems and other structures. They use starch as an energy store.
4 Using glucose, and minerals from the soil, they also produce macromolecules of **proteins** and **fats**.

Enzymes in plant cells act as catalysts, for the reactions in 3 and 4.

A plant: a natural chemical factory at work.

From plants to you

This is how the macromolecules from plants reach you:

1 Animals eat plants and plant seeds. They digest them, and build their own carbohydrates, proteins, and fats from them.

2 You eat animal carbohydrates, proteins, and fats, in animal produce such as eggs, milk, and cheese.

3 You eat them in meat and fish too.

4 You also eat plant parts. For example maize, rice, potatoes and other vegetables, and fruit.

5 And things like bread and pasta that are made from plant products.

6 During digestion, you break the macromolecules in food down to their building blocks.

You then use these for energy, or to build up the carbohydrates, proteins and fats that your own body needs.

We will now look more closely at the carbohydrates, proteins, and fats that plants produce, in this unit and the next one.

Carbohydrates

Carbohydrates contain just carbon, hydrogen and oxygen. Glucose is called a **simple carbohydrate**. It is also called a **monosaccharide**, which means *a single sugar unit*.

The structure of a glucose molecule is shown on the right. Now look how glucose molecules join:

A molecule of glucose, $C_6H_{12}O_6$.

| 1 We can draw a glucose molecule like this, showing only the two groups that react:

HO—☐—OH | 2 Two glucose molecules can join like this, giving maltose, a **disaccharide**:

HO—☐—OH HO—☐—
↓ water molecule eliminated
HO—☐—O—☐—OH | 3 Hundreds or thousands of glucose molecules join like this, to give **starch**, a **complex carbohydrate**. It is also called a **polysaccharide**:

HO—☐—OH HO—☐—OH HO—☐—OH HO—☐—OH
↓ water molecules eliminated
—O—☐—O—☐—O—☐—O—☐— |

In reaction **2**, two molecules join, eliminating a small molecule (water). So this is a condensation reaction. And **3** is a condensation polymerisation.

Cellulose

Cellulose is also a polysaccharide. Its molecules are built from at least 1000 glucose units. But they are joined differently than those in starch, so cellulose has quite different properties.

All the cell walls in plants contain cellulose. So we eat cellulose every time we eat cereals, vegetables, and fruit. We can't digest it, but it helps to clean out our digestive systems. We call it **fibre**.

The importance of carbohydrates in your food

Your body breaks down starch to glucose again. It uses some of this for energy. It also builds some into another complex carbohydrate called **glycogen**, to use as an energy store.

So carbohydrates are an important part of your diet. Rice, wheat, millet, maize, pasta, bread, potatoes, yams, and bananas are all rich in starch. Honey and fruit juices are rich in glucose.

Grass is mainly cellulose, which cows can digest. But humans can't!

Questions
1. All life depends on photosynthesis. Explain why.
2. Explain what it is, and name two examples:
   a a carbohydrate   b a monosaccharide
   c a disaccharide   d a polysaccharide
3. In what ways is cellulose:
   a like starch?   b different from starch?
4. The cellulose in vegetables is good for us. Why?
5. Name three foods you eat, that are rich in starch.

18.7 The macromolecules in food (II)

Proteins – built from amino acids

For proteins, the building blocks are molecules of **amino acids**. Amino acids contain carbon, hydrogen, oxygen, and nitrogen, and some contain sulphur. The general structure of an amino acid molecule is shown on the right. Note the COOH and NH_2 groups.

There are twenty common amino acids. Here are three of them:

$$H-N-\overset{H}{\underset{H}{C}}-\overset{O}{\underset{OH}{C}} \quad H-N-\overset{H}{\underset{CH_3}{C}}-\overset{O}{\underset{OH}{C}} \quad H-N-\overset{H}{\underset{C_3H_7}{C}}-\overset{O}{\underset{OH}{C}}$$

glycine   alanine   cysteine

How amino acids join up to make proteins

This shows four different amino acids combining:

[diagram of four amino acids joining, with water molecules eliminated, and the linking group identified as an **amide** group, as in nylon]

From 60 to 6000 amino acid units can join to make a macromolecule of protein. They include different amino acids, in different orders – so there are a huge number of proteins!

The reaction is a condensation polymerisation, with loss of water molecules. Note the **amide linkage**, as in nylon (page 260).

The importance of proteins in your food

When your body digests food, it breaks the proteins back down to amino acids. These then join up again to make proteins for your body. For example all these things in your body are proteins:

- the enzymes that act as catalysts for reactions in your body cells
- the collagen in your skin, bones, and teeth
- the keratin that forms your hair
- haemoglobin, the red substance in blood, that carries oxygen
- hormones, the chemicals that dictate how you grow and develop.

Your body needs all 20 amino acids to make these proteins. It *can* make 11 of them by itself. But there are 9 **essential amino acids** that it cannot make. To be healthy, you must eat foods that can provide these.

$$H-N-\overset{H}{\underset{R}{C}}-\overset{O}{\underset{OH}{C}}$$

An amino acid is a carboxylic acid with an amino (NH_2) group. R stands for the rest of the molecule.
We show the bonds in the acid group (COOH) straight up, in this unit, just to help you see how the amino acids join.

Part of a protein. The macromolecules may be coiled up because of cross-links in the chains. The genes in cells control which amino acids join up, and in what order.

Rich in proteins

chicken and other meats
fish
cheese
yoghurt
milk
eggs
soya beans
lentils
beans and peas
spinach
nuts
seeds (such as sunflower seeds)

The foods above are rich in proteins. The proteins from animals usually have all 20 amino acids. So do those from soya beans. But in proteins from other plants, some essential amino acids are often missing.

Fats

Foods also contain natural **fats** and **oils** (liquid fats). These are **esters**, which means they are compounds formed from an **alcohol** and an **acid**. (Page 253.)

- The alcohol in this case is always **glycerol**, a natural alcohol with three OH groups. (Its chemical name is propan-1,2,3-triol.)
- The acids are carboxylic acids, usually with long chains of carbon atoms. These natural acids are called **fatty acids**.

There are around forty common natural fatty acids. For example palmitic acid, $C_{15}H_{31}COOH$. It combines with glycerol to give esters found in palm oil.

$$HO-CH_2$$
$$HO-CH$$
$$HO-CH_2$$
glycerol

How fats are formed

This shows the reaction between glycerol and a fatty acid. R stands for the long chain of carbon atoms with hydrogen atoms attached, in the acid:

3 molecules of a fatty acid   and   1 molecule of glycerol   gives   1 macromolecule of a fat

this group is called an **ester** linkage

This is a condensation reaction, with the elimination of water. Each OH group in a glycerol molecule can react with a *different* fatty acid, so you can get many different esters. Note the ester linkage, as in Terylene (page 261).

The importance of fats in your food

In your body, fats and oils in food are broken down to fatty acids and glycerol. Some of these are used for energy. Some are combined into new fats, to make the membranes in your body cells. Some cells also store fat droplets. These cells form a layer under your skin, which keeps you warm.

So you need some fats in your diet. But runny **unsaturated fats** (containing carbon–carbon double bonds) are better for you than the hard, saturated, fats found in meat and cheese. Saturated fats have been linked to heart disease.

Crushing palm fruit to get palm oil. Palm oil is a mixture of esters. Most are from the fatty acid palmitic acid.

Rich in fats

meat
oily fish
butter, cheese, cream
avocados
nuts and seeds
vegetable oils (such as palm oil, olive oil, sunflower oil)
margarine and other spreads

Fish oil and vegetable oils contain unsaturated fats. These are better for you than hard fats.

Questions

1. What is: **a** an amino acid? **b** a protein?
2. Describe in your own words a protein macromolecule.
3. Give three examples of the important roles proteins play in your body.
4. Name six foods that are rich in protein.
5. Show how palmitic acid reacts with propan-1,2,3-triol to give an ester found in palm oil.
6. What happens to fats when you eat them?
7. Compare the reactions that produce carbohydrates, proteins, and fats. What do they have in common?

18.8 Breaking down the macromolecules

What happens during digestion?

You saw earlier how the macromolecules in food are built up by condensation reactions, with the loss of water molecules.

The opposite happens when you eat them. In your mouth, stomach and small intestine, the macromolecules are broken down again, by reacting with water. This is called **hydrolysis**.

Hydrolysis is a reaction in which molecules are broken down by reaction with water.

Hydrolysis in the digestive system

This is what happens in your body, during digestion:

- **Starch** and any disaccharides get broken down to glucose. Your cells then use the glucose to provide energy, in a process called respiration. It is the reverse of photosynthesis:

$$C_6H_{12}O_6 \,(aq) + 6O_2 \,(g) \longrightarrow 6CO_2 \,(g) + 6H_2O \,(l) + \text{energy}$$
glucose + oxygen $\longrightarrow$ carbon dioxide + water + energy

- **Proteins** get broken down to **amino acids** which your body then uses to build up the proteins it needs.
- **Fats** and **oils** (which are esters) get broken down into **glycerol** and **fatty acids**. These are used for energy, or to make new fats for cell membranes, or to be stored.

All the 'breaking down' reactions during digestion are hydrolyses.

The hydrolysis of starch starts in your mouth, where the enzyme *amylase* in saliva start breaking it down. (If you keep chewing bread, it begins to taste sweet – because glucose is forming.)

Example: hydrolysis of an ester during digestion

This shows the hydrolysis of an ester in a vegetable oil, in your digestive system. The green blocks represent long chains of carbon atoms:

an ester in a vegetable oil + water $\xrightarrow{\text{an enzyme as catalyst}}$ fatty acid + glycerol

Compare it with the reaction shown on page 269. What do you notice?

Enzymes as catalysts

Enzymes were used as catalysts, in building the macromolecules in food. During digestion, other enzymes are used as catalysts to break them down again. (Look at the hydrolysis above.) Without enzymes, the reactions would be far too slow. Enzymes called **amylases** act on starch, **lipases** act on fats and oils, and **proteinases** act on proteins.

Getting ready for hydrolysis.

Hydrolysis in the lab

You can also carry out hydrolysis of starch, proteins and fats in the lab. You have to boil them in acid, unless you have enzymes to act as catalysts. (You can buy enzymes.)

Proteins are especially hard to break down, so you may have to boil them for 24 hours in 6M acid solution. Compare this with 3 or 4 hours at body temperature in your digestive system.

The products will be the same as in digestion. So:

- proteins will give a mixture of amino acids.
- fats and oils will give a mixture of fatty acids, plus glycerol.
- starch will give glucose.

But if hydrolysis is not complete, the macromolecules will not all be broken down into their smallest units. So you will get a mixture of molecules of different sizes. For example partial hydrolysis of starch can give glucose, plus maltose (made of two glucose units), maltotriose (three glucose units), and dextrins (many glucose units).

You can use chromatography to identify the products from the hydrolyses, as described in Unit 1.9.

Hydrolysis of esters to make soap

In your digestive system, hydrolysis of the esters in fats and oils is carried out in the presence of acid, giving glycerol and fatty acids.

In industry, the hydrolysis is carried out using **sodium hydroxide**. This gives glycerol and *the sodium salts of the fatty acids*. These salts are used as **soaps**. This shows a typical reaction:

$$\begin{array}{c} R-COO-CH_2 \\ | \\ R-COO-CH \\ | \\ R-COO-CH_2 \end{array} (l) + 3NaOH\,(aq) \longrightarrow \begin{array}{c} HO-CH_2 \\ | \\ HO-CH \\ | \\ HO-CH_2 \end{array} (l) + 3R-COONa\,(aq)$$

an ester in a vegetable oil   glycerol   soap – the sodium salt of a fatty acid

The soap you buy may be made from vegetable oil – for example palm oil or coconut oil – or even from fish oil or animal fat. Chemicals are added to make it smell nice. These are usually artificial esters, like those on page 253.

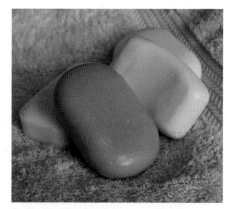

Sodium salts that keep you clean!

Questions

1. **a** What does *hydrolysis* mean?
   b Show that the complete hydrolysis of starch to glucose, in the lab, is the opposite of a condensation polymerisation.
   c If you carry out an incomplete hydrolysis of starch in the lab, you get a mixture of products. Explain.

2. Hydrolysis of a protein in the lab will give a mixture of products. Explain why, and how to identify them.

3. Oils are broken down in your digestive system. And oils are used to make soaps, in industry.
   a What do these two processes have in common?
   b In what way are they different?

Questions on Chapter 18

1. Propene is a hydrocarbon made from oil. Like *propane*, its molecules contain three carbon atoms. Like *ethene*, it has a double bond.
   a. Write down the formula of propene.
   b. How does it differ from propane?
   c. To which group of hydrocarbons does:
      i. propane   ii. propene   belong?
   d. Draw the structural formulae for propane and propene (like the formula in 3 below).
   e. Which of the two is *unsaturated*?
   f. Addition reactions are possible for propene, but not propane. Explain why.
   g. Propene is used as the monomer for making an important plastic.
      i. Which type of polymerisation occurs?
      ii. Show three units in the structure of the long-chain macromolecule that forms.
      iii. How do the bonds in this structure differ from the bonds in a propene molecule?
      iv. Write an equation for the polymerisation reaction, using n.
      v. Suggest a name for the polymer.

2. a. Which of these could be used as monomers for addition polymers? Explain your choice.
      i. ethene, $CH_2=CH_2$
      ii. ethanol, C_2H_5OH
      iii. propane, C_3H_8
      iv. styrene, $C_6H_5CH=CH_2$
      v. chloropropene, $CH_3CH=CHCl$
   b. Suggest a name for each polymer obtained.

3. The polymer 'Teflon' is obtained from the monomer tetrafluoroethene, which has this structure:

   $F_2C=CF_2$

   a. Is this monomer a hydrocarbon? Explain.
   b. Which feature of the monomer makes polymerisation possible?
   c. Which type of polymerisation occurs?
   d. Draw three units in the structure of the macromolecule that forms.
   e. Give the chemical name for this polymer.
   f. Write an equation for the polymerisation reaction, using n.
   g. Teflon is used to coat frying pans, to make them non-stick. Give three properties that make it suitable for this purpose.

4. The polymer poly(dichloroethene) has been used to make 'cling film', for covering food to keep it fresh. This shows the structure of the polymer:

   $$\left(\begin{array}{cc} H & Cl \\ | & | \\ -C - C- \\ | & | \\ H & Cl \end{array} \right)_n$$

   a. What does n represent?
   b. Name the monomer, and draw its structural formula.
   c. Which type of polymerisation takes place?
   d. One property of poly(dichlorothene) is its *low permeability* to moisture and gases.
      i. What does the term in italics mean?
      ii. This property is important in keeping food fresh. Why?
      iii. Give three other *physical* properties a polymer would need, to be suitable for use as 'cling film'.
   e. Poly(dichlorothene) is *non-biodegradable*.
      i. Explain the term in italics.
      ii. Describe two environmental problems caused by the disposal of such plastics.

5. PVA glue, an adhesive, is an addition polymer made from the monomer vinyl acetate. This shows the structure of the monomer:

   $$\begin{array}{c} H \\ \diagdown \\ C=C \\ \diagup \diagdown \\ H COCH_3 \end{array}$$

   (with H on the other C)

   a. What does PVA stand for?
   b. Deduce some properties of this polymer.
   c. Why is it called an *addition polymer*?
   d. Draw a structure showing three units in the macromolecule.
   e. Write an equation for the polymerisation.

6. This diagram represents two units of an addition polymer called polyacrylamide:

   $$\begin{array}{cccc} H & H & H & H \\ | & | & | & | \\ -C- & C- & C- & C- \\ | & | & | & | \\ H & CONH_2 & H & CONH_2 \end{array}$$

   a. Draw the structure of the monomer.
   b. Suggest a name for the monomer.
   c. Is the monomer saturated, or unsaturated?

7 **Polyamides** are polymers made by condensation polymerisation. One polyamide was developed for use in puncture-resistant bicycle tyres. The two monomers for it are:

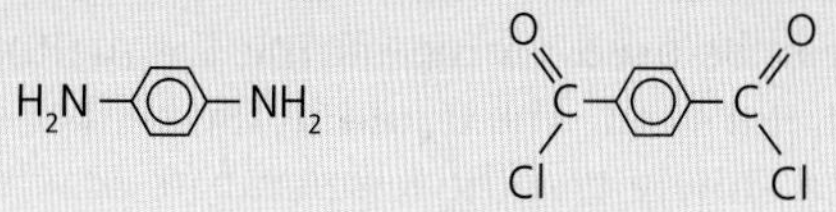

$H_2N-\bigcirc-NH_2$ $\underset{Cl}{\overset{O}{\|}}{C}-\bigcirc-\underset{Cl}{\overset{O}{\|}}{C}$

The hexagon with the circle in the middle stands for a ring of 6 carbon atoms, with 3 double bonds.
   a What is *condensation polymerisation*?
   b Show in detail how the monomers join.
   c Name the other product of the reaction.
   d i In what way is this polymer similar to nylon? (See page 260.)
     ii But its properties are different from those of nylon. Why?

A similar polymer has been developed as a fabric for fireproof clothing. Its structure is:

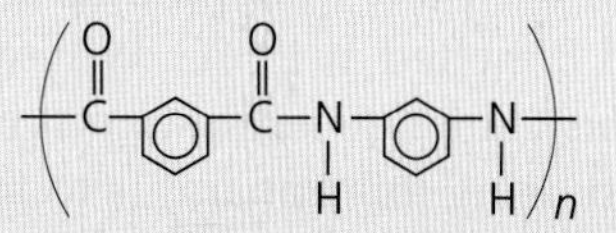

$\left(\overset{O}{\overset{\|}{C}}-\bigcirc-\overset{O}{\overset{\|}{C}}-\underset{H}{N}-\bigcirc-\underset{H}{N} \right)_n$

   e Draw the structures of the two monomers that could be used to make this polymer.

8 Many synthetic polymers contain amide links.
   a Draw the structure of the amide link.
   b Which important natural macromolecules also contain amide links?

The substances in **b** will undergo hydrolysis in the laboratory, in the presence of acid.
   c i What does *hydrolysis* mean?
     ii What are the products of the hydrolysis?
     iii How can the products be separated?

9 One very strong polymer has this structure:

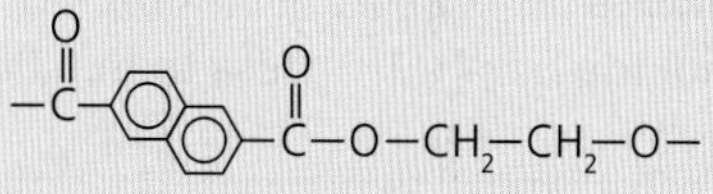

$-\overset{O}{\overset{\|}{C}}-\bigcirc-\bigcirc-\overset{O}{\overset{\|}{C}}-O-CH_2-CH_2-O-$

   a Which type of polymerisation produced it?
   b Which type of linkage joins the monomers?
   c Draw the structures of the two monomers from which this polymer could be made.
   d Compare the structure above with that for Terylene (page 261). What may be responsible for the greater strength of this polymer?
   e i Which *natural* macromolecules have the same linkage as this polymer?
     ii Hydrolysis of these macromolecules, using an alkali, gives a useful product. Name it.

10 Starch is a carbohydrate. It is a natural polymer. This shows part of a starch macromolecule:

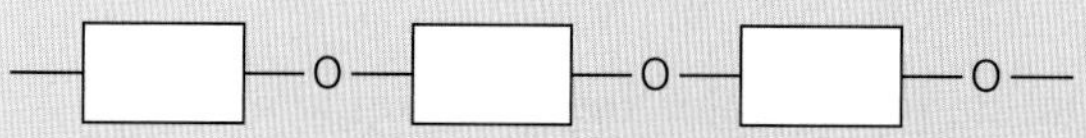

—☐—O—☐—O—☐—O—

   a What is a *macromolecule*?
   b What is a *carbohydrate*?
   c Which type of polymerisation gives starch?
   d What do the blocks represent, above?
   e i Draw a diagram showing the structure of the monomer for starch. (Use a block.)
     ii Name this monomer.
   f Starch is also called a *polysaccharide*. Why?
   g Starch can be broken down by hydrolysis.
     i Describe two ways in which the hydrolysis is carried out. (One occurs in your body.)
     ii One takes place at a far lower temperature than the other. What makes this possible?

11 In the lab, *partial* hydrolysis of starch gives a mixture of colourless products. They can be identified using chromatography. A locating agent is needed.
   a Draw diagrams showing at least two of the products.
   b What is a *locating agent* and why is it needed?
   c Outline the steps in carrying out the chromatography. (Page 23 may help.)

12 Soaps are salts of fatty acids.
   a Name one fatty acid.
   b In which way is a fatty acid different from ethanoic acid? In which way is it similar?
   c Below is one example of a compound found in vegetable oil, and used to make soap.

$H_2C - OOC(C_{17}H_{35})$
$|$
$HC - OOC(C_{15}H_{31})$
$|$
$H_2C - OOC(C_{14}H_{29})$

   i This compound is an *ester*. Explain that term.
   ii To make soap, the oil is usually reacted with a sodium compound. Which one?
   iii Which type of reaction takes place?
   d i The reaction in **c** will give *four* different products. Write down their formulae.
     ii Which ones can be used as soap?
     iii One product is an alcohol. Name it.
     iv In which way is this product similar to ethanol? In which way is it different?
   e Name three vegetable oils used to make soap.
   f Which type of compounds are added to soap, to make it smell good?

Cambridge IGCSE exam questions: core material

1   The states of matter are solid, liquid and gas. The diagram below shows how the molecules are arranged in these three states.

   a   State the name given to the change of state labelled:
       i A   ii B   iii C   (3)
   b   Which one of the following best describes the movement of molecules in the liquid state?
       i   The molecules are not moving from place to place
       ii   The molecules are sliding over each other
       iii The molecules are moving freely   (1)
   c   Which of the changes **A**, **B**, or **C**, is endothermic? Explain your answer.   (2)
   d   Choose from the following list of substances to answer the questions below.
       bromine   chlorine   iron
       mercury   sodium chloride sulphur

   Name a substance which is
       i   a gas at room temperature
       ii   a non-metallic liquid at room temperature
       iii a compound which is a solid at room temperature   (3)
   e   A student set out the apparatus shown in the diagram below.

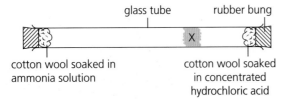

   The white solid is formed because the molecules of hydrogen chloride gas and ammonia gas move at random throughout the tube and eventually react with each other.
       i   State the name given to this random movement of molecules.
       ii   Name the white solid formed at **X**.
       iii Suggest why the white solid is formed towards one end of the tube and not in the middle.   (3)
   f   What type of chemical reaction occurs when ammonia reacts with hydrochloric acid?   (1)

   g   The diagram below shows a simple apparatus that can be used for measuring the melting point of a solid. The liquid in the beaker is heated slowly and the temperature at which the solid B melts is recorded.

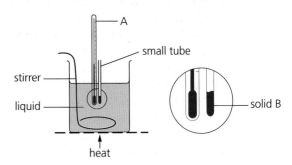

       i   Name the piece of apparatus labelled A.
       ii   Solid B melted at 155°C. Why would water not be a suitable liquid to put in the beaker when using this apparatus to find the melting point of solid B?
       iii Suggest why the liquid needs to be kept stirred.   (3)

   CIE 0620/02 J '03 Q3

2   The table below shows the composition of the mixture of gases coming from a typical car exhaust.

gas	% of the gas in the exhaust fumes
carbon dioxide	9
carbon monoxide	5
oxygen	4
hydrogen	2
hydrocarbons	0.2
nitrogen oxides	0.2
sulphur dioxide	less than 0.003
gas X	79.6

   a   State the name of the gas **X**.   [1]
   b   The carbon dioxide comes from the burning of hydrocarbons, such as octane, in the petrol.
       i   Complete the word equation for the complete combustion of octane.
            octane + ... → carbon dioxide + ...   [2]
       ii   Which **two** chemical elements are present in hydrocarbons?   [1]
       iii To which homologous series of hydrocarbons does octane belong?   [1]
   c   Suggest a reason for the presence of carbon monoxide in the exhaust fumes.   [1]

d Nitrogen oxides are present in small quantities in the exhaust fumes.
  i Complete the following equation for the formation of nitrogen dioxide.
    $N_2(g) + \ldots O_2(g) \rightarrow \ldots NO_2(g)$ [1]
  ii State **one** harmful effect of nitrogen dioxide on organisms. [1]
e Sulphur dioxide is an atmospheric pollutant found only in small amounts in car exhausts.
  i What is the main source of sulphur dioxide pollution of the atmosphere? [1]
  ii Sulphur dioxide is oxidised in the air to sulphur trioxide. The sulphur trioxide may dissolve in rainwater to form a dilute solution of sulphuric acid, H_2SO_4. State the meaning of the term *oxidation*. [1]
  iii Calculate the relative molecular mass of sulphuric acid. [1]
  iv Sulphuric acid reacts with metals such as iron. Complete the following word equation for the reaction of sulphuric acid with iron.
    sulphuric acid + iron → .... + ... [2]
  v What effect does acid rain have on buildings made of stone containing calcium carbonate? [1]

CIE 0620/02 N '04 q4

3 Iron is extracted from the ore, haematite. The iron ore is put in a blast furnace with coke and a current of air is blown through the heated mixture.

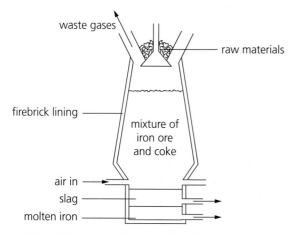

a What do you understand by the term ore? [1]
b What other raw material needs to be added to the blast furnace? Put a ring around the correct answer.
  cement   limewater   limestone   slag [1]

c Near the bottom of the furnace, iron(III) oxide is reduced by carbon.
  $Fe_2O_3 + 3C \rightarrow 2Fe + 3CO$
  i Write a word equation for this reaction. [1]
  ii Explain what is meant by *reduction*. [1]
d The table shows the composition of the waste gases leaving the blast furnace.

gas	percentage of gas in the mixture
carbon dioxide	12
carbon monoxide	24
hydrogen	4
nitrogen	60

  i The hydrogen in the waste gas is formed by the reaction of hot carbon with water vapour. There is no water in the materials added to the top of the furnace. Suggest where this water vapour comes from. [1]
  ii The reaction of hot carbon with water vapour is endothermic. What is meant by the term *endothermic*? [1]
e Iron can be converted into steel, which is more resistant to corrosion.
  i Describe briefly how iron is converted into steel. [2]
  ii State one use of mild steel. [1]
f In some conditions, steel corrodes more quickly than in others. The graphs show the rate of corrosion of a particular type of steel under different controlled conditions.

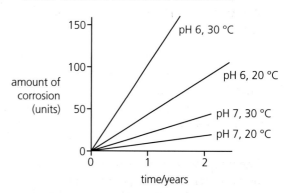

  i How does pH affect the rate of corrosion? [1]
  ii How does temperature affect the rate of corrosion? [1]
     Explain this in terms of moving particles. [2]
  iii The presence of acidic gases in the air may increase the rate of corrosion. State the name and source of one acidic gas found in the air as a result of pollution. [2]

CIE 0620/02 N '02 q 5

4 The structures of some compounds found in plants are shown below.

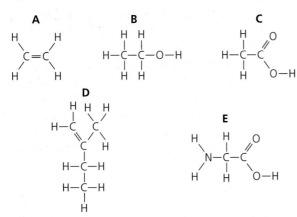

a Which **two** of these compounds are unsaturated hydrocarbons? [1]
b Which **two** of these compounds contain a carboxylic acid functional group? [1]
c Write the molecular formula of compound **D**. [1]
d Draw the structure of the product formed when compound **A** reacts with bromine. Show all atoms and all bonds. [1]
e Strawberry fruits produce compound **A** (ethene) naturally. A scientist left some green strawberry fruits to ripen. The scientist measured the concentration of ethene and carbon dioxide produced over a ten day period. The graph below shows the results.

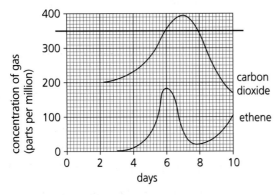

i Between which two days does the rate of ethene production increase most rapidly? [1]
ii What is the name given to the process in which carbon dioxide is produced by living organisms? Put a ring around the correct answer.

  acidification combustion
  neutralisation respiration [1]

iii Carbon dioxide concentration over 350 ppm has an effect on ethene production by the fruits. What effect is this? [1]
iv Ethene gas spreads throughout the fruit by a random movement of molecules. What is the name given to the random movement of molecules?
Put a ring around the correct answer.
  aeration diffusion
  ionization evaporation [1]
v Ethene gas promotes the ripening of strawberry fruits. Ripening of strawberries is slowed down by passing a stream of nitrogen over the fruit. Suggest why this slows down the ripening process. [1]
vi Enzymes are involved in the ripening process. What is an *enzyme*? [2]

f Plants make a variety of coloured pigments. A student extracted red colouring from four different plants, **R, S, T** and **U**. The student put a spot of each colouring on a piece of filter paper. The filter paper was dipped into a solvent and left for 30 minutes. The results are shown below.

i What is name given to the process shown in the diagram? [1]
ii Which plant contained the greatest number of different pigments? [1]
iii Which two plants contained the same pigments? [1]

CIE 0620/02 N '04 Q2

5 A student collected some water from a polluted river. The water contained soluble solids and insoluble clay and had a pH of 5.
a How can the student separate the clay from the rest of the river water? [1]
b The student uses litmus paper to show that the river water is acidic. What will be the result of this test? [1]
c The student then boiled the river water to obtain the soluble solids. The diagram that follows shows how she heated the water.

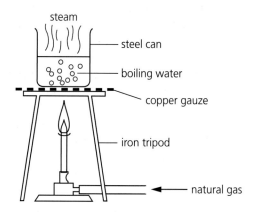

Which of the substances named in the diagram above is
i an alloy, [1]
ii a compound which is liquid at room temperature, [1]
iii an element, [1]
iv a fuel? [1]
d Name the main substance in natural gas. [1]
e What is the normal temperature of boiling water? [1]
f After the student boiled off the water, she analysed the white powder on the inside of the steel can. The table shows her results.

name of ion	formula of ion	mass of ion present /milligrams
calcium	Ca^{2+}	16
carbonate	CO_3^{2-}	35
chloride	Cl^-	8
nitrate	NO_3^-	4
sodium	Na^+	8
sulphate	SO_4^{2-}	6

i Which positive ion had the greatest concentration in the sample of river water? [1]
ii Complete the following equation to show how a sodium ion is formed from a sodium atom.

$Na \rightarrow Na^+ +$ .... [1]

g Instead of using natural gas, the student could have used butane to heat the water. The formula of butane is C_4H_{10}.
i What products are formed when butane burns in excess air? [1]
ii Name the poisonous gas formed when butane undergoes incomplete combustion. [1]

CIE 0620/02 J '05 Q2

6 Bromine is in Group VII of the Periodic Table.
a State the name given to the Gp VII elements. [1]
b Bromine has two **isotopes.** The nucleon (mass) number of bromine-79 is 79 and of bromine-81 is 81.
i What is the meaning of the term *isotopes*?
ii Complete the table to show the numbers of electrons, neutrons and protons in one atom of bromine-79 and bromine-81. [5]

number of	bromine-79	bromine-81
electrons		
neutrons		
protons		

c Bromine is extracted from seawater by treatment with chlorine. When chlorine is bubbled through a solution of potassium bromide, the solution turns orange-red.
i What does this tell you about the reactivity of chlorine compared with bromine?
ii Write a word equation for this reaction. [2]
d In order to get the maximum yield of bromine from seawater, acid is added during the extraction procedure. The graph shows how the yield of bromine changes with pH.

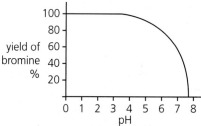

i What is the highest pH at which the yield of bromine is 100%?
ii The pH scale is used to measure acidity. Some pH values are given below.

pH 3   pH 5   pH 7   pH 9   pH 11

From this list of pH values choose
the pH which is most acidic.
the pH of a neutral solution. [3]

e Bromine water can be used to distinguish between ethane and ethene.

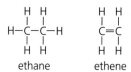

Describe what you would observe when bromine water is added to ethene. [1]

CIE 0620/02 N '03 Q4

7 The diagram below shows a modern landfill site for the disposal of waste materials.

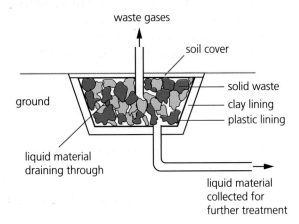

The waste materials are broken down naturally in several stages.

a In the first stage, micro-organisms (mainly bacteria) break down some of the organic material in the waste to carbon dioxide. What is the name given to the process by which organisms use food to produce carbon dioxide? [1]

b In the second stage, the micro-organisms break down organic substances to produce ammonia, hydrogen and more carbon dioxide.
  i Describe a test for hydrogen.
  ii The large volumes of hydrogen produced may be hazardous. Explain why hydrogen may be hazardous when mixed with air.
  iii Ammonia is a base. Describe a test for ammonia. [5]

c In the third stage, ethanoic acid is produced. Draw the structure of ethanoic acid showing all atoms and bonds. [1]

d In the fourth stage, carbon dioxide reacts with hydrogen to form methane and oxygen.
  i Complete the equation for this reaction.

  $CO_2 + \ldots \rightarrow CH_4 + O_2$

  ii State one use of methane.
  iii Methane is a gas. Which **two** of the following statements about gas molecules are true?
      A The molecules are far apart.
      B The molecules are not moving.
      C The molecules are randomly arranged.
      D The molecules are arranged in a regular manner. [4]

e The list below shows some of the substances which are found in the liquid which drains through the waste.

  aluminium   calcium carbonate
  iron   lead
  magnesium   nickel
  sodium sulphate   zinc

  From this list choose
  i a metal used to galvanise iron.
  ii a transition metal.
  iii a metal which is in Group IV of the Periodic Table.
  iv a substance which will release carbon dioxide when an acid is added.
  v a metal which is used to make aircraft bodies. [5]

  CIE 0620/02 N '03 Q2

8 The electrolysis of a concentrated solution of sodium chloride provides us with chemicals.
  a Sodium chloride has an ionic giant structure. Which **one** of the following is a correct description of a property of sodium chloride?
      i sodium chloride has a low melting point
      ii sodium chloride conducts electricity when it is solid
      iii sodium chloride has a high boiling point
      iv sodium chloride is insoluble in water [1]
  b i Explain what is meant by the term *electrolysis*. [1]
     ii At which electrode is hydrogen produced during the electrolysis of aqueous sodium chloride? [1]
     iii Name a suitable substance that can be used for the electrodes. [1]
  c i State the name of the particle which is added to a chlorine atom to make a chloride ion. [1]
     ii Describe a test for chloride ions. [2]
  d If chlorine is allowed to mix with sodium hydroxide, sodium chlorate(I), NaOCl, is formed. Balance the equation for this reaction.

  $Cl_2 + \ldots NaOH \rightarrow NaCl + NaOCl + H_2O$ [1]

  e One tonne (1000 kg) of a commercial solution of sodium hydroxide produced by electrolysis contains the masses of compounds shown in the following table.

Compound	Mass of compound kg/ tonne
sodium hydroxide	510
sodium chloride	10
sodium chlorate(V)	9
water	471
total	1000

i How many kilograms of sodium hydroxide will be present in 5 tonnes of the solution? [1]

ii All the water from one tonne of impure sodium hydroxide is evaporated.
What would the approximate percentage of the remaining impurities be?
Put a ring around the correct answer.

   0.036%   3.6%   36%   96%   [1]

f The hydrogen obtained by electrolysis can be used in the manufacture of margarine. This is a reaction that takes place:

$$\underset{\substack{H\\|\\H-C-\\|\\H}}{}\left[\underset{\substack{H\\|\\C-\\|\\H}}{}\right]_n\underset{\substack{H\\|\\C=\\}}{}\underset{\substack{H\\|\\C-CO_2H\\}}{} + H_2 \xrightarrow{Ni} \underset{\substack{H\\|\\H-C-\\|\\H}}{}\left[\underset{\substack{H\\|\\C-\\|\\H}}{}\right]_n\underset{\substack{H\\|\\C-\\|\\H}}{}\underset{\substack{H\\|\\C-CO_2H\\|\\H}}{}$$

i Complete the following sentences about this reaction using words from the list.
catalyst
inhibitor
monomeric
saturated
unsaturated

Hydrogen gas is bubbled through .... carbon compounds using a nickel .... which speeds up the reaction.

The margarines produced are .... compounds. [3]

ii State **one** other use of hydrogen. [1]
CIE 0620/02 N '04 Q6

9 The electroplating of iron with chromium involves four stages.
  1. The iron object is cleaned with sulphuric acid, then washed with water.
  2. The iron is plated with copper.
  3. It is then plated with nickel to prevent corrosion.
  4. It is then plated with chromium.

a The equation for stage 1 is

   $Fe + H_2SO_4 \rightarrow FeSO_4 + H_2$

   i Write a word equation for this reaction
   ii Describe a test for the gas given off in this reaction. [2]

b The diagram shows how iron is electroplated with copper.

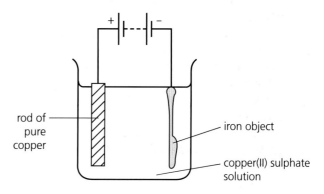

   i Choose a word from the list below which describes the iron object.

   anion   anode   cathode   cation   [1]

   ii What is the purpose of the copper(II) sulphate solution? [1]
   iii Describe what happens during the electroplating to
   the iron object, ...
   the rod of pure copper .... [2]
   iv Describe a test for copper(II) ions. [3]

c Suggest why chromium is used to electroplate articles. [1]

d The information below shows the reactivity of chromium, copper and iron with warm hydrochloric acid.
   chromium – few bubbles of gas produced every second
   copper – no bubbles of gas produced
   iron – many bubbles of gas produced every second

Put these three metals in order of their reactivity with hydrochloric acid. [1]
CIE 0620/02 J '05 Q6

10 The diagram shows part of the Periodic Table:

					He
C	N	O	F	Ne	
		S	Cl	Ar	
			Br	Kr	

a Answer these questions using only the elements shown in the diagram. Write down the symbol for an element which
  i has five electrons in its outer shell, (1)
  ii has diatomic molecules, (1)
  iii reacts with sodium to form sodium bromide, (1)
  iv is a noble gas, (1)
  v has a giant covalent structure, (1)
  vi has a lower proton number than fluorine, (1)
  vii is the most abundant gas in the air. (1)
b Write down a use for each of these elements.
  i argon ii helium iii oxygen (3)
c i Draw a diagram to show the electronic structure of argon. (2)
  ii Why is argon very unreactive? (1)

CIE 0620/02 N '05 Q1

11 The list shows some oxides.
**calcium oxide   magnesium oxide
nitrogen dioxide   sodium oxide   sulphur dioxide**
a From this list choose two oxides which are basic. Give a reason for your answer. (2)
b i Which two oxides contribute to acid rain? (2)
  ii How do each of these oxides get into the atmosphere? (2)
c Calcium oxide is manufactured from calcium carbonate.
  i Complete this word equation:
    calcium carbonate → calcium oxide + ... (1)
  ii What condition is needed for this reaction to take place? (1)
d i Explain why calcium oxide and sodium oxide cannot be reduced by heating with carbon (1)
  ii Copper(II) oxide can be reduced by heating with carbon. Complete the equation for this reaction.
    .... CuO + C → 2 Cu + ... (2)
  iii What do you understand by the term *reduction*? (1)

CIE 0620/02 N '05 Q4

12 Hydrogen peroxide solution, H_2O_2, decomposes slowly in the absence of a catalyst. Oxygen and water are formed.

$$2H_2O_2 (aq) \rightarrow 2H_2O_2 (aq) + O_2 (g)$$

a Draw a diagram of the apparatus you could use to investigate the speed of this reaction. You must label your diagram. (3)
b Catalyst X was added to 50 cm³ of hydrogen peroxide solution at 20°C and the amount of oxygen given off was recorded over a two minute period.
The experiment was repeated with the same amounts of catalyst Y and catalyst Z.
Apart from the type of catalyst, all conditions were kept the same in the three experiments. A graph of the results is shown below.

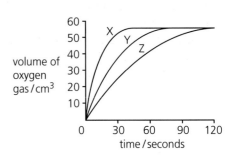

  i What is a catalyst?
  ii Which catalyst, X, Y or Z, produced oxygen gas the fastest? Explain your answer. (2)
  iii Why is the final amount of oxygen gas the same in each experiment? (1)
  iv Many transition metals and their oxides are good catalysts. State two other properties of transition metals which are not shown by other metals. (2)
c The experiment with catalyst Z was repeated at 40°C. All other conditions were kept the same. The speed of the reaction increased. Explain why, using ideas about particles. (2)
d Some enzymes also catalyse the decomposition of hydrogen peroxide.
  i State one difference between an enzyme and an inorganic catalyst such as a transition metal. (1)
  ii Enzymes are responsible for fermentation. Which of the following equations **A, B, C** or **D** describes fermentation?
    A $C_6H_{12}O_6 + 6O_2 \rightarrow 6CO_2 + 6H_2O$
    B $C_2H_4 + H_2O \rightarrow C_2H_5OH$
    C $C_6H_{12}O_6 \rightarrow 6C + 6H_2O$
    D $C_6H_{12}O_6 \rightarrow 2C_2H_5OH + 2CO_2$ (1)

CIE 0620/02 N '05 Q3

13 The diagram shows an electrolysis cell used to extract aluminium.

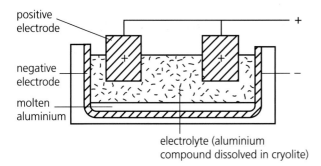

a What compound of aluminium is used for the electrolyte? (1)
b The electrolyte must be molten for the electrolysis to work. Explain why. (1)
c i State the name of the substance used for the electrodes. (1)
 ii To which electrode do the aluminium ions move during electrolysis? Explain why. (2)
d Complete the following sentences about the molten electrolyte using words from this list:
*bauxite   chemical   cryolite   decreased
electrical   haematite   increased   light*
The melting point of the electrolyte is … by adding ……. This means that less ……. energy is needed to melt the electrolyte. (3)
e Aluminium is used in overhead power cables.

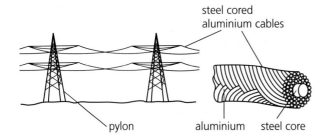

The table shows some properties of three metals which could be used for power cables.

metal	relative electrical conductivity	density/ grams per cm^3	price/ £ per kg	relative strength
aluminium	0.4	2.70	18	9
copper	0.7	8.92	15	30
steel	0.1	7.86	2.7	50

  i Suggest why aluminium is used for overhead power cables rather than copper. (1)
  ii Suggest why steel is not used alone for overhead power cables. (1)
  iii Why is steel used as a core for overhead power cables? (1)
  iv Electrical insulators are used in parts of the pylons which carry the electrical cables. Which one of the following is an electrical insulator? Put a ring around the correct answer.
   aluminium   ceramic   graphite   zinc
f Aluminium has many uses.
  i Why is it used for aircraft bodies? (1)
  ii Describe a test for aluminium ions.
   test …………………..
   result ………………… (3)

CIE 0620/02 N '05 Q6

14 The structures of some organic compounds are shown below.

A: CH_4 structure (H–C–H with 4 H's)
B: 4-carbon alkane
C: 4-carbon alcohol (–O–H)
D: 4-carbon with C=C double bond
E: 2-carbon with COOH group

a Name compound A. (1)
b Which two of the compounds A to E belong to the same homologous series? (1)
c i Which one of the compounds A to E has the same functional group as ethanol? (1)
 ii Draw the structure of ethanol, showing all atoms and bonds. (2)
 iii Describe how ethanol is made in industry from ethene. (2)
d i Which one of the compounds A to E is an unsaturated hydrocarbon? (1)
 ii Describe a chemical test for an unsaturated hydrocarbon.
  test………………….....
  result …………….. (2)
e Compound E is acidic.
 i State the name of compound E. (1)
 ii Describe a test to show that compound E is acidic.
  test……………..
  result …………….. (2)

CIE 0620/02 N '05 Q5

Cambridge IGCSE exam questions: extended material

1 Strontium and zinc are both metals with a valency of 2. Strontium is more reactive than zinc. Its chemistry is similar to that of calcium.

  a i Complete the following table to show the number of protons, electrons and neutrons in each particle.

particle	protons	electrons	neutrons
^{88}Sr			
^{90}Sr			
$^{65}Zn^{2+}$			

  (3)

  ii Explain why ^{88}Sr and ^{90}Sr are isotopes. (1)
  iii Complete the electron distribution of an atom of strontium.
  2 + 8 + 18 + ... + ... (1)

  b The major ore of zinc is zinc blende, ZnS.
  i Describe how zinc is extracted from zinc blende. (2)
  ii Give a use of zinc. (1)

  c The major ore of strontium is its carbonate, $SrCO_3$. Strontium is extracted by the electrolysis of its molten chloride.
  i Name the reagent that will react with the carbonate to form the chloride. (1)
  ii The electrolysis of molten strontium chloride produces strontium metal and chlorine. Write ionic equations for the reactions at the electrodes.

  negative electrode (cathode) ....
  positive electrode (anode) .... (2)

  iii One of the products of the electrolysis of concentrated aqueous strontium chloride is chlorine. Name the other two. (2)

  d Both metals react with water.
  i Write a word equation for the reaction of zinc and water and state the reaction conditions.
  word equation .... (1)
  conditions .... (1)
  ii Write an equation for the reaction of strontium with water and give the reaction condition.
  equation .... (1)
  condition .... (1)

  CIE 0620/03 N '05 Q5

2 Calcium and other minerals are essential for healthy teeth and bones. Tablets can be taken to provide these minerals.

Healthy Bones
Each tablet contains
calcium   magnesium   zinc
copper   boron

  a Boron is a non-metal with a macromolecular structure.
  i What is the valency of boron?
  ii Predict two physical properties of boron.
  iii Name another element and a compound that have macromolecular structures.
  iv Sketch the structure of one of the above macromolecular substances. [7]

  b Describe the reactions, if any, of zinc and copper(II) ions with an excess of aqueous sodium hydroxide.
  i zinc ions
  addition of aqueous sodium hydroxide ....
  excess sodium hydroxide ....
  ii copper(II) ions
  addition of aqueous sodium hydroxide ....
  excess sodium hydroxide .... [4]

  c Each tablet contains the same number of moles of $CaCO_3$ and $MgCO_3$. One tablet reacted with excess hydrochloric acid to produce 0.24 dm^3 of carbon dioxide at r.t.p.

  $CaCO_3 + 2HCl \rightarrow CaCl_2 + CO_2 + H_2O$
  $MgCO_3 + 2HCl \rightarrow MgCl_2 + CO_2 + H_2O$

  i Calculate how many moles of $CaCO_3$ there are in one tablet.
  number of moles CO_2 = ...
  number of moles of $CaCO_3$ and $MgCO_3$ = ...
  number of moles of $CaCO_3$ = ... [3]

  ii Calculate the volume of hydrochloric acid, 1.0 mol/dm^3, needed to react with one tablet.
  number of moles of $CaCO_3$ and $MgCO_3$ in one tablet = ...
  Use your answer to (c)(i).
  number of moles of HCl needed to react with one tablet = ...
  volume of hydrochloric acid, 1.0 mol/dm^3, needed to react with one tablet = ... [2]

  CIE 0620/03 J '03 Q2

3 The first three elements in Period 6 of the Periodic Table of the elements are caesium, barium and lanthanum.
  a How many **more** protons, electrons and neutrons are there in one atom of lanthanum than in one atom of caesium?
  Use the Periodic Table of the elements on page 298 to help you.

  number of protons ....
  number of electrons ....
  number of neutrons .... [3]

  b All three metals can be obtained by the electrolysis of a molten halide. The electrolysis of the aqueous halides does not produce the metal.
    i Complete the equation for the reduction of lanthanum ions at the negative electrode (cathode).

    La^{3+} + ... → ....

    ii Name the three products formed by the electrolysis of aqueous caesium bromide. [4]
  c All **three** metals react with cold water. Complete the word equation for these reactions.

    metal + water → .... + ... [2]

  d Barium chloride is an ionic compound. Draw a diagram that shows the formula of the compound, the charges on the ions, and gives the arrangement of the valency electrons around the negative ion.
    The electron distribution of a barium atom is 2.8.18.18.8.2
    Use x to represent an electron from a barium atom.
    Use O to represent an electron from a chlorine atom. [2]
  e Describe, by means of a simple diagram, the lattice structure of an ionic compound, such as caesium chloride. [2]
  f The reactions of these metals with oxygen are exothermic.

    $2Ba(s) + O_2(g) → 2BaO(s)$

    i Give an example of bond forming in this reaction.
    ii Explain using the idea of bond breaking and forming why this reaction is exothermic. [3]

    CIE 0620/03 J '03 Q5

4 In 1909, Haber discovered that nitrogen and hydrogen would react to form ammonia. The yield of ammonia was 8%.

  $N_2(g) + 3H_2(g) \rightleftharpoons 2NH_3(g)$

  catalyst platinum
  temperature 600 °C
  pressure 200 atm

  The forward reaction is exothermic.

  a Describe how hydrogen is obtained for the modern process. [2]
  b i What is the catalyst in the modern process? [1]
    ii Explain why the modern process, which uses a lower temperature, has a higher yield of 15%. [2]
  c i Complete the following table that describes the bond breaking and forming in the reaction between nitrogen and hydrogen to form ammonia.

bonds	energy changes /kJ	exothermic or endothermic
1 mole of N≡N broken	+945	....
3 moles of .... broken	+1308	....
6 moles of N—H formed	–2328	....

    ii Explain, using the above data, why the forward reaction is exothermic. [2]

    CIE 0620/03 N '05 Q7

5 The salt copper(II) sulphate can be prepared by reacting copper(II) oxide with sulphuric acid. Complete the list of instructions for making copper(II) sulphate using **six** of the words below:

blue	cool	dilute	filter
saturated	sulphate	white	oxide

  Instructions
  1 Add excess copper(II) oxide to .... sulphuric acid in a beaker and boil it.
  2 .... to remove the unreacted copper(II) oxide.
  3 Heat the solution until it is ....
  4 .... the solution to form coloured crystals of copper(II) .... [6]

  CIE 0620/03 N '04 Q2

6 Redox reactions involve the transfer of electrons. Oxidation is the loss of electrons and reduction is the gain.

   a i Describe the colour change observed when acidified potassium manganate(VII) is reduced. [2]
      ii Suggest a suitable reducing agent for this reaction. [1]

   b Chromium(III) chloride is changed into potassium chromate(VI). Is this change oxidation, or reduction? Give a reason for your choice. [2]

   c A piece of paper is coated with a layer of silver(I) chloride. It is used in this experiment.

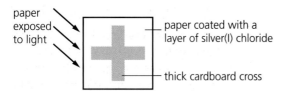

The cross is removed.

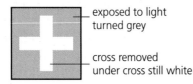

     i Explain why the silver(I) chloride that was not exposed to the light remained white but that which was exposed turned grey. [2]
     ii Write an equation for the reduction of the silver(I) ion. [1]
     iii What difference would using a brighter light make? [1]
     iv What is an important application of this reaction? [1]

   d The diagrams show an example of electrolysis and of a cell.

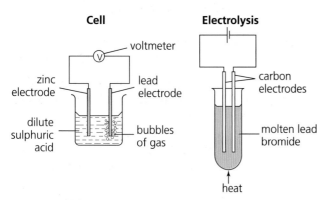

     i One of the above reactions is exothermic. State, giving a reason, which one. [1]
     ii How could you tell from the cell experiment which is the more reactive, zinc or lead? [1]
     iii Name the products of the electrolysis of molten lead bromide.
         product at negative electrode .... [1]
         product at positive electrode .... [1]

   e To compare the reactivity of lead, manganese, silver and iron, each metal was added to a solution containing the positive ion of a different metal. The results of this experiment are given in the table below.

	Pb^{2+}	Mn^{2+}	Ag^+	Fe^{2+}
Pb		no reaction	reaction	no reaction
Mn	reaction		reaction	reaction
Ag	no reaction	no reaction		no reaction
Fe	reaction	no reaction	reaction	

     i Write the four metals in order of reactivity. [2]
     ii Which metal most readily forms positive ions? [1]
     iii Which ion is the best oxidising agent? [1]

CIE 0620/03 Specimen paper Q3

7 Reversible reactions can come to equilibrium. They have both a forward and a backward reaction.

   a When water is added to an acidic solution of bismuth(III) chloride, a white precipitate forms and the mixture slowly goes cloudy.

$$BiCl_3(aq) + H_2O(l) \underset{backward}{\overset{forward}{\rightleftharpoons}} BiOCl(s) + 2HCl(aq)$$
colourless   white

     i Explain why the rate of the forward reaction decreases with time. [2]
     ii Why does the rate of the backward reaction increase with time? [1]
     iii After some time why does the appearance of the mixture remain unchanged? [2]
     iv When a few drops of concentrated hydrochloric acid are added to the cloudy mixture, it changes to a colourless solution. Suggest an explanation. [2]

   b Both of the following reactions are reversible.

reaction 1 $N_2(g) + O_2(g) \rightleftharpoons 2NO(g)$
reaction 2 $2NO(g) + O_2(g) \rightleftharpoons 2NO_2(g)$

i Suggest a reason why an increase in pressure does not affect the position of equilibrium for reaction 1. [1]
ii What effect would an increase in pressure have on the position of equilibrium for reaction 2? Give a reason for your answer. [2]

CIE 0620/03 N '05 Q3

8 The following apparatus was used to measure the rate of the reaction between zinc and iodine.

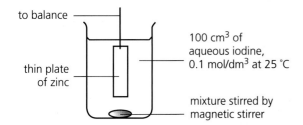

The mass of the zinc plate was measured every minute until the reaction was complete.

a Write an ionic equation for the redox reaction that occurred between zinc atoms and iodine molecules. [2]
b Describe how you could show, by adding aqueous sodium hydroxide and aqueous ammonia, that a solution contained zinc ions.

result with sodium hydroxide ....
excess sodium hydroxide ....
result with aqueous ammonia..............................
excess aqueous ammonia.............................. [3]

c From the results, two graphs were plotted.

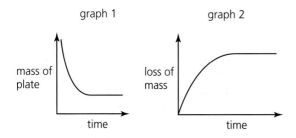

i Which reagent, iodine or zinc, was in excess? Give a reason for your choice. [1]
ii Describe how the shape of graph 1 would change if 100cm³ of 0.05 mol/dm³ iodine had been used. [2]
iii On graph 2, sketch the shape if the reaction is carried out using 100 cm³ of mol/dm³ iodine at 35 °C instead of at 25° C. [2]

CIE 0620/03 J '05 Q2

9 a Copper has the structure of a typical metal. It has a lattice of positive ions and a 'sea' of mobile electrons. The lattice can accommodate ions of a different metal.

Give a **different** use of copper that depends on each of the following.
i the ability of the ions in the lattice to move past each other [1]
ii the presence of mobile electrons [1]
iii the ability to accommodate ions of a different metal in the lattice [1]

b Aqueous copper(II) sulphate solution can be electrolysed using carbon electrodes. The ions present in the solution are as follows.

Cu^{2+} (aq), SO_4^{2-} (aq), H^+(aq), OH^-(aq)

i Write an ionic equation for the reaction at the negative electrode (cathode). [1]
ii A colourless gas was given off at the positive electrode (anode) and the solution changes from blue to colourless. Explain these observations. [2]

c Aqueous copper(II) sulphate can be electrolysed using copper electrodes. The reaction at the negative electrode is the same but the positive electrode becomes smaller and the solution remains blue.
i Write a word equation for the reaction at the positive electrode. [1]
ii Explain why the colour of the solution does not change. [2]
iii What is the large scale use of this electrolysis? [1]

CIE 0620/03 J '04 Q5

10 One use of the polymer, polyacrylonitrile, is to make carbon fibres. The monomer, acrylonitrile, is made by the following reaction.

$2CH_3—CH=CH_2 + 2NH_3 + 3O_2 \rightarrow 2CH_2=CH—CN + 6H_2O$

a Propene is made by the thermal cracking of the naphtha fraction of petroleum. This is a mixture of alkanes, C_4 to C_{10}.
i Name the technique used to obtain naphtha from petroleum. [1]
ii Predict the formula for the C_{10} alkane.
   C_{10} .... [1]
iii Write a symbol equation for the cracking of hexane (C_6H_{14}) to form propene. [1]

b Ammonia is manufactured by the Haber Process.
   forward reaction is exothermic

   $N_2(g) + 3H_2(g) \rightleftharpoons 2NH_3(g)$
   450 °C
   200 atmospheres pressure

   i Explain why a high pressure increases the % of ammonia in the equilibrium mixture. [2]
   ii At 300 C, the yield of ammonia would be greater. Why is this temperature not used? [1]
c The repeat unit of the polymer polyacrylonitrile is shown below. Add two more units to the diagram.

   —C(H)(H)—C(CN)(H)—
   [2]

   CIE 0620/03 Specimen paper Q4

11 Enzymes are biological catalysts. They are used both in research laboratories and in industry.
a Enzymes called proteases can hydrolyse proteins to amino acids. The amino acids can be separated and identified by chromatography. This diagram shows a typical chromatogram.

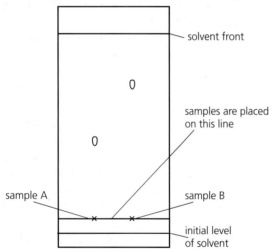

   i The R_f value of a sample =
      $\dfrac{\text{distance travelled by sample}}{\text{distance travelled by solvent front}}$

   Some R_f values for amino acids are:
   glutamic acid = 0.4   glycine = 0.5
   alanine = 0.7   leucine = 0.9
   Identify the two amino acids on the chromatogram.
   A is .... B is .... [2]
   ii Explain why the chromatogram must be exposed to a locating agent before R_f values can be measured. [1]
   iii R_f values give one way to identify amino acids on a chromatogram. Suggest another. [1]
   iv Nylon has the same linkage as proteins. Draw the structural formula of nylon.
b Enzymes called carbohydrases can hydrolyse complex carbohydrates to simple sugars which can be represented as HO—☐—OH. Draw the structure of a complex carbohydrate. [2]

   CIE 0620/03 J '05 Q5

12 Manganese is a transition element. It has more than one valency and the metal and its compounds are catalysts.
a i Predict three other properties of manganese that are typical of transition elements. [3]
   ii Complete the electron distribution of manganese by inserting one number.
      2 + 8 + ... + 2   [1]
b It has several oxides. Three are shown below:
   manganese(II) oxide, which is basic.
   manganese(III) oxide, which is amphoteric.
   manganese(IV) oxide, which is acidic.
   i Complete the word equation.
      manganese(II) oxide + hydrochloric acid
      → .... + ... [2]
   ii Which, if any, of these oxides will react with sodium hydroxide? [1]
c Aqueous hydrogen peroxide decomposes to form water and oxygen. This reaction is catalysed by manganese(IV) oxide:

   $2H_2O_2(aq) \rightarrow 2H_2O(l) + O_2(g)$

   The following experiments were carried out to investigate the rate of this reaction.
   A 0.1 g sample of manganese (IV) oxide was added to 20 cm³ of 0.2 M hydrogen peroxide solution. The volume of oxygen produced was measured every minute. The results of this experiment are shown on the graph.

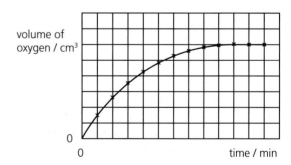

i How does the rate of reaction vary with time? Explain why the rate varies. [3]
ii The experiment was repeated at the same temperature, using 0.1 g of manganese(IV) oxide and 20 cm³ of 0.4 M hydrogen peroxide. Sketch the curve for this experiment on the same grid. [2]
iii How would the shape of the graph differ if only half the mass of catalyst had been used in these experiments? [2]

CIE 0620/03 N '02 Q2

13 The Carisbad caverns in New Mexico are very large underground caves. Although the walls of these caves are coated with gypsum (hydrated calcium sulphate), the caves have been formed in limestone.

a It is believed that the caves were formed by sulphuric acid reacting with the limestone.
  i Complete the word equation.
  calcium carbonate + sulphuric acid →
  calcium sulphate + ...+ .... [1]
  ii Describe how you could test the water entering the cave to show that it contained sulphate ions. [2]
  iii How could you show that the water entering the cave has a high concentration of hydrogen ions? [1]

b Hydrogen sulphide gas which was escaping from nearby petroleum deposits was being oxidised to sulphuric acid.
  i Complete the equation for this reaction forming sulphuric acid.
  H_2S + ... O_2 → .... [2]
  ii Explain why all the hydrogen sulphide should be removed from the petroleum before it is used as a fuel. [1]
  iii Draw a diagram to show the arrangement of the valency electrons in one molecule of the covalent compound hydrogen sulphide. Use O to represent an electron from a sulphur atom. Use x to represent an electron from a hydrogen atom. [2]

c Sulphuric acid is manufactured by the Contact Process. Sulphur dioxide is oxidised to sulphur trioxide by oxygen.
$2SO_2$ + O_2 → $2SO_3$
  i Name the catalyst used for this reaction. [1]
  ii What temperature is used for it? [1]
  iii Describe how sulphur trioxide is changed into sulphuric acid. [1]

d Gypsum is hydrated calcium sulphate, $CaSO_4.xH_2O$. It contains 20.9% water by mass. Calculate x. M_r: $CaSO_4$, 136; H_2O, 18

79.1 g of $CaSO_4$ = ... moles
20.9 g of H_2O = ... moles
x = ... [3]

CIE 0620/03 J '05 Q4

14 The elements in Period 3 and some of their common oxidation states are shown below.

Element	Na	Mg	Al	Si	P	S	Cl	Ar
Oxidation state	+1	+2	+3	+4	–3	–2	–1	0

a i Why do the oxidation states increase from sodium to silicon? [1]
  ii After Group(IV) the oxidation states are negative and decrease across the period. Explain why. [2]
b Predict the formulae of these compounds:

  aluminium sulphide ....
  silicon phosphide .... [2]

c Choose a different element from Period 3 that matches each description.
  i It has a similar structure to diamond. [1]
  ii It reacts violently with cold water to form a solution with pH = 14. [1]
  iii It has a gaseous oxide of the type XO_2, which is acidic. [1]
d The only oxidation state of argon is zero. Why is argon used to fill light bulbs? [1]
e Draw a diagram that shows the arrangement of the valency electrons in the ionic compound sodium phosphide. Use O to represent an electron from sodium. Use x to represent an electron from phosphorus. [3]
f Sodium reacts with sulphur to form sodium sulphide: $2Na + S \rightarrow Na_2S$
An 11.5 g sample of sodium is reacted with 10 g of sulphur. All of the sodium reacted but there was an excess of sulphur.
Calculate the mass of sulphur left unreacted.
  i Number of moles of sodium atoms reacted = ...
  [2 moles of Na react with 1 mole of S]
  ii Number of moles of sulphur atoms that reacted = ... g
  iii Mass of sulphur reacted = ... g
  iv Mass of sulphur left unreacted = ... g [4]

CIE 0620/03 N '02 Q3

Working with gases in the lab

Preparing gases in the lab

The usual way to make a gas is to displace it from a solid or solution, using apparatus like this. The table below gives some examples.

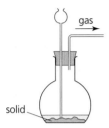
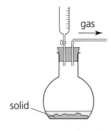

Apparatus for preparing gases. The dropping funnel in the right-hand apparatus is used for adding concentrated acids.

To make ...	Place in flask	Add	Reaction
carbon dioxide	calcium carbonate (marble chips)	dilute hydrochloric acid	$CaCO_3\,(s) + 2HCl\,(aq) \longrightarrow CaCl_2\,(aq) + H_2O\,(l) + CO_2\,(g)$
chlorine	manganese(IV) oxide (as an oxidizing agent)	conc hydrochloric acid	$2HCl\,(aq) + [O] \longrightarrow H_2O\,(l) + Cl_2\,(g)$
hydrogen	pieces of zinc	dilute hydrochloric acid	$Zn\,(s) + 2HCl\,(aq) \longrightarrow ZnCl_2\,(aq) + H_2\,(g)$
oxygen	manganese(IV) oxide (as a catalyst)	hydrogen peroxide	$2H_2O_2\,(aq) \longrightarrow 2H_2O\,(l) + O_2\,(g)$

But to make ammonia, heat any ammonium compound with a strong alkali.

Collecting the gases you have prepared

The table below shows four ways of collecting a gas you have prepared. The method depends on whether the gas is heavier or lighter than air, whether you need it dry, and what you want to do with it.

Method	upward displacement of air	downward displacement of air	over water	gas syringe
Use when ...	the gas is heavier than air	the gas is lighter than air	the gas is sparingly soluble in water	you want to measure the volume
Apparatus	gas jar	gas jar	gas jar / water	gas syringe
Examples	carbon dioxide, CO_2 chlorine, Cl_2 sulphur dioxide, SO_2 hydrogen chloride, HCl	ammonia, NH_3 hydrogen, H_2	carbon dioxide, CO_2 hydrogen, H_2 oxygen, O_2	any gas

Tests for gases

You have a sample of gas. You think you know what it is, but you're not sure. So you need to do a test. Below are some tests for common gases. Each is based on particular properties of the gas, including its appearance, and sometimes its smell.

Gas	Description and test details
Ammonia, NH_3 Properties Test Result	 Ammonia is a colourless alkaline gas with a strong sharp smell. Carefully smell the gas, and hold damp indicator paper in it. Recognizable odour, and indicator paper turns blue.
Carbon dioxide, CO_2 Properties Test Result	 Carbon dioxide is a colourless, weakly acidic gas. It reacts with **lime water** (a solution of calcium hydroxide in water) to give a white precipitate of calcium carbonate: $CO_2\ (g) + Ca(OH)_2\ (aq) \longrightarrow CaCO_3\ (s) + H_2O\ (l)$ Bubble the gas through lime water. Lime water turns cloudy or milky.
Chlorine, Cl_2 Properties Test Result	 Chlorine is a green poisonous gas which bleaches dyes. Hold damp indicator paper in the gas, *in a fume cupboard*. Indicator paper turns white.
Hydrogen, H_2 Properties Test Result	 Hydrogen is a colourless gas which combines violently with oxygen when lit. Collect the gas in a tube and hold a lighted splint to it. The gas burns with a squeaky pop.
Oxygen, O_2 Properties Test Result	 Oxygen is a colourless gas. Fuels burn much more readily in it than in air. Collect the gas in a test tube and hold a glowing splint to it. The splint immediately bursts into flame.

Testing for ions in the lab

You have a solid or solution. You think you know what it is, but you're not sure. It's time for some detective work! Below are tests you can do.

Testing for halide ions (Cl⁻, Br⁻, I⁻)

- Take a small amount of the solution.
- Add an equal volume of dilute nitric acid.
- Then add silver nitrate solution.
- Silver halides are insoluble. So if halide ions are present a precipitate will form, as shown in this table.

Precipitate	Indicates presence of …	Reaction taking place
white	chloride ions, Cl⁻	$Ag^+ (aq) + Cl^- (aq) \longrightarrow AgCl (s)$
cream	bromide ions, Br⁻	$Ag^+ (aq) + Br^- (aq) \longrightarrow AgBr (s)$
yellow	iodide ions, I⁻	$Ag^+ (aq) + I^- (aq) \longrightarrow AgI (s)$

Note that iodide ions can also be identified using acidified lead(II) nitrate solution. A deep yellow precipitate of lead(II) iodide forms.

Testing for sulphate ions (SO_4^{2-})

- Take a small amount of the solution.
- Add an equal volume of dilute hydrochloric acid.
- Then add barium nitrate solution.
- Barium sulphate is insoluble. So if sulphate ions are present a precipitate will form, as shown in this table:

Precipitate	Indicates presence of …	Reaction taking place
white	sulphate ions, SO_4^{2-}	$Ba^{2+}(aq) + SO_4^{2-} (aq) \longrightarrow BaSO_4(s)$

Testing for metal ions (Cu^{2+}, Fe^{2+}, Fe^{3+}, Al^{3+}, Zn^{2+}, Ca^{2+})

- Take a small amount of the solution.
- Add a few drops of dilute sodium hydroxide solution.
- If a *coloured* precipitate forms it indicates the presence of Cu^{2+}, Fe^{2+}, or Fe^{3+} ions. These form insoluble coloured hydroxides, as shown in this table:

Precipitate	Indicates presence of …	Reaction taking place
pale blue	copper ions, Cu^{2+}	$Cu^{2+} (aq) + 2OH^- (aq) \longrightarrow Cu(OH)_2 (s)$
green	iron(II) ions, Fe^{2+}	$Fe^{2+} (aq) + 2OH^- (aq) \longrightarrow Fe(OH)_2 (s)$
red-brown	iron(III) ions, Fe^{3+}	$Fe^{3+} (aq) + 3OH^- (aq) \longrightarrow Fe(OH)_3 (s)$

- Note that the pale blue precipitate of copper(II) hydroxide will dissolve in excess ammonia solution, giving a deep blue solution.
- If a *white* precipitate forms it suggests that Al^{3+}, Zn^{2+}, or Ca^{2+} ions are present. These form insoluble white hydroxides.
- To find out which of those three ions it is, divide the solution containing the precipitate into two equal volumes.

- To one, add double the volume of sodium hydroxide solution. To the other, add double the volume of ammonium hydroxide solution. Check the result against this table:

If the precipitate ...	It indicates the presence of ...
dissolves again in sodium hydroxide solution, giving a colourless solution	Al^{3+} ions
dissolves in both sodium hydroxide and ammonium hydroxide solutions, giving a colourless solution	Zn^{2+} ions
dissolves in neither	Ca^{2+} ions

Carbonate ions (CO_3^{2-})

- Take a small amount of the unknown solid or solution.
- Add a little dilute hydrochloric acid.
- If the mixture bubbles and gives off a gas that turns lime water milky, the unknown substance contained carbonate ions.

The reaction that takes place is:
$$2H^+ (aq) + CO_3^{2-} (aq) \longrightarrow CO_2 (g) + H_2O (l)$$
from the acid   carbon dioxide
                                    (turns limewater milky)

Ammonium ions (NH_3^+)

- Take a small amount of the unknown solid or solution.
- Add a little dilute sodium hydroxide solution and heat gently.
- If ammonia gas is given off, the unknown substance contained ammonium ions. (Ammonia gas has a strong sharp smell, and turns red litmus blue.)

The reaction that takes place is:
$$NH_4^+ (aq) + OH^- (aq) \longrightarrow NH_3 (g) + H_2O (l)$$
           from sodium   ammonia
           hydroxide

Nitrate ions (NO_3^-)

- Take a small amount of the unknown solid or solution.
- Add a little sodium hydroxide solution.
- Add some aluminium foil. Heat gently.
- If ammonia gas is given off, the unknown substance contained nitrate ions.

The reaction that takes place is:
$$8Al (s) + 3NO_3^- (aq) + 5OH^- (aq) + 2H_2O (l) \longrightarrow 3NH_3 (g) + 8AlO_2^- (aq)$$
                  from sodium
                  hydroxide

Safety first!

Rules for working safely in the lab

These rules will help you avoid accidents in the lab.

- Follow instructions exactly.
- Always wear safety glasses when working in the lab. If chemicals get in your eyes they could blind you.
- Tie hair back.
- Don't fool around with Bunsen burners.
- Stand up when working with liquids.
- Don't splash chemicals around.
- Mop up spills carefully with a cloth for that purpose.
- If you have to smell a substance, do so with care. DO NOT smell poisonous gases such as chlorine.
- DO NOT EVER taste anything in the lab.
- Handle chemicals with respect : especially acids, alkalis, and poisonous and flammable substances.
- Use a fume cupboard for experiments involving poisonous or highly reactive substances.
- Handle glassware with care.
- Find out NOW what to do in case of fire, or chemicals in your eye, or other kinds of accident. Talk the safety drill over with your teacher.
- Know the symbols for chemical hazards. They're given below. Watch out for them on bottles and other chemical containers.

Symbols for chemical hazards

Toxic

These substances can kill. They may act when you swallow them, or breathe them in, or absorb them through your skin.

Examples: chlorine gas
mercury

Harmful

These are similar to toxic substances, but less dangerous.

Examples: copper(II) sulphate
lead oxide

Highly flammable

These substances catch fire easily, so pose a fire risk.

Examples: hydrogen
ethanol

Oxidising

These provide oxygen, which allows other substance to burn more fiercely.

Examples: hydrogen peroxide
potassium manganate(VII)

Corrosive

These attack and destroy living tissue, including eyes and skin.

Examples: conc sulphuric acid
bromine

Irritant

These are not corrosive but can cause reddening or blistering of skin.

Examples: calcium chloride
zinc sulphate

Answers to numerical questions

page 15 3 b 20 times
page 29 6 C 6p 6e 6n; O 8p 8e 8n; Mg 12p 12e 12n; Al 13p 13e 14n; Cu 29p 29e 35n
page 34 4 b i 60 ii 34 iii 0 iv 10 v 146
page 35 12 a i 38 ii 40
page 73 8 b The correct values for rubidium are: melting point 39 °C, boiling point 688 °C d i 5 ii 37 iii 1
page 77 4 a 32 b 254 c 16 d 71 e 58 f 46 g 132
page 79 1 60% 2 hydrogen 5%, oxygen 60% 3 80% 4 a 15g b 3g
page 81 4 a 1g b 127g c 35.5g d 71g 5 a 32g b 64g 6 138g 7 a 9 moles b 3 moles 8 a 6.02×10^{23} b 35.5g
page 83 1 a 4 b 4g 3 FeS 4 SO_3
page 85 4 CH 5 C_2H_4 6 a C_7H_{16} b C_7H_{16} 7 P_4O_6
page 87 1 a 1 mole b 1 mole 2 a 2 mol/dm^3 b 1.5 mol/dm^3 3 a 0.5 dm^3 (500 cm^3) b 0.005 dm^3 (or 5 cm^3) 4 a 20g b 0.5g 5 a 0.5 mol/dm^3 b 0.25 mol/dm^3
page 88 1 a 64g b 48g c 48g d 60g e 355g f 1.4g g 4 g 2 a 2g b 4g c 32g d 35.5g e 248g f 1024g g 144g 3 a 1 mole b 2 moles c 1 mole d 2 moles e 0.2 moles f 0.1 moles g 0.4 moles h 0.2 moles i 2 moles 4 a 80g of sulphur b 80g of oxygen c 8 moles of chlorine atoms d 1 mole of oxygen molecules 5 a 90g b 250g c 34g d 30g e 8g f 15.8g g 325g 7 The missing numbers are: a 40, 16, 1, 56 b 1.6, 5.6 c 0.16 d 6, 3 e 71.4% 8 a 64g b 4 moles c 2 moles d MnO_2 e 632.2g 9 a 106.5 b 3 moles c 1mole d $AlCl_3$ e 0.1 mol/dm^3
page 89 10 a N_2H_2 b C_2N_2 c N_2O_4 d $C_6H_{12}O_6$ 11 b CH_2 c A is C_3H_6 B is C_6H_{12} 12 a Zn_3P_2 b 24.1% 13 a P_2O_3 b 41.3g c P_4O_{10} d P_4O_6 14 a 1 dm^3 of 2M sodium chloride b 1 dm^3 of 1M sodium chloride c 100 cm^3 of 2M sodium chloride d 1 dm^3 of 1M sodium hydroxide 15 a 45.5 cm^3 b 41.7 cm^3 c 62.5 cm^3 16 c The missing figures are: group 4, 0.19g; group 5, 0.20g e 0.16g f 16g g 2 moles h 1 mole i Cu_2O
page 95 1 b 2 moles c i 32g ii 8g 2 b $CuCO_3$, 124g; CuO, 80g; CO_2, 44g c i 11g ii 20g
page 97 3 24 dm^3 4 a 168 dm^3 b 12 dm^3 c 0.0024 dm^3 (24 cm^3) 5 a 12 dm^3 b 2.4 dm^3 6 a 12 dm^3 b 12 dm^3 7 a 12 dm^3 b 6 dm^3
page 99 2 76.6% 3 63% 4 172g 5 88%
page 100 5 a 217g b 20.1g of mercury, 1.6g of oxygen c 94.5 % 6 a 32g b sulphur c 11g of iron(II) sulphide and 6g of sulphur d 17.5g 7 b 160g c 2000 moles d 2 moles e 4000 moles f 224kg
page 101 8 a 48 dm^3 b 12 dm^3 c 0.24 dm^3 d 7.2 dm^3 e 2.4 dm^3 f 0.072 dm^3 or 72 cm^3 9 b i 0.25 moles of molecules ii 0.125 moles of molecules iii 0.1 moles of atoms iv 25 moles of molecules v 0.05 moles of molecules vi 0.02 moles of molecules c i 0.5g ii 4g iii 2g iv 700g v 2.2g vi 1.28g 10 b i 2 moles ii 168g c i 1 mole ii 24 dm^3 or 24000 cm^3 d i 12 dm^3 or 12000 cm^3 ii 1.2 dm^3 or 1200 cm^3 11 b 0.5 moles c i 11.2g ii 8.8g iii 4.8 dm^3 or 4800 cm^3 12 a 0.5 moles b 25 cm^3 c i 75 cm^3 ii 50 cm^3 13 a i 4 moles ii 19 moles b 4.75 moles c 114 dm^3 d 227g e 502 dm^3 14 a 0.052 moles b 4.375g c 87.5% 15 a 1.4g b 0.025 moles c 0.025 moles d Fe^{2+} e 0.6 dm^3
page 129 2 50 cm^3 3 1.6 mol/dm^3
page 131 8 c 0.0015 moles d 0.0005 moles e 3 9 b 0.014 moles c 0.007 moles d 0.742g e 1.258g f 0.07 moles g 10 moles
page 147 3 a i 29 cm^3 ii 39 cm^3 b i 0.65 minutes ii 1.5 minutes c i 5 cm^3 of hydrogen per minute ii 0 cm^3 of hydrogen per minute
page 149 1 a i 60 cm^3 ii 60 cm^3
page 151 1 a experiment 1, 0.55g; experiment 2, 0.95g b experiment 1, 0.55g; experiment 2, 0.95g c experiment 1, 0.33g per minute; experiment 2, 0.5g per minute
page 158 2 c i 14 cm^3 ii 9 cm^3 iii 8 cm^3 e 40 cm^3 f 5 minutes g 8 cm^3 per minute
page 163 3 486 kJ/mol
page 170 2 a fall of 4 °C for NH_4NO_3, rise of 20 °C for $CaCl_2$ d i fall of 8 °C for NH_4NO_3, rise of 40 °C for $CaCl_2$ ii fall of 2 °C for NH_4NO_3, rise of 10 °C for $CaCl_2$ iii fall of 4 °C for NH_4NO_3, rise of 20 °C for $CaCl_2$ 3 b 4 c 416 kJ/mol d iii 330 kJ/mol 4 d i 2148 kJ/mol ii 2354 kJ/mol iii −206 kJ/mol
page 171 5 d 55.6 kJ/mol 6 c i 2220 kJ/mol ii 2801 kJ/mol d −581 kJ/mol
page 182 2 b iv the metal is iron (density 7.9 g/cm^3)
page 183 11 c i 2.0 volts ii 0.2 volts
page 200 5 e 1000g of zinc ore
page 210 3 d 33 cm^3 e oxygen 33 %, nitrogen 67 % f oxygen 21 %, nitrogen 79 % 5 d 20.8%
page 219 6 a 21.2 %
page 221 10 c 26.2 %

Examination questions
core
6 b ii bromine-79: 35e 44n 35p; bromine-81: 35e 46n 35p
8 e i 2550 kg ii 3.6%
extended
1 a i ^{88}Sr 38p 38e 50n; ^{90}Sr 38p 38e 52n; ^{65}Zn^{2+} 30p 28e 35n 2 c i 0.01 moles, 0.01 moles, 0.005 moles ii 0.01 moles, 0.02 moles, 20 cm^3
11 a i A glutamic acid, B alanine 13 d 0.582 moles, 1.16 moles, $x = 2$ 14 f i 0.5 moles ii 0.25 moles iii 8 g iv 2 g

Glossary

A

acidic solution has a pH less than 7; an acidic solution contains H^+ ions

acid rain rain that is acidic because gases such as sulphur dioxide are dissolved in it (from burning fossil fuels)

activation energy the energy needed to break the bonds in the reactants, before a reaction can take place

addition polymerisation where small molecules join to form a very large molecule, by adding on at double bonds

addition reaction where a molecule adds onto an alkene, and the C = C double bond of the alkene changes to a single bond

air the mixture of gases that surrounds us

alkali a soluble base; for example sodium hydroxide

alkali metals the Group 1 elements of the Periodic Table

alkaline earth metals the Group 2 elements of the Periodic Table

alkaline solution has a pH above 7; alkaline solutions contain OH^- ions

alkanes a family of saturated hydrocarbons with the general formula C_nH_{2n+2}; 'saturated' means they have only single C–C bonds

alkenes a family of unsaturated hydrocarbons with general formula C_nH_{2n}; their molecules contain a carbon= carbon double bond

allotropes different forms of the same element; diamond and graphite are allotropes of carbon

alloy a mixture where at least one other substance is added to a metal, to improve its properties; the other substance is often a metal too (but not always)

amphoteric can be both acidic and basic in its reactions; for example aluminium oxide is an amphoteric oxide

anode the positive electrode of a cell

aquifer underground rocks containing a large volume of water; it can be pumped out to give a water supply

atmosphere the mixture of gases around the Earth; down here at the Earth's surface, we call it air

atomic number the number of protons in the atoms of an element; is is also called the proton number

atoms elements are made up of atoms, which contain protons, neutrons, and electrons

Avogadro constant the number of particles in one mole of an element or compound; it is 6.02×10^{23}

B

back reaction the reaction in which the product breaks down again, in a reversible reaction

bacteria tiny organisms, some of which can cause disease; others break down dead plant and animal material

balanced equation a chemical equation in which the number of each type of atom is the same on both sides of the arrow

base a chemical that neutralises an acid, to form a salt and water

batch process a batch of reactants is mixed, and the products are removed when the reaction is over

battery a portable electrical cell; for example a torch battery

biodegradable will decay naturally in the soil, with the help of bacteria

biopolymer a polymer made by bacteria

blast furnace the chemical plant in which iron is extracted from its ore, iron(III) oxide

boiling the change from a liquid to a gas, which takes place at the boiling point

boiling point the temperature at which a substance boils

bond energy the energy needed to break a bond, or released when the bond is formed; it is given in kilojoules (kJ) per mole

bonding how the atoms are held together in an element or compound; there are three types of bonds: ionic, covalent, and metallic

brine the industrial name for a concentrated solution of sodium chloride in water; it can be made by dissolving rock salt

brittle breaks up easily when struck

burette a piece of laboratory equipment for delivering a measured volume of liquid

burning an exothermic chemical reaction in which the reactant combines with oxygen to form an oxide

C

carbon dating a way to find the age of something (for example ancient bones, or cloth, or wood) by measuring the radiation from its carbon-14 atoms

cast iron iron from the blast furnace that is run into moulds to harden; it contains a high % of carbon, which makes it brittle

catalyst a substance that speeds up a chemical reaction, without being used up in the process

catalytic converter a device in a car exhaust, in which catalysts are used to convert harmful gases to harmless ones

catalytic cracking where large molecules of hydrocarbons are split up into smaller ones, with the help of a catalyst

cathode the negative electrode

cell (biological) the building blocks for animals and plants

cell (electrical) a device that converts chemical energy to electrical energy

cement a substance used in building, made from limestone and clay

centrifuge to separate a solid from a liquid by rapid spinning; the solid falls to the bottom

ceramic a hard, ureactive material that can withstand high temperatures, made by baking clay in a kiln

chalk a rock made of calcium carbonate

change of state a change in the physical state of a substance – for example from solid to liquid, or liquid to gas

chemical change a change in which a new chemical substance forms

chemical equation uses chemical symbols to describe a chemical reaction in a short way

chemical reaction a process in which chemical change takes place

chromatogram the paper showing the separated coloured substances, after chromatography has been carried out

chromatography a way to separate the substances in a mixture, using a solvent and filter paper (or other support); the substances separate because they travel over the paper at different speeds

coagulant a substance that will make small particles stick together; coagulants are used in cleaning up water, ready for piping to homes

coke a form of carbon made by heating coal; it is used in the blast furnace,

combination reaction where two or more substances chemically combine to give a single substance

combustible can catch fire and burn very easily

combustion another name for burning

compound a substance in which two or more elements are chemically combined

compound fertiliser it provides nitrogen, potassium, and phosphorus for plants

compound ion an ion containing more than one element; for example the nitrate ion NO_3^-

concentration tells you how much of one substance is dissolved in another

condensation the physical change in which a gas turns into a liquid on cooling

condensation polymerisation where molecules join to make very large molecules, by eliminating small molecules (such as water molecules)

condenser a piece of laboratory equipment used to cool a gas rapidly, and turn it into a liquid

conductor a substance that allows heat or electricity to pass through it easily

Contact process the industrial process for making sulphuric acid

continuous flow process a process in which reactants flow into the system non-stop, and the products flow out

corrosion where a substance is attacked by air or water, from the surface inwards; the corrosion of iron is called rusting

covalent bond the chemical bond formed when two atoms share electrons

covalent compound a compound made of atoms joined by covalent bonds

cracking the chemical process in which long-chain hydrocarbon molecules are broken down to shorter, more useful molecules

cross-linking the chemical bonds between the long-chain molecules in some polymers, that hold the chains together

crude oil the fossil fuel formed over millions of years from the remains of tiny sea plants and animals

crystallisation the process in which crystals form, as a saturated solution cools

D

decomposition reaction where a substance breaks down to give two or more products

degradeable will break down naturally (for example through the action of bacteria)

denature to destroy the structure and properties of an enzyme by heating, or a change in pH

density tells you how 'heavy' something is; the density of a substance is its mass per unit volume; for water it is 1 g/cm^3

diatomic a substance is called diatomic if its molecules contain two atoms joined by a covalent bond

diffusion the process in which particles mix by colliding randomly with each other, and bouncing off in all directions

displacement reaction a reaction in which a more reactive element takes the place of a less reactive one, in a compound

dissolving the process in which a soluble substance forms a solution

distillation separating a liquid from a mixture by boiling it off, then condensing it

double bond a covalent bond in which two atoms share two pairs of electrons

ductile can be drawn out into a wire; for example copper is ductile

dynamic equilibrium where forward and back reactions take place at the same rate, so there is no *overall* change

E

economic process a process that is profitable to carry out

electrodes the electrical conductors used to carry current into and out of an electrolyte; they could be graphite rods, or strips of platinum, for example

electrolysis the process of breaking down a compound by passing an electric current through it

electrolyte the liquid through which the current is passed, in electrolysis; the current is carried by ions in the electrolyte

electronic configuration how the electrons in an atom are arranged

electrons the particles with a charge of 1-, and almost no mass, in an atom

electron shells the different energy levels where electrons are found

electroplating coating one metal with another, using electricity

element a substance that cannot be split into anything simpler, in a chemical reaction

empirical formula shows the simplest ratio in which the atoms in a compound are combined

endothermic describes a chemical reaction that takes in energy from the surroundings

environment the surroundings

enzymes proteins made by living cells, that act as biological catalysts

equation it represents a chemical reaction (reactants to products) using words or symbols

equilibrium the state where the forward and back reactions are taking place at the same rate, in a reversible reaction; so there is no *overall* change

ester a compound formed when an alcohol reacts with a carboxylic acid; esters often smell of fruit or flowers

evaporation the physical change where a liquid turns to a gas, at a temperature below its boiling point

exothermic describes a chemical reaction that gives energy out to the surroundings

extract to remove a metal from its ore

F

fermentation the process in which the enzymes in yeast break down sugars, to form alcohol (ethanol) and carbon dioxide

fertilisers substances added to soil to help crops grow well

filtering separating solids from liquids by pouring through filter paper

filtrate the liquid obtained from filtration (from which the solid has been removed)

flammable burns easily

formula uses symbols and numbers to tell you what elements are in a compound, and the ratio in which they are combined

forward reaction the reaction in which the product is made, in a reversible reaction

fossil fuels fuels formed over millions of years from the remains of living things

fractional distillation a method used to separate two or more liquids that have different boiling points

fractions different parts of a mixture, with different ranges of boiling point, that are obtained by fractional distillation

freezing the physical change from a liquid to a solid, that happens at the melting point

fuel a substance that is burned to release energy

fuel cell a cell in which a chemical reaction provides electricity, to light homes, power cars and so on

functional group the group of atoms in a family of organic compounds, that dictates most of its reactions; it is C=C for the alkene family, and –OH for the alcohol family

G

galvanised iron iron coated with zinc, to prevent it from rusting

giant structure where millions of atoms or ions are held by a network of strong bonds; metals, diamond and ionic solids such as sodium chloride are all giant structures

global warming the rise in average temperatures that is now taking place around the world; experts say that carbon dioxide from burning fossil fuels is the main cause

greenhouse gas a gas such as carbon dioxide that traps heat around the Earth, preventing it from escaping into space; too much carbon dioxide in the atmosphere means the Earth is getting warmer

ground water the water trapped in rocks below the ground

group a column of the Periodic Table; elements in a group have similar properties

H

Haber process the industrial process for making ammonia from nitrogen and hydrogen

halogens the Group 7 elements of the Periodic Table

heating curve a graph showing how the temperature of a substance changes on heating, while it goes from solid to liquid to gas

homologous series a family of organic compounds, that share the same general formula and have similar properties

hydrocarbon a compound containing only carbon and hydrogen; the alkanes and alkenes are families of hydrocarbons

hydrogenation adding hydrogen to the double bonds of an unsaturated compound, to make it saturated

hydrogen fuel cell it uses hydrogen from a tank, and oxygen from the air, to give an electric current

hydrolysis the breaking down of a compound by reaction with water

I

immiscible liquids liquids that do not mix (for example cooking oil and water)

incomplete combustion the burning of fuels in a limited supply of oxygen; it gives carbon (soot) and carbon monoxide

indicator a chemical that shows by its colour whether a substance is acidic or alkaline

inert does not react (except under extreme conditions)

inert electrode is not itself changed during electrolysis; all it does is conduct a current

insoluble does not dissolve in a solvent

insulator a poor conductor of heat or electricity

intermolecular forces forces between molecules

ion a charged atom or group of atoms formed by the gain or loss of electrons

ionic bond the bond formed between ions of opposite charge

ionic compound a compound made up of ions, joined by ionic bonds

ionic equation shows only the ions that actually take part in a reaction, and ignores the other ions that are present; the other ions are called spectator ions

ionosphere the top layer of the atmosphere, that starts about 80 km above us; it mainly contains charged particles

isomers compounds that have the same formula, but a different arrangement of atoms

isotopes atoms of the same element, that have a different numbers of neutrons

L

lattice a regular arrangement of particles

lime water a solution of the slightly soluble compound calcium hydroxide, which is used to test for carbon dioxide

locating agent a substance used to show up colourless substances on a chromatogram; it reacts with them to give coloured substances

M

macromolecule a very large molecule; for example a molecule in a polymer

malleable can be hammered into shape

mass number the total number of protons and neutrons in an atom; it is another term for nucleon number

mass spectrometer an instrument used to find the masses of atoms and molecules

melting the physical change from a solid to a liquid

melting point the temperature at which a solid substance melts

mesosphere the layer of the atmosphere that extends from about 50 km to 80 km above us; there is not much gas in the atmosphere at that height

metal an element to the left of the zig-zag line on the Periodic Table, that shows metallic properties

metallic bond the bond that holds the atoms together in a metal

metalloid an element that has properties of both a metal and a non-metal; the metalloids lie along the zig-zag line that separates the metals and non-metals, in the Periodic Table

minerals compounds that occur naturally in the Earth; rocks are made up of different minerals

mixture contains two or more substances that are not chemically combined

mobile phase the solution that travels over the filter paper or other support, in chromatography

molar solution contains one mole of a substance in 1 dm^3 (1 litre) of water

molecular substance a substance made up of molecules

molecule a unit of two or more atoms held together by covalent bonds

monatomic describes an element made up of single atoms; for example neon

monomers small molecules that join together to form polymers

N

native describes a metal that is found in the Earth as the element

negative electrode another name for the cathode

negative ion an ion with a negative charge

neutral (electrical) has no charge

neutral (oxide) is neither acidic nor basic; carbon monoxide is a neutral oxide

neutral (solutions) neither acidic nor alkaline; neutral solutions have a pH of 7

neutralisation the chemical reaction between an acid and a base, which produces a salt and water

neutron a particle with no charge and a mass of 1 unit, found in the nucleus of an atom

nitrogenous fertiliser it provides nitrogen for plants, in the form of nitrate ions or ammonium ions

noble gases the Group 0 elements of the Periodic Table; they are called 'noble' because they are so unreactive

non-metal an element to the right of the zig-zag line in the Periodic Table, that does not show metallic properties

non-renewable resource a resource such as crude oil that we are using up, and which will run out one day

non-toxic not harmful to health

nucleon number the total number of protons and neutrons in an atom of an element; it is also called the mass number

nucleus the centre part of the atom, made up of protons and neutrons

O

ore rock containing a metal, or metal compounds, from which the metal is extracted

organic chemistry the study of organic compounds

organic compound a compound containing carbon, that originated in a living thing; crude oil is a mixture of organic compounds

oxidation a chemical reaction in which a substance gains oxygen, loses hydrogen, or loses electrons

oxidation state every atom in a formula can be given a number that describes its oxidation state; this is zero for the atoms in an element

oxide a compound formed between oxygen and an other element

ozone a gas with the formula O_3

ozone layer the layer of ozone up in the atmosphere, which protects us from harmful UV radiation from the sun

P

percentage composition it tells you which elements are in a compound, and what % of each is present by mass

period a row of the Periodic Table

periodicity the repeating pattern of properies when elements are arranged in order of their atomic number; this shows up clearly in the Periodic Table

Periodic Table the table showing the elements in order of increasing atomic number; similar elements are arranged in columns called groups

pH scale a scale that tells you how acidic or alkaline a solution is

photochemical reaction a reaction that depends on light energy to start it off

photodegradeable can be broken down by light

photosynthesis the process in which plants convert carbon dioxide and water to glucose

physical change a change in which no new chemical substance is made; melting and boiling are physical changes

physical properties properties such as density and melting point (that are not about chemical behaviour)

pipette a piece of laboratory equipment used to deliver a known volume of liquid, accurately

plastics a term used for synthetic polymers (made in factories, rather than in nature)

pollutant a substance that causes harm if it gets into the air or water

pollution when harmful substances are released into the environment

polymer a compound containing very large molecules, formed by polymerisation

polymerisation a chemical reaction in which many small molecules join to form very large molecules; the product is called a polymer

positive ion an ion with a positive charge

precipitate an insoluble chemical produced during a chemical reaction

precipitation reaction a reaction in which a precipitate forms

product a chemical made in a chemical reaction

protein a polymer made up of many different amino acid units joined together

proton a particle with a charge of 1+ and a mass of 1 unit, found in the nucleus of an atom

proton number the number of protons in the atoms of an element; it is also called the atomic number

pure there is only one substance in it

Q

quicklime another name for calcium oxide

R

radioactive isotopes (radioisotopes) unstable atoms that break down, giving out radiation

random motion the zig-zag path a particle follows as it collides with other particles and bounces away again

rate of reaction how fast a reaction is

reactant a starting chemical for a chemical reaction

reactive tends to react easily

reactivity how readily a substance reacts

reactivity series the metals listed in order of their reactivity

recycling reusing resources such as scrap metal, glass, paper and plastics

redox reaction any reaction in which electrons are transferred; one substance is oxidised (it loses electrons) and another is reduced (it gains electrons)

reducing agent a substance which brings about the reduction of another substance

reduction when a substance loses oxygen, gains hydrogen, or gains electrons

refining (metals) the process of purifying a metal; copper is refined using electrolysis

refining (oil) the process of separating crude oil into groups of compounds with molecules of roughly similar size; it is carried out by fractional distillation

relative atomic mass (A_r) the average mass of the atoms of an element, relative to the mass of an atom of carbon-12

relative formula mass (M_r) the mass of a formula unit of a substance; it is found by adding the relative atomic masses of the atoms in the formula (eg in the formula NaCl)

relative molecular mass the mass of a molecule, found by adding the relative atomic masses of the atoms in it

renewable resource a resource that will not run out; for example water, air, sunlight

residue the solid you obtain when you separate a solid from a liquid by filtering

respiration the reaction between glucose and oxygen that takes place in your body cells, and gives you energy

reversible reaction a reaction that can go both ways: a product can form, then break down again; the symbol $\rightleftharpoons$ is used to show a reversible reaction

rusting the name given to the corrosion of iron; oxygen and water attack the iron and rust forms

S

sacrificial protection letting one metal corrode in order to protect another metal

salt a metal compound formed when an acid is neutralised by a base

saturated compound an organic compound in which all the bonds between carbon atoms are single covalent bonds

saturated solution no more of the solute will dissolve in it, at that temperature

separating funnel a piece of equipment for separating two immiscible liquids

single bond the bond formed when two atoms share just one pair of electrons

slaked lime another name for calcium hydroxide

solubility the amount of solute that will dissolve in 100 grams of a solvent, at a given temperature

soluble will dissolve in a solvent

solute the substance you dissolve in the solvent, to make a solution

solution a mixture obtained when a solute is dissolved in a solvent

solvent the liquid in which a solute is dissolved, to make a solution

sonorous makes a ringing noise when hit

spectator ions ions that are present in a reaction mixture, but do not actually take part in the reaction

stable does not tend to react easily; it is unreactive

state symbols these are added to an equation to show the physical states of the reactants and products (g = gas, l = liquid, s = solid, aq = aqueous)

stationary phase the filter paper or other support that the solution travels over, in chromatography

stratosphere the layer of the atmosphere that extends from about 30 to 50 km above us

T

thermal decomposition the breaking down of a compound by heating it

thermite process the redox reaction between iron oxide and aluminium, which produces molten iron

titration a laboratory technique for finding the concentration of a solution

toxic poisonous

transition elements the elements that occupy the wide middle block of the Periodic Table; they are all metals and include iron, tin, copper and lead

trend a gradual change; the groups within the Periodic Table show trends in their properties; for example as you go down group 1, reactivity increases

triple bond the bond formed when two atoms share three pairs of electrons; a nitrogen molecule has a triple bond

U

universal indicator a paper or liquid you can use to find the pH of a solution; it changes colour across the whole range of pH

unreactive does not react easily

unsaturated compound an organic compound with at least one double bond between carbon atoms

V

valency a number that tells you how many electrons an atom gains, loses or shares, in forming a compound

valency electrons the electrons in the outer shell of an atom

variable valency an element shows variable valency if its atoms can lose different numbers of electrons, in forming compounds

viscosity a measure of how runny a liquid is; the more runny it is, the lower its viscosity

viscous thick and sticky

volatile forms a vapour easily

W

water of crystallization water that is locked into a compound; for example in crystals of copper(II) sulphate: $CuSO_4.5H_2O$

Y

yield the amount of a product that you obtain from a reaction; it is often given as a % of the maximum yield possible, according to the equation

The Periodic Table and atomic masses

Group																	
	1	2															1
1																	$^{1}_{1}$H hydrogen
2	$^{7}_{3}$Li lithium	$^{9}_{4}$Be beryllium															
3	$^{23}_{11}$Na sodium	$^{24}_{12}$Mg magnesium				The transition elements											
4	$^{39}_{19}$K potassium	$^{40}_{20}$Ca calcium	$^{45}_{21}$Sc scandium	$^{48}_{22}$Ti titanium	$^{51}_{23}$V vanadium	$^{52}_{24}$Cr chromium	$^{55}_{25}$Mn manganese	$^{56}_{26}$Fe iron	$^{59}_{27}$Co cobalt	$^{59}_{28}$Ni nickel	$^{64}_{29}$Cu copper	$^{65}_{30}$Zn zinc					
5	$^{85}_{37}$Rb rubidium	$^{88}_{38}$Sr strontium	$^{89}_{39}$Y yurium	$^{91}_{40}$Zr zirconium	$^{93}_{41}$Nb niobium	$^{96}_{42}$Mo molybdenum	$^{99}_{43}$Tc technetium	$^{101}_{44}$Ru ruthenium	$^{103}_{45}$Rh rhodium	$^{106}_{46}$Pd palladium	$^{108}_{47}$Ag silver	$^{112}_{48}$Cd cadmium					
6	$^{133}_{55}$Cs caesium	$^{137}_{56}$Ba barium	$^{139}_{57}$La lanthanium	$^{178.5}_{72}$Hf hafnium	$^{181}_{73}$Ta tantalum	$^{184}_{74}$W tungsten	$^{186}_{75}$Re rhenium	$^{190}_{76}$Os osmium	$^{192}_{77}$Ir iridium	$^{195}_{78}$Pt platinum	$^{197}_{79}$Au gold	$^{201}_{80}$Hg mercury					
7	$^{223}_{87}$Fr francium	$^{226}_{88}$Ra radium	$^{227}_{89}$Ac actinuim														

Lanthanides

$^{140}_{58}$Ce cerium	$^{141}_{59}$Pr praseodymium	$^{144}_{60}$Nd neodymium	$^{147}_{61}$Pm promethium	$^{150}_{62}$Sm samarium	$^{152}_{63}$Eu europium	$^{157}_{64}$Gd gadolinium	$^{159}_{65}$Tb terbium

Actinides

$^{232}_{90}$Th thorium	$^{231}_{91}$Pa protactinium	$^{238}_{92}$U uranium	$^{237}_{93}$Np neptunium	$^{244}_{94}$Pu plutonium	$^{243}_{95}$Am americium	$^{247}_{96}$Cm curium	$^{247}_{97}$Bk berkelium

Relative atomic masses based on internationally agreed figures

Element	Symbol	Atomic number	Relative atomic mass	Element	Symbol	Atomic number	Relative atomic mass
actinium	Ac	89		erbium	Er	68	167.26
aluminium	Al	13	26.9815	europium	Eu	63	151.96
americium	Am	95		fermium	Fm	100	
antimony	Sb	51	121.75	fluorine	F	9	18.9984
argon	Ar	18	39.948	francium	Fr	87	
arsenic	As	33	74.9216	gadolinium	Gd	64	157.25
astatine	At	85		gallium	Ga	31	69.72
barium	Ba	56	137.34	germanium	Ge	32	72.59
berkelium	Bk	97		gold	Au	79	196.967
beryllium	Be	4	9.0122	hafnium	Hf	72	178.49
bismuth	Bi	83	208.980	helium	He	2	4.0026
boron	B	5	10.811	holmium	Ho	67	164.930
bromine	Br	35	79.909	hydrogen	H	1	1.00797
cadmium	Cd	48	112.40	indium	In	49	114.82
caesium	Cs	55	132.905	iodine	I	53	126.9044
calcium	Ca	20	40.08	iridium	Ir	77	192.2
californium	Cf	98		iron	Fe	26	55.847
carbon	C	6	12.01115	krypton	Kr	36	83.80
cerium	Ce	58	140.12	lanthanum	La	57	138.91
chlorine	Cl	17	35.453	lawrencium	Lw	103	
chromium	Cr	24	51.996	lead	Pb	82	207.19
cobalt	Co	27	58.9332	lithium	Li	3	6.939
copper	Cu	29	63.54	lutetium	Lu	71	174.97
curium	Cm	96		magnesium	Mg	12	24.312
dysprosium	Dy	66	162.50	manganese	Mn	25	54.9380
einsteinium	Es	99		mendelevium	Md	101	

Group

3	4	5	6	7	0
					4/2 He helium
11/5 B boron	12/6 C carbon	14/7 N nitrogen	16/8 O oxygen	19/9 F fluorine	20/10 Ne noen
27/13 Al aluminium	28/14 Si silicon	31/15 P phosphorus	32/16 S sulphur	35.5/17 Cl chlorine	40/18 Ar argon
70/31 Ga gallium	73/32 Ge germanium	75/33 As arsenic	79/34 Se selenium	80/35 Br bromine	84/36 Kr krypton
115/49 In indium	119/50 Sn tin	122/51 Sb antimony	128/52 Te tellurium	127/53 I iodine	131/54 Xe xenon
204/81 Tl thallium	207/82 Pb lead	209/83 Bi bismuth	210/84 Po polonium	210/85 At astatine	222/86 Rn radon

162/66 Dy dysprosium	165/67 Ho holmium	167/68 Er erbium	169/69 Tm thutium	173/70 Yb ytterbium	175/71 Lu lutetium
251/98 Cf californium	252/99 Es einsteinium	257/100 Fm fermium	258/101 Md mendelevium	259/102 No nobelium	262/103 Lw lawrencium

Approximate atomic masses for calculations

Element	Symbol	Atomic mass (A_r) for calculations
aluminium	Al	27
bromine	Br	80
calcium	Ca	40
carbon	C	12
chlorine	Cl	35.5
copper	Cu	64
helium	He	4
hydrogen	H	1
iodine	I	127
iron	Fe	56
lead	Pb	207
lithium	Li	7
magnesium	Mg	24
manganese	Mn	55
neon	Ne	20
nitrogen	N	14
oxygen	O	16
phosphorus	P	31
potassium	K	39
silicon	Si	28
silver	Ag	108
sodium	Na	23
sulphur	S	32
zinc	Zn	65

Element	Symbol	Atomic number	Relative atomic mass
mercury	Hg	80	200.59
molybdenum	Mo	42	95.94
neodymium	Nd	60	144.24
neon	Ne	10	20.179
neptunium	Np	93	
nickel	Ni	28	58.71
niobium	Nb	41	92.906
nitrogen	N	7	14.0067
nobelium	No	102	
osmium	Os	76	190.2
oxygen	O	8	15.9994
palladium	Pd	46	106.4
phosphorus	P	15	30.9738
platinum	Pt	78	195.09
plutonium	Pu	94	
polonium	Po	84	
potassium	K	19	39.102
praseodymium	Pr	59	140.907
promethium	Pm	61	
protactinium	Pa	91	
radium	Ra	88	
radon	Rn	86	
rhenium	Re	75	186.2
rhodium	Rh	45	102.905
rubidium	Rb	37	85.47
ruthenium	Ru	44	101.07

Element	Symbol	Atomic number	Relative atomic mass
samarium	Sm	62	150.35
scandium	Sc	21	44.956
selenium	Se	34	78.96
silicon	Si	14	28.086
silver	Ag	47	107.868
sodium	Na	11	22.9898
strontium	Sr	38	87.62
sulphur	S	16	32.064
tantalum	Ta	73	180.948
technetium	Tc	43	
tellurium	Te	52	127.60
terbium	Tb	65	158.924
thallium	Tl	81	204.37
thorium	Th	90	232.038
thulium	Tm	69	168.934
tin	Sn	50	118.69
titanium	Ti	22	47.90
tungsten	W	74	183.85
uranium	U	92	238.03
vanadium	V	23	50.942
xenon	Xe	54	131.30
ytterbium	Yb	70	173.04
yttrium	Y	39	88.905
zinc	Zn	30	65.37
zirconium	Zr	40	91.22

Index

Where several page numbers are given and one is **bold**, look that one up first.

A
acetone 15
acetylene 205
acidic oxides 225
acidity 116
    and pH number 115
acid rain **123**, 227
acids 114–119
    and pH 116
    as proton donors 120, **121**
    carboxylic 252–253
    in the home 122
    reactions of 118–119
    strong and weak 116
activation energy 154, **163**
addition polymerisation 257–259
air 202–205
alanine 268
alcoholic drinks **249**, 251
alcohols 248–249
alkali metals 58, **60–61**, 62-63
alkaline earth metals 58
alkalis 114–119
alkanes 244–245
alkenes 246–247
allotropes
    of carbon 51
    of oxygen 202
    of sulphur 226
alloys 194
alumina (aluminium oxide) 190
aluminium
    alloys of 195
    corrosion in 197
    extraction of 190–191
    in reactivity series 178
    ore 185, 187, **190**
    properties of 191
    recycling of 199
    uses of 190, 192, **193**
aluminium ions, test for 290
aluminium oxide 190
amide link **260**, 268
amino acids 23, **268**, 270, 271
ammonia 215–217
    as an alkali 117
    bonding in 46
    Haber process for 216
    in making fertilisers 219
    laboratory preparation 215
    manufacture 168–169, **216–217**
    test for 289
ammonium compounds
    chloride 166
    nitrate 219
ammonium ion, test for 291
amphoteric oxides 225
anhydrous salts 229
anode 134
anodising 197
aqueous solutions 15
A_r 77
argon 36, **66**, 67
artificial elements 59
artificial fertilisers 218
atmosphere 202
atoms 26, **28**
atomic number 28
Avogadro's Law 96
Avogadro's number 80
aquifer 208

B
back reaction 166
bacteria 208
baking soda 122
balancing equations 92–93
barium sulphate 126
bases 118–119
    alkalis (soluble bases) 114–119
    and acids 118–121
    as proton acceptors 121
    outside the lab 122
    reactions of 119
basic oxides 224
batteries 132, **181**
bauxite (aluminium ore) 185, **190**
biological catalysts 156–157
biological detergents 157
bitumen 241
blast furnace 188–189
bleaching
    by chlorine 230
    by sulphur dioxide 227
boiling (change of state) 11
boiling points 8
    and purity 17
    for ionic compounds 42–43
    for molecular solids 48
bond energies 162–163
bonding
    between ions 38–39
    in metals 52
    in molecules 44
brass 195
bread making 157
breathalyser test 105

brine **138**, 230
    electrolysis of 138–139
bromide ion, test for 290
bronze 195
burning of fuels 223
    energy from 164–165
butane 244
butanoic acid 252
but-1-ene 246

C
calamine lotion 122
calcium, reactivity of 174, **178**
calcium compounds
    carbonate 234
    hydroxide 114, **174**, 235
    oxide 234
calcium ions, test for 290
calculations from equations 94–95
cancer treatment by radioisotopes 31
carbohydrates 266–267
carbon 232–233
    allotropes of 51
    and metal extraction 186
    in steel **189**, 194, 195
    isotopes of 30
    reactivity compared to metals **177**, 178
carbon-12 **30**, 76
carbonates 179, **233**
carbon dioxide 225, **232**
    and global warming 207
    bonding in 47
    in photosynthesis 223, **266**
    properties of 232
    uses of 232
carbonic acid 123
carbon monoxide 206, 233
    and metal extraction 186
carboxyl functional group 252
carboxylic acids 252–253
car exhausts 155
car fuel cells 213
cast iron 189
catalysts 154
    and activation energy 154
    biological (enzymes) **156–157**, 268, 270
    in car exhausts 155
    transition metals as **69**, 155
catalytic cracking 243
cathode 134
cell (producing electricity) 181
cellulose 267
cement 235
centrifuging 18

chain length
    and boiling point 244
    and homologous series 244
    in polythene 258
changes of state 8–11
charcoal 232
charge (in iron extraction) 188
charge (on atomic particles) 28
chemical change 90–91
chemical equations 36, **92–93**
chemical properties 172
chemical reactions 91–93
chlorine 230–231
    bonding in 45
    manufacture from brine 138
    reaction with sodium 36
    uses 139
chlorine water 230
chloromethane 245
chlorophyll 266
chromatogram **21**, 22
chromatography 21–23
chromium, extraction of 180
chromium plating 141
citric acid 114
cleaning (using acids and alkalis) 122
coal **206**, 226
cobalt(II) chloride 166
coke 188
collision theory (for reaction rate) 152–153
combination reactions 102
combustion 102, 232
competition among metals 176–178
compound ions 41
compounds 27
    and mixtures 90
    covalent 46–477
    forming 36
    ionic 42–43
    names and formulae 74–75
    percentage composition of 78–79
compression of gases 12
concentration of a solution 86–87
    finding by titration 128–129
concrete 226, **235**
condensation 11
condensation polymerisation 257, **260–261**
condenser 20
conductivity of metals 53
conductors 132–133
contact process 228–229
copper **68**, 69
    in reactivity series 180
    refining of 140
    uses of 192, **193**
copper ion, test for 290
copper(II) oxide 104, **224**
copper(II) sulphate **124**, 166
    electrolysis of solution 140

corrosion 196–197
covalent bond 44–45
covalent compounds 46–47
covalent giant structures 50–51
cracking of hydrocarbons 242–243
crude oil 238
crystallisation 19
cysteine 268

D
decay (radioactivity) 30
decomposition reactions 102
dehydrating agent 229
denaturing (enzymes) 157
density 172
detergents (biological) 157
diamond 51
diaphragm cell 138
diatomic 44
dichloromethane 245
dichromate ion 111
diesel (fraction in oil) 241
diffusion **7**, 13
displacement reactions **103**, 176, 230, 231
dissolving 14
distillation 20
distilled water 20
double bonds 45
dry ice 232
ductile (of metals) **53**, 172
dynamic equilibrium 167

E
Earth's crust 184
electricity 132
electrode 134
electrolysis 134–141
    in extraction of aluminium 191
    in extraction of metals 186
    of brine 138
    of copper(II) sulphate solutions 140–1
    of solutions 136–137
    of molten compounds 134–135
electrolyte 134
electronic configuration 33
electrons **28**, 32
electron shells 32–33
electron transfer 107
electroplating 141
elements **26–27**, 58-59
empirical formula 82–85
endothermic reactions 91, **161**
energy changes during reactions 160–163
energy from fuels 164–165
energy level diagram **160**, 161
enzymes 156–157
equations 36, **92–93**
    calculations using 94–95
    half equations 106
    ionic equations 120

equilibrium (in reversible reactions) 167
ester link 261
esters 253
ethane 244
ethanoates 252
ethanoic acid 252
ethanol 27, 165, **248–251**
ethene **246**, 247
evaporation 11
exothermic reactions 91, 150
extraction of metals 186–187

F
fats 269
fatty acids 269
fermentation 157, **250**
fertilisers 218
filtrate 18
filtration 18
fizzy drinks 232
flexible-fuel cars 251
fluoride (in water supply) 209
formulae 27
    empirical 82–85
    of compounds 74–75
    of ionic compounds 40–41
formula mass 77
forward reaction 166
fossil fuels 206
fountain experiment 215
fractional distillation 20
fractions (in distillation) 240
fractions from oil refining 241
freezing 8
freezing point 8
fuel cells 213
fuel oil 241
fuels 164–165
functional group **246**, 248, 252

G
galena (ore) 226
galvanising 180
gas chromatography 22
gases **8**, 10, 12–13
    and Avogadro's Law 96
    calculations involving 97
    compression of 12
    diffusion in 13
    effect of temperature on 12
    pressure in 12
gasoline (petrol) 241
giant covalent structures 50–51
global warming 207
glucose 267
    and photosynthesis 266
glycerol 269
glycine 268
gold 173
graphite 51, **232**
greenhouse gases 207

ground water 208
groups in Periodic Table 26, **58**
    group 0 (noble gases) 66–67
    group 1 (alkali metals) 62–63
    group 2 (alkaline earth metals) 58
    group 7 (halogens), 64–65
gypsum 235

H
Haber process 216–217
haematite (iron ore) 188
haemoglobin 206, 233
half-equations 106
halides 64
halogens 58, **64–65**, 231
heating curve 9
helium 36, **66**, 67
homologous series 244
hydrocarbons **238**, 242, 247
hydrochloric acid **114**, 116, 175, 230
    electrolysis of dilute 136
    electrolysis of concentrated 136
hydrogen 212–213
    and redox reactions 105
    as a fuel 165, **213**
    bonding in 44
    in Periodic Table 59
    reactivity compared with metals 175, **177**
    test for 212
    uses of 139
hydrogen chloride 116, **230**
    bonding in 47
hydrogen ions, in acids **116**, 120
hydrogen sulphide 226
hydrolysis
    in digestion 270
    in the lab 271
    of esters 271
hydroxide ions (and alkalinity) 117
hydroxides 117, **174**, 179
hypochlorous acid 230

I
impurities 16–17
incomplete combustion 245
indicators 115
indigestion 122
insect stings 122
insoluble salts 126–127
insulators 132
iodide ion, test for 290
iodine 64
    bonding and structure in 48
    displacement of **230**, 231
ionic bonds 38–39
ionic compounds 42–43
    and electrolysis 134
    formulae of 40–41
    names of 40–41
    properties of 42–43

ionosphere 202
ions 37
    compound ions 41
    mass of 77
iron **68**, 69, 184
    alloys of **194**, 195
    corrosion (rusting) of **196**, 197
    extraction of 188–189
    in reactivity series 178
    ore 187
    properties of **68**, 69, 173, 194
iron ions, test for 290
iron(III) oxide 188, **224**
iron sulphide 90
isomers 245, 247
isotopes **30–31**, 76

K
kerosene 241
kilojoule (kJ) 160
krypton 67

L
lattice **42**, 48, 52
Lavoisier 203
lead 181, **192**, 206, 207
    reactivity of 178
    uses of 181, 192, **193**
lead bromide, electrolysis of 134–135
Le Chatelier's principle 168
light (and photochemical reactions) 151
limestone 189, **234–235**
lime water 114, **232**
liquids 8–11
lithium **60**, 62
lithium chloride, decomposition of 107
litmus 114
locating agent 23
lubricating fraction 241

M
macromolecules **50**, 256, 266
magnesium 39
magnesium chloride 39
magnesium oxide 83, **224**
magnesium sulphate (gypsum) 235
malleable (of metals) **53**, 172
manganese (and variable valency) 110
manganese(IV) oxide 156
mass number *see* nucleon number
mass of atoms, molecules and ions 76–77
    calculations involving 95
mass spectrometer 76, **84**
melting 8–10
melting points **8**, 9
    and purity 17
    of ionic compounds 42, **43**
    of metals 52
    of molecular solids 48
    of giant covalent structures 50
mesosphere 202

metallic bonds 52
metalloids (semi-metals) 70
metal oxides 224
metals 172
    bonding in 52–53
    extraction from ores 186–187
    in Earth's crust 184
    in Periodic Table 59
    in reactivity series 178
    ores 185
    properties of 53, **172**
    reactions with oxygen 224
    reactivity of 174–178
    recycling of 198–199
    stability of compounds 179
    transition metals 68–69
    uses of 192–195
metal salts 118, 119
methane 244
    as fuel 223
    bonding in 46
methanoic acid 252
methyl orange 115
mild steel 194
mining of metal ores 185, 198–199
mixtures 14
    compared with compounds 90
mobile phase (in chromatography) 22
molar gas volume 96
molar solution 86
mole 80
molecular formulae 44, **84–85**
molecular solids 48
molecular substances **48–49**, 52, 133
molecules 7, **44**, 48
    masses of 76–77
monomer **258**, 258, 260
monosaccharides 267
M_r 77

N
names for compounds 40–41, **74–75**
naphtha **241**, 243
natural gas 164, 165, **216**
natural polymers 257
neon 36, **66**, 67
neutralisation reactions 113, 118, **120–1**
neutral liquids (and pH) 115
neutral oxides 225
neutrons 28, 29
nickel 68
ninhydrin 23
nitrate ion, test for 291
nitrates **118**, 179, 219
nitric acid 219
nitrogen 214
    bonding in 45
    for plants **218**, 219
    production in industry 204
    uses 205

nitrogen oxides as pollutants 206
noble gases 58, **66**
    electronic configuration in 33
    in Periodic Table 58
    uses of **67**, 205
non-conductor 132
non-metals 212–235
    compared with metals 172–173
    in Periodic Table 59
    reactions with oxygen 225
non-renewable resource 239
nuclear fuels 165
nucleon number 29
nucleus of atoms 26, **28**
nylon **260**, 263

O

oil 238–239
    refining of 240–241
    sulphur extraction from 226
ores **185**, 186, 187
organic chemistry 233
organic compounds **233**, 238
oxidation 104
    and electron transfer 106–107
oxidation state 108
oxides 224–225
oxidising agents **110**, 111, 231
oxy-acetylene torch 205
oxygen 222–223
    in air **202**, 203
    bonding in 45
    ionic compounds with metals 39
    uses 205
oxygen mask 205
oxygen tent 205
oxygen, test for **156**, 209
ozone layer 202

P

paint (against rust) 196
palmitic acid 269
paper chromatography 21
paraffin (kerosene) 241
particles in solids, liquids and
    gases 6–7, **10–11**
pentane 244
percentage composition of
    compounds 78–79
percentage purity 98, **99**
percentage yield 98
Periodic Table **58–59**, 70–71
petrol (gasoline) 241
pH scale 115
phenolphthalein 115
phosphorus (for plants) 218
phosphorus pentoxide 225
photographic film 127
photosynthesis 151, **266**
physical change 91
physical properties 172

plastics 262, 264–265
platinum, as catalyst **154**, 155
pollution 206–207
polyamide 260
polychloroethene (PVC) 263
polyester 261
polymerisation **256**, 258–261
polymers **256–257**, 258–263
polypropene 263
polysaccharide 267
polystyrene 263
polythene **256**, 258
    and chain lengths 258
    and pollution 264–265
    changing properties of 262
    uses of 263
potassium (as group 1 metal) 60–63
    reactivity 174, **178**
potassium compounds
    bromide, electrolysis of 137
    dichromate(VI) 111
    iodide, electrolysis of 135
    iodide, as reducing agent 111
    hydroxide 114, 117, **174**
    manganate(VII) 7, **110**
potassium for plants 218
power stations 206, 207
precipitation 103, **126–127**
pressure in gases 12
propane 244
propane-1,2,3-triol 269
propanoic acid 252
propanone 15
propene 246
proteins 266, **268**, 270, 271
proton donors and acceptors 121
proton number 28
protons **28**, 29
purity 16
    and melting and boiling points 17
    percentage 98, **99**
PVC (polychloroethene) **259**, 263
pyrite 90

Q

quartz **50**, 51
quicklime 234

R

radiation **30**, 31
radioactivity 30–31
radioisotopes **30**, 31,
random motion of particles 6
rates of reaction **145**, 146–151
    and catalysts 154–155
    and collisions 152–153
    and concentration **148**, 152
    and light 151
    and surface area **150–151**, 153
    and temperature **149**, 153
    measuring **145**, 146–147

reactants 92
reactions, types of 102–103
    energy changes in 160–161
    reversible 166–167
reactivity
    of group 1 metals 62–6
    of halogens **65**, 231
    of group 0 66
    of metals 174–179
reactivity series of metals 178
recycling 198
    aluminium 199
    metals 198–199
    plastics 265
    steel 199
redox reactions 104–109
    and electron transfer 106–107
    and oxidation state 108–109
reducing agents **110**, 111
reduction 104–107
    in metal extraction 186
refining of oil 20, **240**
refrigerators 149
relative atomic mass 76
relative formula mass 77
relative molecular mass 77
residue from filtration 18
respiration 223
reversible reactions **166–167**, 228
rhombic sulphur 226
rock salt 230
rotary kilns 234, 235
rubber (vulcanizing of) 226
rusting 196–197

S

sacrificial protection **180**, 197
salt (sodium chloride) 36
    electrolysis of solutions 136–138
salts 118–119
    making in laboratory 124–127
    solubility of some 126
sand 50
saturated compounds (alkanes) 246
saturated solutions 15
semi-conductors 71
separation methods 18–21
    centrifuging 18
    chromatography 21–23
    crystallisation 19
    filtering 18
    fractional distillation 20
    paper chromatography 21
    simple distillation 20
shells (electron) 32
silicon dioxide (silica) **50**, 51, 74
silver 192, **193**
    and electroplating 141
    in photography 127
    reactivity of 174, 175, **178**
    uses of 193

silver bromide **102**, 126
silver chloride **102**, 126
single bonds 45
slag 189
slaked lime 235
smoke particles 6
soap 271
sodium
    as alkali metal 60–63
    extraction of 186
    properties of 60–61
sodium chloride **36**, 38
    bonding in 38
    electrolysis of solutions 136–138
    structure of compound 38
sodium hydroxide 60
    as alkali 114, **117**
    production in industry 138
    uses of 139
soil acidity **122**, 234
solidifying 11
solids **8**, 11
solutions 14–15
    concentration of 86–87
    electrolysis of 136–137
solvents 14–15, 18
sonorous 172
stability of metal compounds 179
stainless steel 69, **194**, 195
standard solution 128
starch **267**, 270, 271
state symbols 92
stationary phase (in chromatography) 22
steel 194, 195
    manufacture of 195
    recycling of 199
stratosphere 202
styrene 259
substitution reaction 245

sulphates 229
    solubility of 126
sulphur 226–227
sulphur dioxide 206, 225, **227**, 228
sulphur trioxide 228
sulphuric acid 228–229
    manufacture of 228
    properties of 229
sulphurous acid 227
surface area (rates of reaction) **150–151**, 153
symbols for elements 26–27
synthesis reactions 102
synthetic polymers **256**, 262–263

T
Teflon **259**, 263
Terylene **261**, 263
test for unsaturation 247
tests for gases 289
tests for ions 290–291
tests for water 166
tetrachloroethane 245
thermal decomposition **179**, 234, 242
thermite process 180
tin plating 197
titanium 192, 193
titrations 125, **128–129**
tracers 31
transition elements 59, **68–69**
transition metals as catalysts 155
trichloromethane 245
troposphere 202

U
universal indicator 115
unsaturated (organic compounds) 246
    test for unsaturation 247
unsaturated oils 268

V
valency 71, 74
valency electrons **33**, 58
vinegar 129
vinyl chloride 259
viscosity 241
volatile liquids **14**, 240
voltage (cells) 181
vulcanization 226

W
water 208
    as solvent 15
    bonding in 47
    changes of state for 10
    formula of 74
    heating curve for 9
    reaction with metals 174
    supply 208–209
    tests for 166
    uses of 208
water vapour 8

X
xenon 67

Y
yeast **157**, 250
yield 98
    in making ammonia 217

Z
zinc 192, **193**
    extraction of 187
    reactivity of 174–177, **178**
zinc blende (ore) 187
zinc chloride 135
zinc sulphate 124